Finance Construction-4, *Tim Asikin, Steve Asikin, Indra Senihardja*

FINANCE

Construction-4

Corporate IFRS-GAAP (B/S-I/S) Engineering

Technologies No. 3,001-4,000 of 111,111 Laws

by

**Tim Asikin, Steve Asikin,
Indra Senihardja**

0
Corporate IFRS-GAAP (B/S-I/S), ISBN 13: **978-1977659187**, ISBN 10: **1977659187**

FINANCE CONSTRUCTION-4
Corporate IFRS-GAAP (B/S-I/S) Engineering
Technology No. 3,001-4,000 of 111,111 Laws
By Tim Asikin, Steve Asikin, Indra Senihardja

Creative Team: Steve Asikin
        & Asikin Business Quant, New York
An Amazon Paperback
First Published in United States 2018
By Amazon Createspace
This 1st edition published in 2018
AMZN, PO. Box 81226, Seattle
WA 98108-1226, United States

Copyright © 2018 by Steve Asikin,
gundul.ganesha@yahoo.com, indra.senihardja@yahoo.com
This book is sold subject to the condition
That it shall not, by the way of trade or otherwise,
Be lent, resold, hired out, or otherwise circulated
Without the publisher's prior consent in any form
Of binding and cover or other than that in which it
Is published and without a similar condition
Including this condition being imposed
On the subsequent purchaser.

ISBN-13: 978-1977659187
ISBN-10: 1977659187

Finance Construction-4, *Tim Asikin, Steve Asikin, Indra Senihardja*

PREFACE

This one is the only Real Finance Engineering Book, written by Engineering & Business Experts with strong Accounting Maths. It comes from a very long world wide experience, analyzing more than 100,000 Corporate Balancesheet and Income Statement reports, in conservative IFRS (International Finance Reporting Standard) and GAAP (Generally Accepted Accounting Principles) of 9,000 days works as Financing Banker, plus Business CEO expert & fresh Engineering views.

This is a manual for healthy Corporate Governance assurance dan practical business management. It is best for Government Authorities, Accounting Regulators, Business Owners & Small Public Investors (all majority & minority share holders as well).

This is not just a tool for making debt instrument unrelated to daily Corporate Finance. This only good for sincere Directors & Executive, and truly bad for the lazy and /or incapable one.

The nicest thing about it, is that only requires simple Junior High-school Algebra. We just need to be consistent with its less number treatment of constellations. Its saddest part is that many professors and doctors are no longer able to understand Junior High-school Algebra, and hiding in Accounting jargon. Please go back to its Introduction for glossary, at anywhere inside the book, so then we can still understand what are those mean with simple Algebraic Rule there. We try to avoid some sentences multiplied, added, eliminated divided, rooted or powered by another sentences, that make many traditional finance book successful (delivering few simple things, seen as difficult and scary), but less useful.

Finance Construction-4, Tim Asikin, Steve Asikin, Indra Senihardja

TABLE OF CONTENTS

Preface, p.002
Table of Contents, p.003
INTRODUCTION: The Math Fin Law Terminologies, p.004

Chapter-01: Growth of **S**= SALES, in Optimum Math Fin Law,
Chapter-02: Manageable **V**= VARIABLE COST, Optimization,
Chapter-03: **M**= MARGIN of CONTRIBUTION, Optimization,
Chapter-04: Manageable **F**= FIXED COST, Optimization,
Chapter-05: **O**= OPERATIONAL SURPLUS, Optimization,
Chapter-06: Manageable **I**= INTEREST COST, Optimization,
Chapter-07: **B**= BEFORE TAX INCOME, Optimization,
Chapter-08: Manageable **T**= TAX, Planning Optimization,
Chapter-09: **A**= AFTER TAX INCOME, Optimization,
Chapter-10: Manageable **D**= DIVIDEND, Optimization.
Chapter-11: Manageable **Y**= YIELDING TAX, Optimization,
Chapter-12: **H**= HOME TAKEN DIVIDEND, Optimization, p.6
Chapter-13: **R**= RETAINED EARNINGS, Optimization, p.236
Chapter-14: **E**= EQUITY, or Capital Optimization, p.509
Chapter-15: **L**= LIABILITIES, or Debt Optimization,
Chapter-16: **W**= WEALTH, or Total Assets Optimization,
Chapter-17:**P**= PROCURED TRADE INVENTORY, Optimization,
Chapter-18: **X**= XPRESS or Current Debt Optimization,
Chapter-19: **Q**= QUOTED LONGTERM, Debt Optimization,
Chapter-20: **G**= GOODS or Trade Payable Optimization,
Chapter-21: **Z**= ZERO TRADE or Current Debt Optimization,
Chapter-22: **J**= JOB or Trade Account Receivable Optimization,
Chapter-23: **C**= CURRENT ASSETS, Optimization,
Chapter-24: **K**=KIND of CASH, Optimization.
Chapter-25: **N**=NON CURRENT ASSETS, Optimization, p.006

Finance Construction-4, *Tim Asikin, Steve Asikin, Indra Senihardja*

INTRODUCTION:
The Math Fin Law Terminologies

Income Statement

S	V				
	M	F			
		O	I		
			B	T	
				A	D	Y
						H
				R		

Balance Sheet

K	C	W	L	X	G
J					Z
P				Q	
N			E	U	
				R	

Please carefully look that all of 26 characters (**A** to **Z**) used once, except the **R**, that in Accountancy must be appeared in both diagrams at same identical value, with glossary (of well known abbreviations) in the next page. Only 2 (two) of them are historical as **U=E'** and **S'** means carried forward legal numbers from previous year.

The 13 (thirteen) common measures (**c, d, f, g, i, j, l, p, q, s, t, v,** and **y**), are put up front as desired criteria along the way, so then later analyst could be surprised.

In Balance Sheet, we found that **W= C+N= K+J+P+N= L+E= X+Q+U+R= G+Z+Q+U+R** as well, so then we put **W** in the middle of its T-Account balance.

In Income Statement, same applied as **S= V+F+I +T+Y+H+R= V+M= V+F+ O= V+F+I+B= V+F+I+T+A= V+F+I+B= V+F+I+T+D+R.**

Finance Construction-4, *Tim Asikin, Steve Asikin, Indra Senihardja*

A= After Tax Income;
B= Before Tax Income,
C= Current Assets, **c**= C/X= Current Ratio,
D= Dividend Paid, **d**= D/A=Dividend Portion or Payout,
E= Equity or Capital, **E'**= Equity of Past Year,
F= Fixed Cost, **f**= F/S=Fixed Portion,
G= Goods or Trade Payables,
    g= 360G/V= Goods Payable Days,
H= Home Taken Dividend,
I= Interest Expense, **i**= I/S= Interest Portion,
J= Jobs or Account Receivable,
    j= 360J/S= Jobs or Trade Account ReceivableDays,
K= Kind of Cash,
L= Liabilities or Debt, **l**= L/E= Leverage or Gearing Ratio,
M= Margin of Contribution,
N= Non Current Assets,
O= Operational Surplus,
P= Procured Inventories,
    p= 360P/V= Procured Inventory Days,
Q= Quoted Longterm Liabilities,
    q= [C-P]/X= Quick or Acid Test Ratio,
R= Retained Earnings,
S= Sales or Revenues, **S'**= Sales of Past Year,
    s= [S/S']-1= Sales Growth,
T= Tax Paid, **t**= T/B= Tax Rate,
U= Utilized or Starting Capital,
V= Variable Cost, **v**= V/S= Variable Portion,
W= Wealth or Total Assets,
X= Xpress or Current Debt,
Y= Yielding Tax of Dividend,
    y= Y/D= Yielding Tax Rate of Dividend,
Z= Zero Trade Current Debt.

CHAPTER-12:
H= HOME TAKEN DIVIDEND Optimization

Based on 12/31/15 Pfizer Annual Report
(http://www.nasdaq.com/symbol/pfe/financials?query=income-statement)

Continuing the Laws No.1-3,000
at Finance Construction 1-3

Law-3001:

If both (**H**= Home Taken Dividend= 1,659 millions USD), (**i**= I/S= Interest Portion= 1.00%), (**y**= Y/D= Yielding Tax Rate of Dividend= 30.00%), (**d**= D/A= Dividend Portion or Payout= 30.00%), (**$**= Sales or Revenues= 51,294 millions USD), (**f**= F/S= Fixed Portion= 58.00%), (**s**= [S/S']-1= Sales Growth= 5.00%), (**v**= V/S= Variable Portion= 19.00%), and (**t**= T/B= Tax Rate= 30.00%), are known, then it's (**$'**= Sales Past) must be:

$' = ($[1-**f**]-**H**/{**d**[1-**y**][1-**t**]})/{[1+**s**][**v**+**i**]}
  = (51,294[1-0.5800]-1,659/{0.3000
     [1-0.3000][1-0.3000]})
       /{[1+0.0500][0.1900+0.0100]}
  = 48,851 millions USD

Law-3002:

If both (**H**= Home Taken Dividend= 1,659 millions USD), (**i**= I/S= Interest Portion= 1.00%), (**y**= Y/D= Yielding Tax Rate of Dividend= 30.00%), (**d**= D/A= Dividend Portion or Payout= 30.00%), (**$**= Sales or Revenues= 51,294 millions USD), (**f**= F/S= Fixed Portion= 58.00%), (**$'**= Sales of Past Year= 48,851 millions USD), (**v**= V/S= Variable Portion= 19.00%), and (**t**= T/B= Tax Rate= 30.00%), are known, then it's (**s**= Sales Growth planned) is:

$$s = (\$[1-f]-H/\{d[1-y][1-t]\})/\{\$'[v+i]\}-1$$
$$= (51{,}294[1-0.5800]-1{,}659/\{0.3000$$
$$[1-0.3000][1-0.3000]\})/\{48{,}851$$
$$[0.1900+0.0100]\}-1$$
$$= \underline{5.00\%}$$

Law-3003:

If both (**H**= Home Taken Dividend= 1,659 millions USD), (**i**= I/S= Interest Portion= 1.00%), (**y**= Y/D= Yielding Tax Rate of Dividend= 30.00%), (**d**= D/A= Dividend Portion or Payout= 30.00%), (**$**= Sales or Revenues= 51,294 millions USD), (**f**= F/S= Fixed Portion= 58.00%), (**$'**= Sales of Past Year= 48,851 millions USD), (**s**= [S/S']-1= Sales Growth= 5.00%), and (**t**= T/B= Tax Rate= 30.00%), are known, then it's (**v**= Variable Portion planned) is:

$$v = (\$[1-f]-H/\{d[1-y][1-t]\})/\{\$'[1+s]\}-i$$
$$= (51{,}294[1-0.5800]-1{,}659/\{0.3000$$
$$[1-0.3000][1-0.3000]\})/\{48{,}851$$
$$[1+0.0500]\}-0.0100$$
$$= \underline{19.00\%}$$

Law-3004:

If both (**H**= Home Taken Dividend= 1,659 millions USD), (**y**= Y/D= Yielding Tax Rate of Dividend= 30.00%), (**d**= D/A= Dividend Portion or Payout= 30.00%), (**$**= Sales or Revenues= 51,294 millions USD), (**f**= F/S= Fixed Portion= 58.00%), (**$'**= Sales of Past Year= 48,851 millions USD), (**s**= [S/S']-1= Sales Growth= 5.00%), (**v**= V/S= Variable Portion= 19.00%), and (**t**= T/B= Tax Rate= 30.00%), are known, then it's (**i**= Interest Portion planned) is:

$$i = (\$[1-f]-H/\{d[1-y][1-t]\})/\{\$'[1+s]\}-v$$
$$= (51{,}294[1-0.5800]-1{,}659/\{0.3000 \\ [1-0.3000][1-0.3000]\})/\{48{,}851 \\ [1+0.0500]\}-0.1900$$
$$= 1.00\%$$

Finance Construction-4, *Tim Asikin, Steve Asikin, Indra Senihardja*

Law-3005:
If both (**I**= Interest Expense= 513 millions USD), (**y**= Y/D= Yielding Tax Rate of Dividend= 30.00%), (**d**= D/A= Dividend Portion or Payout= 30.00%), (**$**= Sales or Revenues= 51,294 millions USD), (**f**= F/S= Fixed Portion= 58.00%), (**$'**= Sales of Past Year= 48,851 millions USD), (**s**= [S/S']-1= Sales Growth= 5.00%), (**v**= V/S= Variable Portion= 19.00%), and (**t**= T/B= Tax Rate= 30.00%), are known, then it's (**H**= Home Taken Dividend planned) is:

$$H = d[1-y][1-t]\{\$-\$'v[1+s]-\$'f[1+s]-I\}$$
$$= d[1-y][1-t]\{\$-\$'[1+s][v+f]-I\}$$
$$= 0.3000[1-0.3000][1-0.3000]\{51,294$$
$$-48,851[1+0.0500][0.1900$$
$$+0.5800]-513\}$$
$$= 1,659 \text{ millions USD}$$

Law-3010:
If both (l= Interest Expense= 513 millions USD), (d= D/A= Dividend Portion or Payout= 30.00%), (H= Home Taken Dividend= 1,659 millions USD), (t= T/B= Tax Rate= 30.00%), (f= F/S= Fixed Portion= 58.00%), ($= Sales or Revenues= 51,294 millions USD), (s= [S/S']-1= Sales Growth= 5.00%), (v= V/S= Variable Portion= 19.00%), and (y= Y/D= Yielding Tax Rate of Dividend= 30.00%), are known, then it's ($'= Sales Past) must be:

$$\$' = (\$-l-H/\{d[1-y][1-t]\})/\{[1+s][v+f]\}$$
$$= (51{,}294 - 513 - 1{,}659/\{[1-0.3000][1-0.3000]\})/\{[1+0.0500][0.1900+0.5800]\}$$
$$= 48{,}851 \text{ millions USD}$$

Law-3011:

If both (**I**= Interest Expense= 513 millions USD), (**d**= D/A= Dividend Portion or Payout= 30.00%), (**H**= Home Taken Dividend= 1,659 millions USD), (**t**= T/B= Tax Rate= 30.00%), (**f**= F/S= Fixed Portion= 58.00%), (**S**= Sales or Revenues= 51,294 millions USD), (**S'**= Sales of Past Year= 48,851 millions USD), (**v**= V/S= Variable Portion= 19.00%), and (**y**= Y/D= Yielding Tax Rate of Dividend= 30.00%), are known, then it's (**s**= Sales Growth planned) is:

$$s = (S-I-H/\{d[1-y][1-t]\})/\{S'[v+f]\} - 1$$
$$= (51,294-513-1,659/\{[1-0.3000][1-0.3000]\})/\{48,851[0.1900+0.5800]\} - 1$$
$$= 5.00\%$$

Law-3012:
If both (**I**= Interest Expense= 513 millions USD), (**d**= D/A= Dividend Portion or Payout= 30.00%), (**H**= Home Taken Dividend= 1,659 millions USD), (**t**= T/B= Tax Rate= 30.00%), (**f**= F/S= Fixed Portion= 58.00%), (**$**= Sales or Revenues= 51,294 millions USD), (**$'**= Sales of Past Year= 48,851 millions USD), (**s**= [S/S']-1= Sales Growth= 5.00%), and (**y**= Y/D=Yielding Tax Rate of Dividend= 30.00%), are known, then it's (**v**= Variable Portion planned) is:

$$v = (\$-I-H/\{d[1-y][1-t]\})/\{\$'[1+s]\}-f$$
$$= (51,294-513-1,659/\{[1-0.3000][1-0.3000]\})/\{48,851[1+0.0500]\}-0.5800$$
$$= 19.00\%$$

Law-3013:

If both (**I**= Interest Expense= 513 millions USD), (**d**= D/A= Dividend Portion or Payout= 30.00%), (**H**= Home Taken Dividend= 1,659 millions USD), (**t**= T/B= Tax Rate= 30.00%), (**v**= V/S= Variable Portion= 19.00%), (**S**= Sales or Revenues= 51,294 millions USD), (**S'**= Sales of Past Year= 48,851 millions USD), (**s**= [S/S']-1= Sales Growth= 5.00%), and (**y**= Y/D= Yielding Tax Rate of Dividend= 30.00%), are known, then it's (**f**= Fixed Portion planned) is:

$$f = (S-I-H/\{d[1-y][1-t]\})/\{S'[1+s]\}-v$$
$$= (51{,}294-513-1{,}659/\{[1-0.3000][1-0.3000]\})/\{48{,}851[1+0.0500]\}-0.1900$$
$$= \underline{58.00\%}$$

Law-3014:
If both (**f**= F/S= Fixed Portion= 58.00%), (**d**= D/A= Dividend Portion or Payout= 30.00%), (**H**= Home Taken Dividend= 1,659 millions USD), (**t**= T/B= Tax Rate= 30.00%), (**v**= V/S= Variable Portion= 19.00%), (**$**= Sales or Revenues= 51,294 millions USD), (**$'**= Sales of Past Year= 48,851 millions USD), (**s**= [S/S']-1= Sales Growth= 5.00%), and (**y**= Y/D= Yielding Tax Rate of Dividend= 30.00%), are known, then it's (**I**= Interest Expense planned) is:

$$I = \$-\$'[1+s][v+f]-H/\{d[1-y][1-t]\}$$
$$= 51{,}294-48{,}851[1+0.0500][0.1900+0.5800]-1{,}659/\{0.3000[1-0.3000][1-0.3000]\}$$
$$= \underline{513} \text{ millions USD}$$

Law-3015:

If both (**f**= F/S= Fixed Portion= 58.00%), (**d**= D/A= Dividend Portion or Payout= 30.00%), (**i**= I/S= Interest Portion= 1.00%), (**t**= T/B= Tax Rate= 30.00%), (**v**= V/S= Variable Portion= 19.00%), (**$**= Sales or Revenues= 51,294 millions USD), (**$'**= Sales of Past Year= 48,851 millions USD), (**s**= [S/S']-1= Sales Growth= 5.00%), and (**y**= Y/D= Yielding Tax Rate of Dividend= 30.00%), are known, then it's (**H**= Home Taken Dividend planned) is:

$$H = d[1-y][1-t]\{\$-\$'v[1+s]-\$'f[1+s]-\$i\}$$
$$= d[1-y][1-t]\{\$[1-i]-\$'[1+s][v+f]\}$$
$$= 0.3000[1-0.3000][1-0.3000]\{51,294$$
$$[1-0.0100]-48,851[1+0.0500]$$
$$[0.1900+0.5800]\}$$
$$= \underline{1,659} \text{ millions USD}$$

Law-3016:

If both (**f**= F/S= Fixed Portion= 58.00%), (**H**= Home Taken Dividend= 1,659 millions USD), (**i**= I/S= Interest Portion= 1.00%), (**t**= T/B= Tax Rate= 30.00%), (**v**= V/S= Variable Portion= 19.00%), (**$**= Sales or Revenues= 51,294 millions USD), (**$'**= Sales of Past Year= 48,851 millions USD), (**s**= [S/S']-1= Sales Growth= 5.00%), and (**y**= Y/D= Yielding Tax Rate of Dividend= 30.00%), are known, then it's (**d**= Dividend Portion or Payout planned) is:

$$\begin{aligned}
\mathbf{d} &= \mathbf{H}/([1-\mathbf{y}][1-\mathbf{t}]\{\$-\$'\mathbf{v}[1+\mathbf{s}]-\$'\mathbf{f}[1+\mathbf{s}]-\$\mathbf{i}\}) \\
&= \mathbf{H}/([1-\mathbf{y}][1-\mathbf{t}]\{\$[1-\mathbf{i}]-\$'[1+\mathbf{s}][\mathbf{v}+\mathbf{f}]\}) \\
&= 1{,}659/([1-0.3000][1-0.3000]\{51{,}294 \\
&\qquad [1-0.0100]-48{,}851[1+0.0500] \\
&\qquad [0.1900+0.5800]\}) \\
&= 30.00\%
\end{aligned}$$

Law-3017:

If both (**f**= F/S= Fixed Portion= 58.00%), (**H**= Home Taken Dividend= 1,659 millions USD), (**i**= I/S= Interest Portion= 1.00%), (**t**= T/B= Tax Rate= 30.00%), (**v**= V/S= Variable Portion= 19.00%), (**$**= Sales or Revenues= 51,294 millions USD), (**$'**= Sales of Past Year= 48,851 millions USD), (**s**= [S/S']-1= Sales Growth= 5.00%), and (**d**= D/A= Dividend Portion or Payout= 30.00%), are known, then it's (**y**= Yielding Tax Rate of Dividend planned) is:

$$y = 1 - H/(d[1-t]\{\$-\$'v[1+s]-\$'f[1+s]-\$i\})$$
$$= 1 - H/(d[1-t]\{\$[1-i]-\$'[1+s][v+f]\})$$
$$= 1 - 1{,}659/(0.3000[1-0.3000]\{51{,}294[1-0.0100]-48{,}851[1+0.0500][0.1900+0.5800]\})$$
$$= 30.00\%$$

<u>Law-3018</u>:

If both (**f**= F/S= Fixed Portion= <u>58.00</u>%), (**H**= Home Taken Dividend= <u>1,659</u> millions USD), (**i**= I/S= Interest Portion= <u>1.00</u>%), (**y**= Y/D= Yielding Tax Rate of Dividend= <u>30.00</u>%), (**v**= V/S= Variable Portion= <u>19.00</u>%), (**$**= Sales or Revenues= <u>51,294</u> millions USD), (**$'**= Sales of Past Year= <u>48,851</u> millions USD), (**s**= [S/S']-1= Sales Growth= <u>5.00</u>%), and (**d**= D/A= Dividend Portion or Payout= 30.00%), are known, then it's (**t**= Tax Rate planned) is:

$$t = 1 - H/(d[1-y]\{\$ - \$'v[1+s] - \$'f[1+s] - \$i\})$$
$$= 1 - H/(d[1-y]\{\$[1-i] - \$'[1+s][v+f]\})$$
$$= 1 - 1,659/(0.3000[1-0.3000]\{51,294$$
$$[1-0.0100] - 48,851[1+0.0500]$$
$$[0.1900+0.5800]\})$$

$$= \underline{30.00\%}$$

Law-3019:

If both (**f**= F/S= Fixed Portion= 58.00%), (**H**= Home Taken Dividend= 1,659 millions USD), (**i**= I/S= Interest Portion= 1.00%), (**y**= Y/D= Yielding Tax Rate of Dividend= 30.00%), (**v**= V/S= Variable Portion= 19.00%), (**t**= T/B= Tax Rate= 30.00%), (**S'**= Sales of Past Year= 48,851 millions USD), (**s**= [S/S']-1= Sales Growth= 5.00%), and (**d**= D/A= Dividend Portion or Payout= 30.00%), are known, then it's (**S**= Sales or Revenues planned) is:

$$S = (S'[1+s][v+f] + H/\{d[1-y][1-t]\})/[1-i]$$
$$= (48,851[1+0.0500][0.1900+0.5800]$$
$$+ 1,659/\{0.3000[1-0.3000]$$
$$[1-0.3000]\})/ [1-0.0100]$$
$$= \underline{51,294} \text{ millions USD}$$

Law-3020:

If both (**f**= F/S= Fixed Portion= 58.00%), (**H**= Home Taken Dividend= 1,659 millions USD), (**S**= Sales or Revenues= 51,294 millions USD), (**y**= Y/D= Yielding Tax Rate of Dividend= 30.00%), (**v**= V/S= Variable Portion= 19.00%), (**t**= T/B= Tax Rate= 30.00%), (**S'**= Sales of Past Year= 48,851 millions USD), (**s**= [S/S']-1= Sales Growth= 5.00%), and (**d**= D/A= Dividend Portion or Payout= 30.00%), are known, then it's (**i**= Interest Portion planned) is:

$$i = 1 - (S'[1+s][v+f] + H/\{d[1-y][1-t]\})/S$$
$$= 1 - (48,851[1+0.0500][0.1900+0.5800]$$
$$+ 1,659/\{0.3000[1-0.3000]$$
$$[1-0.3000]\})/51,294$$
$$= \underline{1.00\%}$$

Law-3021:

If both (**f**= F/S= Fixed Portion= 58.00%), (**H**= Home Taken Dividend= 1,659 millions USD), (**$**= Sales or Revenues= 51,294 millions USD), (**y**= Y/D= Yielding Tax Rate of Dividend= 30.00%), (**v**= V/S= Variable Portion= 19.00%), (**t**= T/B= Tax Rate= 30.00%), (**i**= I/S= Interest Portion= 1.00%), (**s**= [S/S']-1= Sales Growth= 5.00%), and (**d**= D/A= Dividend Portion or Payout= 30.00%), are known, then it's (**$'**= Sales Past) must be:

$' = ($[1-**i**]-**H**/{**d**[1-**y**][1-**t**]})/{[1+**s**][**v**+**f**]}
= (51,294[1-0.0100]-1,659/{0.3000
    [1-0.3000][1-0.3000]})
    /{[1+0.0500][0.1900+0.5800]}
= 48,851 millions USD

Law-3022:

If both (**f**= F/S= Fixed Portion= 58.00%), (**H**= Home Taken Dividend= 1,659 millions USD), (**$**= Sales or Revenues= 51,294 millions USD), (**y**= Y/D= Yielding Tax Rate of Dividend= 30.00%), (**v**= V/S= Variable Portion= 19.00%), (**t**= T/B= Tax Rate= 30.00%), (**i**= I/S= Interest Portion= 1.00%), (**$'**= Sales of Past Year= 48,851 millions USD), and (**d**= D/A= Dividend Portion or Payout= 30.00%), are known, then it's (**s**= Sales Growth planned) is:

s= ($[1-**i**]-**H**/{**d**[1-**y**][1-**t**]})/{**$'**[**v**+**f**]}-1
= (51,294[1-0.0100]-1,659/{0.3000
    [1-0.3000][1-0.3000]})
    /{48,851[0.1900+0.5800]}-1
= 5.00%

Finance Construction-4, *Tim Asikin, Steve Asikin, Indra Senihardja*

Law-3023:

If both (**f**= F/S= Fixed Portion= 58.00%), (**H**= Home Taken Dividend= 1,659 millions USD), (**$**= Sales or Revenues= 51,294 millions USD), (**y**= Y/D= Yielding Tax Rate of Dividend= 30.00%), (**s**= [S/S']-1= Sales Growth= 5.00%), (**t**= T/B= Tax Rate= 30.00%), (**i**= I/S= Interest Portion= 1.00%), (**$'**= Sales of Past Year= 48,851 millions USD), and (**d**= D/A= Dividend Portion or Payout= 30.00%), are known, then it's (**v**= Variable Portion planned) is:

$\quad$ **v**= (**$**[1-**i**]-**H**/{**d**[1-**y**][1-**t**]})/{**$'**[1+**s**]}-**f**
$\quad\quad$ = (51,294[1-0.0100]-1,659/{0.3000
$\quad\quad\quad\quad$ [1-0.3000][1-0.3000]})
$\quad\quad\quad\quad$ /{48,851[1+0.0500]}-0.5800
$\quad\quad$ = 19.00%

Law-3024:

If both (**v**= V/S= Variable Portion= 19.00%), (**H**= Home Taken Dividend= 1,659 millions USD), (**$**= Sales or Revenues= 51,294 millions USD), (**y**= Y/D= Yielding Tax Rate of Dividend= 30.00%), (**s**= [S/S']-1= Sales Growth= 5.00%), (**t**= T/B= Tax Rate= 30.00%), (**i**= I/S= Interest Portion= 1.00%), (**$'**= Sales of Past Year= 48,851 millions USD), and (**d**= D/A= Dividend Portion or Payout= 30.00%), are known, then it's (**f**= Fixed Portion planned) is:

$\quad$ **f**= (**$**[1-**i**]-**H**/{**d**[1-**y**][1-**t**]})/{**$'**[1+**s**]}-**v**
$\quad\quad$ = (51,294[1-0.0100]-1,659/{0.3000
$\quad\quad\quad\quad$ [1-0.3000][1-0.3000]})
$\quad\quad\quad\quad$ /{48,851[1+0.0500]}-0.1900
$\quad\quad$ = 58.00%

<u>Law-3025</u>:

If both (**v**= V/S= Variable Portion= <u>19.00%</u>), (**f**= F/S= Fixed Portion= <u>58.00%</u>), (**$**= Sales or Revenues= <u>51,294</u> millions USD), (**y**= Y/D= Yielding Tax Rate of Dividend= <u>30.00%</u>), (**s**= [S/S']-1= Sales Growth= <u>5.00%</u>), (**t**= T/B= Tax Rate= <u>30.00%</u>), (**i**= I/S= Interest Portion= <u>1.00%</u>), (**$'**= Sales of Past Year= <u>48,851</u> millions USD), and (**d**= D/A= Dividend Portion or Payout= 30.00%), are known, then it's (**H**= Home Taken Dividend planned) is:

$$\begin{aligned}\mathbf{H} &= \mathbf{d}[1-\mathbf{y}][1-\mathbf{t}]\{\$-\$'\mathbf{v}[1+\mathbf{s}]-\$'\mathbf{f}[1+\mathbf{s}]-\$'\mathbf{i}[1+\mathbf{s}]\}\\ &= \mathbf{d}[1-\mathbf{y}][1-\mathbf{t}]\{\$-\$'[1+\mathbf{s}][\mathbf{v}+\mathbf{f}+\mathbf{i}]\}\\ &= 0.3000[1-0.3000][1-0.3000]\{51,294\\ &\quad -48,851[1+0.0500][0.1900\\ &\quad +0.5800+0.0100]\}\\ &= \underline{1,659} \text{ millions USD}\end{aligned}$$

Law-3026:

If both (v= V/S= Variable Portion= 19.00%), (f= F/S= Fixed Portion= 58.00%), ($\$$= Sales or Revenues= 51,294 millions USD), (y= Y/D= Yielding Tax Rate of Dividend= 30.00%), (s= [S/S']-1= Sales Growth= 5.00%), (t= T/B= Tax Rate= 30.00%), (i= I/S= Interest Portion= 1.00%), ($\$'$= Sales of Past Year= 48,851 millions USD),and (H= Home Taken Dividend= 1,659 millions USD),are known, then it's (d= Dividend Portion or Payout planned) is:

d = H/([1-y][1-t]{$\$$-$\$'v$[1+$s$]-$\$'f$[1+$s$]-$\$'i$[1+$s$]})
 = H/([1-y][1-t]{$\$$-$\$'$[1+$s$][$v$+$f$+$i$]})
 = 1,659/([1-0.3000][1-0.3000]{51,294
  -48,851[1+0.0500][1-0.1900
  -0.5800-0.0100]})
 = 30.00%

Law-3027:

If both (v= V/S= Variable Portion= 19.00%), (f= F/S= Fixed Portion= 58.00%), (S= Sales or Revenues= 51,294 millions USD), (d= D/A= Dividend Portion or Payout= 30.00%), (s= [S/S']-1= Sales Growth= 5.00%), (t= T/B= Tax Rate= 30.00%), (i= I/S= Interest Portion= 1.00%), (S'= Sales of Past Year= 48,851 millions USD), and (H= Home Taken Dividend= 1,659 millions USD), are known, then it's (y= Yielding Tax Rate of Dividend planned), is:

$$y = 1 - H/(d[1-t]\{S - S'v[1+s] - S'f[1+s] - S'i[1+s]\})$$
$$= 1 - H/(d[1-t]\{S - S'[1+s][v+f+i]\})$$
$$= 1 - 1{,}659/(0.3000[1-0.3000]\{51{,}294$$
$$-48{,}851[1+0.0500][1-0.1900$$
$$-0.5800-0.0100]\})$$
$$= 30.00\%$$

Finance Construction-4, *Tim Asikin, Steve Asikin, Indra Senihardja*

Law-3028:

If both (**v**= V/S= Variable Portion= 19.00%), (**f**= F/S= Fixed Portion= 58.00%), (**$**= Sales or Revenues= 51,294 millions USD), (**d**= D/A= Dividend Portion or Payout= 30.00%), (**s**= [S/S']-1= Sales Growth= 5.00%), (**y**= Y/D= Yielding Tax Rate of Dividend= 30.00%), (**i**= I/S= Interest Portion= 1.00%), (**$'**= Sales of Past Year= 48,851 millions USD), and (**H**= Home Taken Dividend= 1,659 millions USD), are known, then it's (**t**= Tax Rate planned) is:

t = 1-**H**/(**d**[1-**y**]{**$**-**$'v**[1+**s**]-**$'f**[1+**s**]-**$'i**[1+**s**]})
　 = 1-**H**/(**d**[1-**y**]{**$**-**$'**[1+**s**][**v**+**f**+**i**]})
　 = 1-1,659/(0.3000[1-0.3000]{51,294
　　　 -48,851[1+0.0500][0.1900
　　　 +0.5800+0.0100]})
　 = 30.00%

Finance Construction-4, *Tim Asikin, Steve Asikin, Indra Senihardja*

Law-3029:

If both (v= V/S= Variable Portion= 19.00%), (f= F/S= Fixed Portion= 58.00%), (t= T/B= Tax Rate= 30.00%), (d= D/A= Dividend Portion or Payout= 30.00%), (s= [S/S']-1= Sales Growth= 5.00%), (y= Y/D= Yielding Tax Rate of Dividend= 30.00%), (i= I/S= Interest Portion= 1.00%), ($\$'$= Sales of Past Year= 48,851 millions USD), and (H= Home Taken Dividend= 1,659 millions USD), are known, then it's ($\$$= Sales or Revenues planned) is:

$$\$ = \$'[1+s][v+f+i]+H/\{d[1-y][1-t]\}$$
$$= 48,851[1+0.0500][0.1900+0.5800$$
$$+0.0100]+1,659/\{0.3000$$
$$[1-0.3000][1-0.3000]\}$$
$$= \underline{51,294} \text{ millions USD}$$

Law-3030:

If both (v= V/S= Variable Portion= 19.00%), (f= F/S= Fixed Portion= 58.00%), (t= T/B= Tax Rate= 30.00%), (d= D/A= Dividend Portion or Payout= 30.00%), (s= [S/S']-1= Sales Growth= 5.00%), (y= Y/D= Yielding Tax Rate of Dividend= 30.00%), (i= I/S= Interest Portion= 1.00%), ($\$$= Sales or Revenues= 51,294 millions USD), and (H= Home Taken Dividend= 1,659 millions USD),are known, then it's ($\$'$= Sales Past) must be:

$$\$' = (\$-H/\{d[1-y][1-t]\})/\{[1+s][v+f+i]\}$$
$$= (51,294-1,659/\{0.3000[1-0.3000]$$
$$[1-0.3000]\}) /\{[1+0.0500]$$
$$[0.1900+0.5800+0.0100]\}$$
$$= \underline{48,851} \text{ millions USD}$$

Law-3031:

If both (**v**= V/S= Variable Portion= 19.00%), (**f**= F/S= Fixed Portion= 58.00%), (**t**= T/B= Tax Rate= 30.00%), (**d**= D/A= Dividend Portion or Payout= 30.00%), (**$'**= Sales of Past Year= 48,851 millions USD), (**y**= Y/D= Yielding Tax Rate of Dividend= 30.00%), (**i**= I/S= Interest Portion= 1.00%), (**$**= Sales or Revenues= 51,294 millions USD), and (**H**= Home Taken Dividend= 1,659 millions USD), are known, then it's (**s**= Sales Growth planned) is:

$s = (\$-H/\{d[1-y][1-t]\})/\{\$'[v+f+i]\}-1$
$\quad = (51,294-1,659/\{0.3000[1-0.3000][1-0.3000]\})/\{48,851[0.1900+0.5800+0.0100]\}-1$
$\quad = 5.00\%$

Law-3032:

If both (s= [S/S']-1= Sales Growth= 5.00%), (f= F/S= Fixed Portion= 58.00%), (t= T/B= Tax Rate= 30.00%), (d= D/A= Dividend Portion or Payout= 30.00%), (S'= Sales of Past Year= 48,851 millions USD), (y= Y/D= Yielding Tax Rate of Dividend= 30.00%), (i= I/S= Interest Portion= 1.00%), (S= Sales or Revenues= 51,294 millions USD), and (H= Home Taken Dividend= 1,659 millions USD), are known, then it's (v= Variable Portion planned) is:

v= (S-H/{d[1-y][1-t]})/{S'[1+s]}-f-i

= 1-0.5800-0.0100-(51,294-1,659
/{0.3000[1-0.3000][1-0.3000]})
/{48,851[1+0.0500]}-0.5800
-0.0100

= 19.00%

Law-3033:

If both (**s**= [S/S']-1= Sales Growth= 5.00%), (**v**= V/S= Variable Portion= 19.00%), (**t**= T/B= Tax Rate= 30.00%), (**d**= D/A= Dividend Portion or Payout= 30.00%), (**$'**= Sales of Past Year= 48,851 millions USD), (**y**= Y/D= Yielding Tax Rate of Dividend= 30.00%), (**i**= I/S= Interest Portion= 1.00%), (**$**= Sales or Revenues= 51,294 millions USD), and (**H**= Home Taken Dividend= 1,659 millions USD), are known, then it's (**v**= Variable Portion planned) is:

v= (**$**-**H**/{**d**[1-**y**][1-**t**]})/{**$'**[1+**s**]}-**v**-**i**
  = (1-0.5800-0.0100-(51,294-1,659
       /{0.3000[1-0.3000]
       [1-0.3000]})/{48,851
       [1+0.0500]}-0.1900-0.0100
  = 58.00%

<u>Law-3034</u>:

If both (**s**= [S/S']-1= Sales Growth= <u>5.00</u>%), (**v**= V/S= Variable Portion= <u>19.00</u>%), (**t**= T/B= Tax Rate= <u>30.00</u>%), (**d**= D/A= Dividend Portion or Payout= 30.00%), (**S'**= Sales of Past Year= <u>48,851</u> millions USD), (**y**= Y/D= Yielding Tax Rate of Dividend= <u>30.00</u>%), (**f**= F/S= Fixed Portion= <u>58.00</u>%), (**S**= Sales or Revenues= <u>51,294</u> millions USD), and (**H**= Home Taken Dividend= <u>1,659</u> millions USD), are known, then it's (**i**= Interest Portion planned) is:

i= (**S**-**H**/{**d**[1-**y**][1-**t**]})/{**S'**[1+**s**]}-**v**-**f**
    = (1-0.5800-0.0100-(51,294-1,659
        /{0.3000[1-0.3000]
        [1-0.3000]}) /{48,851
        [1+0.0500]}-0.1900-0.5800
    = <u>1.00</u>%

Law-3035:
If both (s= [S/S']-1= Sales Growth= 5.00%), (V= Variable Cost= 9,746 millions USD), (t= T/B= Tax Rate= 30.00%), (d= D/A= Dividend Portion or Payout= 30.00%), (S'= Sales of Past Year= 48,851 millions USD), (y= Y/D= Yielding Tax Rate of Dividend= 30.00%), (F= Fixed Cost= 29,750 millions USD), and (I= Interest Expense= 513 millions USD), are known, then it's (H= Home Taken Dividend planned) is:

$H = d[1-y][1-t]\{S'[1+s]-V-F-I\}$
  = 0.3000[1-0.3000][1-0.3000]{48,851
      [1+0.0500]-9,746-29,750-513}
  = 1,659 millions USD

Law-3036:
If both (s= [S/S']-1= Sales Growth= 5.00%), (V= Variable Cost= 9,746 millions USD), (t= T/B= Tax Rate= 30.00%), (H= Home Taken Dividend= 1,659 millions USD), (S'= Sales of Past Year= 48,851 millions USD), (y= Y/D= Yielding Tax Rate of Dividend= 30.00%), (F= Fixed Cost= 29,750 millions USD), and (I= Interest Expense= 513 millions USD), are known, then it's (d= Dividend Portion or Payout planned) is:

$d = H/([1-y][1-t]\{S'[1+s]-V-F-I\}$
  = 1,659/([1-0.3000][1-0.3000]{48,851
      [1+0.0500]-9,746-29,750-513})
  = 30.00%

Law-3037:

If both (s= [S/S']-1= Sales Growth= 5.00%), (V= Variable Cost= 9,746 millions USD), (t= T/B= Tax Rate= 30.00%), (H= Home Taken Dividend= 1,659 millions USD), (S'= Sales of Past Year= 48,851 millions USD), (d= D/A= Dividend Portion or Payout= 30.00%), (F= Fixed Cost= 29,750 millions USD), and (I= Interest Expense= 513 millions USD), are known, then it's (y= Yielding Tax Rate of Dividend planned) is:

y= 1-H/(d[1-t]{S'[1+s]-V-F-I}
 = 1-1,659/(0.3000[1-0.3000]{48,851
  [1+0.0500]-9,746-29,750-513})
 = 30.00%

Law-3038:

If both (s= [S/S']-1= Sales Growth= 5.00%), (V= Variable Cost= 9,746 millions USD), (y= Y/D= Yielding Tax Rate of Dividend= 30.00%), (H= Home Taken Dividend= 1,659 millions USD), (S'= Sales of Past Year= 48,851 millions USD), (d= D/A= Dividend Portion or Payout= 30.00%), (F= Fixed Cost= 29,750 millions USD), and (I= Interest Expense= 513 millions USD), are known, then it's (t= Tax Rate planned) is:

t= 1-H/(d[1-t]{S'[1+s]-V-F-I}
 = 1-1,659/(0.3000[1-0.3000]{48,851
  [1+0.0500]-9,746-29,750-513})
 = 30.00%

Law-3039:

If both (**s**= [S/S']-1= Sales Growth= 5.00%), (**V**= Variable Cost= 9,746 millions USD), (**y**= Y/D= Yielding Tax Rate of Dividend= 30.00%), (**H**= Home Taken Dividend= 1,659 millions USD), (**t**= T/B= Tax Rate= 30.00%), (**d**= D/A= Dividend Portion or Payout= 30.00%), (**F**= Fixed Cost= 29,750 millions USD), and (**I**= Interest Expense= 513 millions USD), are known, then it's (**S'**= Sales Past) must be:

$$S' = (V+F+I+H/\{d[1-y][1-t]\})/[1+s]$$
$$= (9{,}746+29{,}750+513+1{,}659/(0.3000$$
$$[1-0.3000][1-0.3000]\})$$
$$/[1+0.0500]$$
$$= 48{,}851 \text{ millions USD}$$

Law-3040:

If both (**S'**= Sales of Past Year= 48,851 millions USD), (**V**= Variable Cost= 9,746 millions USD), (**y**= Y/D= Yielding Tax Rate of Dividend= 30.00%), (**H**= Home Taken Dividend= 1,659 millions USD), (**t**= T/B= Tax Rate= 30.00%), (**d**= D/A= Dividend Portion or Payout= 30.00%), (**F**= Fixed Cost= 29,750 millions USD), and (**I**= Interest Expense= 513 millions USD), are known, then it's (**s**= Sales Growth planned) is:

$$s = (V+F+I+H/\{d[1-y][1-t]\})/S' - 1$$
$$= (9{,}746+29{,}750+513+1{,}659/(0.3000$$
$$[1-0.3000][1-0.3000]\})$$
$$/48{,}851-1$$
$$= 5.00\%$$

Law-3041:

If both ($'= Sales of Past Year= 48,851 millions USD), (s= [S/S']-1= Sales Growth= 5.00%), (y= Y/D= Yielding Tax Rate of Dividend= 30.00%), (H= Home Taken Dividend= 1,659 millions USD), (t= T/B= Tax Rate= 30.00%), (d= D/A= Dividend Portion or Payout= 30.00%), (F= Fixed Cost= 29,750 millions USD), and (I= Interest Expense= 513 millions USD), are known, then it's (V= Variable Cost planned) is:

$$V = \$'[1+s] - F - I - H/\{d[1-y][1-t]\}$$
$$= 48,851[1+0.0500] - 29,750 - 513$$
$$\quad - 1,659/\{0.3000[1-0.3000][1-0.3000]\}$$
$$= 9,746 \text{ millions USD}$$

Law-3042:

If both ($'= Sales of Past Year= 48,851 millions USD), (s= [S/S']-1= Sales Growth= 5.00%), (y= Y/D= Yielding Tax Rate of Dividend= 30.00%), (H= Home Taken Dividend= 1,659 millions USD), (t= T/B= Tax Rate= 30.00%), (d= D/A= Dividend Portion or Payout= 30.00%), (V= Variable Cost= 9,746 millions USD), and (I= Interest Expense= 513 millions USD), are known, then it's (F= Fixed Cost planned) is:

$$F = \$'[1+s] - V - I - H/\{d[1-y][1-t]\}$$
$$= 48,851[1+0.0500] - 9,746 - 513 - 1,659$$
$$\quad /\{0.3000[1-0.3000][1-0.3000]\}$$
$$= 29,750 \text{ millions USD}$$

Law-3043:

If both ($'$= Sales of Past Year= 48,851 millions USD), (s= [S/S']-1= Sales Growth= 5.00%), (y= Y/D= Yielding Tax Rate of Dividend= 30.00%), (H= Home Taken Dividend= 1,659 millions USD), (t= T/B= Tax Rate= 30.00%), (d= D/A=Dividend Portion or Payout= 30.00%), (V= Variable Cost= 9,746 millions USD), and (F= Fixed Cost= 29,750 millions USD), are known, then it's (I= Interest Expense planned) is:

$I = S'[1+s] - V - F - H / \{d[1-y][1-t]\}$
  = 48,851[1+0.0500] - 9,746 - 29,750
        -1,659/{0.3000[1-0.3000]
        [1-0.3000]}
  = 513 millions USD

Law-3044:

If both (S'= Sales of Past Year= 48,851 millions USD), (s= [S/S']-1= Sales Growth= 5.00%), (y= Y/D= Yielding Tax Rate of Dividend= 30.00%), (i= I/S= Interest Portion= 1.00%), (S= Sales or Revenues= 51,294 millions USD), (t= T/B= Tax Rate= 30.00%), (d= D/A= Dividend Portion or Payout= 30.00%), (V= Variable Cost= 9,746 millions USD), and (F= Fixed Cost= 29,750 millions USD), are known, then it's (H= Home Taken Dividend planned) is:

$H = d[1-y][1-t]\{S'[1+s]-V-F-Si\}$
  = 0.3000[1-0.3000][1-0.3000]{48,851[1+0.0500]-9,746-29,750-51,294*0.0100}
  = 1,659 millions USD

Law-3045:

If both ($'$= Sales of Past Year= 48,851 millions USD), (s= [S/S']-1= Sales Growth= 5.00%), (y= Y/D= Yielding Tax Rate of Dividend= 30.00%), (i= I/S= Interest Portion= 1.00%), ($\$$= Sales or Revenues= 51,294 millions USD), (t= T/B= Tax Rate= 30.00%), (H= Home Taken Dividend= 1,659 millions USD), (V= Variable Cost= 9,746 millions USD), and (F= Fixed Cost= 29,750 millions USD), are known, then it's (d= Dividend Portion or Payout planned) is:

$d = H/([1-y][1-t]\{S'[1+s]-V-F-\$i\})$
$= 1,659/([1-0.3000][1-0.3000]\{48,851[1+0.0500]-9,746-29,750-51,294*0.0100\})$
$= 30.00\%$

Law-3046:

If both (S'= Sales of Past Year= 48,851 millions USD), (s= [S/S']-1= Sales Growth= 5.00%), (d= D/A= Dividend Portion or Payout= 30.00%), (i= I/S= Interest Portion= 1.00%), (S= Sales or Revenues= 51,294 millions USD), (t= T/B= Tax Rate= 30.00%), (H= Home Taken Dividend= 1,659 millions USD), (V= Variable Cost= 9,746 millions USD), and (F= Fixed Cost= 29,750 millions USD), are known, then it's (y= Yielding Tax Rate of Dividend planned) is:

y= 1-H/(d[1-t]{S'[1+s]-V-F-Si})
   = 1-1,659/(0.3000[1-0.3000]{48,851
        [1+0.0500]-9,746-29,750
        -51,294*0.0100})
   = 30.00%

Law-3047:

If both (S'= Sales of Past Year= 48,851 millions USD), (s= [S/S']-1= Sales Growth= 5.00%), (d= D/A=Dividend Portion or Payout= 30.00%), (i= I/S= Interest Portion= 1.00%), (S= Sales or Revenues= 51,294 millions USD), (y= Y/D= Yielding Tax Rate of Dividend= 30.00%), (H= Home Taken Dividend= 1,659 millions USD), (V= Variable Cost= 9,746 millions USD), and (F= Fixed Cost= 29,750 millions USD), are known, then it's (t= Tax Rate planned) is:

t = 1-H/(d[1-y]{S'[1+s]-V-F-Si})
= 1-1,659/(0.3000[1-0.3000]{48,851
 [1+0.0500]-9,746-29,750
 -51,294*0.0100})
= 30.00%

Law-3048:

If both (t= T/B= Tax Rate= 30.00%), (s= [S/S']-1= Sales Growth= 5.00%), (d= D/A= Dividend Portion or Payout= 30.00%), (i= I/S= Interest Portion= 1.00%), (S= Sales or Revenues= 51,294 millions USD), (y= Y/D= Yielding Tax Rate of Dividend= 30.00%), (H= Home Taken Dividend= 1,659 millions USD), (V= Variable Cost= 9,746 millions USD), and (F= Fixed Cost= 29,750 millions USD), are known, then it's (S'= Sales Past) must be:

S'= (V+F+Si+H/{d[1-y][1-t]})/[1+s]
= (9,746+29,750+51,294*0.0100
 +1,659/{0.3000[1-0.3000]
 [1-0.3000]})/[1+0.0500]
= 48,851 millions USD

Law-3049:

If both (**t**= T/B= Tax Rate= 30.00%), (**S'**= Sales of Past Year= 48,851 millions USD), (**d**= D/A= Dividend Portion or Payout= 30.00%), (**i**= I/S= Interest Portion= 1.00%), (**S**= Sales or Revenues= 51,294 millions USD), (**y**= Y/D= Yielding Tax Rate of Dividend= 30.00%), (**H**= Home Taken Dividend= 1,659 millions USD), (**V**= Variable Cost= 9,746 millions USD), and (**F**= Fixed Cost= 29,750 millions USD), are known, then it's (**s**= Sales Growth planned) is:

$$s = (V+F+Si+H/\{d[1-y][1-t]\})/S' - 1$$
$$= (9,746+29,750+51,294*0.0100$$
$$+1,659/\{0.3000[1-0.3000]$$
$$[1-0.3000]\})/48,851 - 1$$
$$= 5.00\%$$

Finance Construction-4, *Tim Asikin, Steve Asikin, Indra Senihardja*

Law-3050:

If both (t= T/B= Tax Rate= 30.00%), (S'= Sales of Past Year= 48,851 millions USD), (d= D/A= Dividend Portion or Payout= 30.00%), (i= I/S= Interest Portion= 1.00%), (S= Sales or Revenues= 51,294 millions USD), (y= Y/D= Yielding Tax Rate of Dividend= 30.00%), (H= Home Taken Dividend= 1,659 millions USD), (s= [S/S']-1= Sales Growth= 5.00%), and (F= Fixed Cost= 29,750 millions USD), are known, then it's (V= Variable Cost planned) is:

$V = S'[1+s] - F - Si - H/\{d[1-y][1-t]\}$

$= 48,851[1+0.0500] - 29,750 - 51,294$
$\quad *0.0100 - 1,659/\{0.3000$
$\quad [1-0.3000][1-0.3000]\}$

$= 9,746$ millions USD

Law-3051:

If both (**t**= T/B= Tax Rate= 30.00%), (**S'**= Sales of Past Year= 48,851 millions USD), (**d**= D/A= Dividend Portion or Payout= 30.00%), (**i**= I/S= Interest Portion= 1.00%), (**S**= Sales or Revenues= 51,294 millions USD), (**y**= Y/D= Yielding Tax Rate of Dividend= 30.00%), (**H**= Home Taken Dividend= 1,659 millions USD), (**s**= [S/S']-1= Sales Growth= 5.00%), and (**V**= Variable Cost= 9,746 millions USD), are known, then it's (**F**= Fixed Cost planned) is:

$$F = S'[1+s] - V - Si - H/\{d[1-y][1-t]\}$$
$$= 48,851[1+0.0500] - 9,746 - 51,294$$
$$*0.0100 - 1,659/\{0.3000$$
$$[1-0.3000][1-0.3000]\}$$
$$= \underline{29,750} \text{ millions USD}$$

Law-3052:

If both (**t**= T/B= Tax Rate= 30.00%), (**$'**= Sales of Past Year= 48,851 millions USD), (**d**= D/A= Dividend Portion or Payout= 30.00%), (**i**= I/S= Interest Portion= 1.00%), (**F**= Fixed Cost= 29,750 millions USD), (**y**= Y/D= Yielding Tax Rate of Dividend= 30.00%), (**H**= Home Taken Dividend= 1,659 millions USD), (**s**= [S/S']-1= Sales Growth= 5.00%), and (**V**= Variable Cost= 9,746 millions USD), are known, then it's (**$**= Sales or Revenues planned) is:

$$\$ = (\$'[1+s]-V-F-H/\{d[1-y][1-t]\})/i$$
$$= 48,851[1+0.0500]-9,746-29,750$$
$$-1,659/\{0.3000[1-0.3000]$$
$$[1-0.3000]\})/0.0100$$
$$= 51,294 \text{ millions USD}$$

Finance Construction-4, *Tim Asikin, Steve Asikin, Indra Senihardja*

Law-3053:

If both (t= T/B= Tax Rate= 30.00%), (S'= Sales of Past Year= 48,851 millions USD), (d= D/A= Dividend Portion or Payout= 30.00%), (S= Sales or Revenues= 51,294 millions USD),(F= Fixed Cost= 29,750 millions USD), (y= Y/D= Yielding Tax Rate of Dividend= 30.00%), (H= Home Taken Dividend= 1,659 millions USD), (s= [S/S']-1= Sales Growth= 5.00%), and (V= Variable Cost= 9,746 millions USD), are known, then it's (i= Interest Portion planned) is:

i= (S'[1+s]-V-F-H/{d[1-y][1-t]})/S
  = 48,851[1+0.0500]-9,746-29,750
      -1,659/{0.3000[1-0.3000]
      [1-0.3000]})/51,294
  = 1.00%

Finance Construction-4, *Tim Asikin, Steve Asikin, Indra Senihardja*

Law-3054:

If both (**t**= T/B= Tax Rate= 30.00%), (**$'**= Sales of Past Year= 48,851 millions USD), (**d**= D/A= Dividend Portion or Payout= 30.00%), (**F**= Fixed Cost= 29,750 millions USD), (**y**= Y/D= Yielding Tax Rate of Dividend= 30.00%), (**i**= I/S= Interest Portion= 1.00%), (**s**= [S/S']-1= Sales Growth= 5.00%), and (**V**= Variable Cost= 9,746 millions USD), are known, then it's (**H**= Home Taken Dividend planned) is:

$$H = d[1-y][1-t]\{\$'[1+s]-V-F-\$'i[1+s]\}$$
$$= d[1-y][1-t]\{\$'[1+s][1-i]-V-F\}$$
$$= 0.3000[1-0.3000][1-0.3000]\{48,851$$
$$[1+0.0500][1-0.0100]-9,746$$
$$-29,750\}$$
$$= \underline{1,659} \text{ millions USD}$$

Finance Construction-4, *Tim Asikin, Steve Asikin, Indra Senihardja*

<u>Law-3055</u>:

If both (**t**= T/B= Tax Rate= <u>30.00</u>%), (**S'**= Sales of Past Year= <u>48,851</u> millions USD), (**H**= Home Taken Dividend= <u>1,659</u> millions USD), (**F**= Fixed Cost= <u>29,750</u> millions USD), (**y**= Y/D= Yielding Tax Rate of Dividend= <u>30.00</u>%), (**i**= I/S= Interest Portion= <u>1.00</u>%), (**s**= [S/S']-1= Sales Growth= <u>5.00</u>%), and (**V**= Variable Cost= <u>9,746</u> millions USD), are known, then it's (**d**= Dividend Portion or Payout planned) is:

$$\begin{aligned}
\mathbf{d} &= \mathbf{H}/([1-\mathbf{y}][1-\mathbf{t}]\{\mathbf{S'}[1+\mathbf{s}]-\mathbf{V}-\mathbf{F}-\mathbf{S'i}[1+\mathbf{s}]\}) \\
&= \mathbf{H}/([1-\mathbf{y}][1-\mathbf{t}]\{\mathbf{S'}[1+\mathbf{s}][1-\mathbf{i}]-\mathbf{V}-\mathbf{F}\}) \\
&= 1,659/([1-0.3000][1-0.3000]\{48,851 \\
&\qquad [1+0.0500][1-0.0100]-9,746 \\
&\qquad -29,750\}) \\
&= \underline{30.00}\%
\end{aligned}$$

Finance Construction-4, *Tim Asikin, Steve Asikin, Indra Senihardja*

Law-3056:

If both (t= T/B= Tax Rate= 30.00%), (S'= Sales of Past Year= 48,851 millions USD), (H= Home Taken Dividend= 1,659 millions USD), (F= Fixed Cost= 29,750 millions USD), (d= D/A= Dividend Portion or Payout= 30.00%), (i= I/S= Interest Portion= 1.00%), (s= [S/S']-1= Sales Growth= 5.00%), and (V= Variable Cost= 9,746 millions USD), are known, then it's (y= Yielding Tax Rate of Dividend planned) is:

$$y = 1-H/(d[1-t]\{S'[1+s]-V-F-S'i[1+s]\})$$
$$= 1-H/(d[1-t]\{S'[1+s][1-i]-V-F\})$$
$$= 1-1{,}659/(0.3000[1-0.3000]\{48{,}851 [1+0.0500][1-0.0100]-9{,}746 -29{,}750\})$$
$$= 30.00\%$$

Law-3057:

If both (**y**= Y/D= Yielding Tax Rate of Dividend= 30.00%), (**S'**= Sales of Past Year= 48,851 millions USD), (**H**= Home Taken Dividend= 1,659 millions USD), (**F**= Fixed Cost= 29,750 millions USD), (**d**= D/A= Dividend Portion or Payout= 30.00%), (**i**= I/S= Interest Portion= 1.00%), (**s**= [S/S']-1= Sales Growth= 5.00%), and (**V**= Variable Cost= 9,746 millions USD), are known, then it's (**t**= Tax Rate planned) is:

$$t = 1-H/(d[1-y]\{S'[1+s][1-i]-V-F\})$$
$$= 1-1,659/(0.3000\,[1-0.3000]\{48,851\,[1+0.0500][1-0.0100]-9,746\,-29,750\})$$
$$= 30.00\%$$

Law-3058:

If both (**y**= Y/D= Yielding Tax Rate of Dividend= 30.00%), (**t**= T/B= Tax Rate= 30.00%), (**H**= Home Taken Dividend= 1,659 millions USD), (**F**= Fixed Cost= 29,750 millions USD), (**d**= D/A= Dividend Portion or Payout= 30.00%), (**i**= I/S= Interest Portion= 1.00%), (**s**= [S/S']-1= Sales Growth= 5.00%), and (**V**= Variable Cost= 9,746 millions USD), are known, then it's (**S'**= Sales Past) must be:

$$S' = (H/\{d[1-y][1-t]\}-V-F)/\{[1+s][1-i]\}$$
$$= (1,659/\{0.3000[1-0.3000]\,[1-0.3000]\})+9,746+29,750\})\,/\{[1+0.0500][1-0.0100]\}$$
$$= 48,851 \text{ millions USD}$$

Law-3059:

If both (**y**= Y/D= Yielding Tax Rate of Dividend= 30.00%), (**t**= T/B= Tax Rate= 30.00%), (**H**= Home Taken Dividend= 1,659 millions USD), (**F**= Fixed Cost= 29,750 millions USD), (**d**= D/A= Dividend Portion or Payout= 30.00%), (**i**= I/S= Interest Portion= 1.00%), (**$'**= Sales of Past Year= 48,851 millions USD), and (**V**= Variable Cost= 9,746 millions USD), are known, then it's (**$**= Sales Growth planned) is:

$$s= (H/\{d[1-y][1-t]\}-V-F)/\{\$'[1-i]\}-1$$
$$= (1,659/\{0.3000[1-0.3000]$$
$$[1-0.3000]\})+9,746+29,750\})$$
$$/\{48,851[1-0.0100]\}-1$$
$$= 5.00\%$$

Law-3060:

If both (**y**= Y/D= Yielding Tax Rate of Dividend= 30.00%), (**t**= T/B= Tax Rate= 30.00%), (**H**= Home Taken Dividend= 1,659 millions USD), (**F**= Fixed Cost= 29,750 millions USD), (**d**= D/A= Dividend Portion or Payout= 30.00%), (**s**= [S/S']-1= Sales Growth= 5.00%), (**$'**= Sales of Past Year= 48,851 millions USD), and (**V**= Variable Cost= 9,746 millions USD), are known, then it's (**i**= Interest Portion planned) is:

$$i= 1-(H/\{d[1-y][1-t]\}-V-F)/\{\$'[1+s]\}$$
$$= (1,659/\{0.3000[1-0.3000]$$
$$[1-0.3000]\})+9,746+29,750\})$$
$$/\{48,851[1+0.0500]\}$$
$$= 1.00\%$$

Law-3061:

If both (**y**= Y/D= Yielding Tax Rate of Dividend= 30.00%), (**t**= T/B= Tax Rate= 30.00%), (**H**= Home Taken Dividend= 1,659 millions USD),(**F**= Fixed Cost= 29,750 millions USD), (**d**= D/A= Dividend Portion or Payout= 30.00%), (**s**= [S/S']-1= Sales Growth= 5.00%), (**S'**= Sales of Past Year= 48,851 millions USD), and (**i**= I/S= Interest Portion= 1.00%), are known, then it's (**V**= Variable Cost planned) is:

$$V = S'[1+s][1-i] - F - H/\{d[1-y][1-t]\}$$
$$= 48,851[1+0.0500][1-0.0100] - 29,750$$
$$-1,659/\{0.3000[1-0.3000]$$
$$[1-0.3000]\}$$
$$= 9,746 \text{ millions USD}$$

Law-3062:

If both (**y**= Y/D= Yielding Tax Rate of Dividend= 30.00%), (**t**= T/B= Tax Rate= 30.00%), (**H**= Home Taken Dividend= 1,659 millions USD), (**V**= Variable Cost= 9,746 millions USD), (**d**= D/A= Dividend Portion or Payout= 30.00%), (**s**= [S/S']-1= Sales Growth= 5.00%), (**S'**= Sales of Past Year= 48,851 millions USD), and (**i**= I/S= Interest Portion= 1.00%), are known, then it's (**F**= Fixed Cost planned) is:

$$F = S'[1+s][1-i] - V - H/\{d[1-y][1-t]\}$$
$$= 48,851[1+0.0500][1-0.0100] - 9,746$$
$$-1,659/\{0.3000[1-0.3000]$$
$$[1-0.3000]\}$$
$$= 29,750 \text{ millions USD}$$

Finance Construction-4, *Tim Asikin, Steve Asikin, Indra Senihardja*

Law-3063:

If both (y= Y/D= Yielding Tax Rate of Dividend= 30.00%), (t= T/B= Tax Rate= 30.00%), (f= F/S= Fixed Portion= 58.00%), (V= Variable Cost= 9,746 millions USD), (d= D/A= Dividend Portion or Payout= 30.00%), (s= [S/S']-1= Sales Growth= 5.00%), ($\$'$= Sales of Past Year= 48,851 millions USD), ($\$$= Sales or Revenues= 51,294 millions USD), and (I= Interest Expense= 513 millions USD), are known, then it's (H= Home Taken Dividend planned) is:

$H = d[1-y][1-t]\{\$'[1+s]-V-\$f-I\}$
 = 0.3000[1-0.3000][1-0.3000]{48,851
   [1+0.0500]-9,746-51,294
   *0.5800-513}
 = 1,659 millions USD

Law-3064:

If both (**y**= Y/D= Yielding Tax Rate of Dividend= <u>30.00</u>%), (**t**= T/B= Tax Rate= <u>30.00</u>%), (**f**= F/S= Fixed Portion= <u>58.00</u>%), (**V**= Variable Cost= <u>9,746</u> millions USD), (**H**= Home Taken Dividend= <u>1,659</u> millions USD), (**s**= [S/S']-1= Sales Growth= <u>5.00</u>%), (**S'**= Sales of Past Year= <u>48,851</u> millions USD), (**S**= Sales or Revenues= <u>51,294</u> millions USD), and (**I**= Interest Expense= <u>513</u> millions USD), are known, then it's (**d**= Dividend Portion or Payout planned) is:

$$d = H/([1-y][1-t]\{S'[1+s]-V-Sf-I\})$$
$$= 1,659/([1-0.3000][1-0.3000]\{48,851[1+0.0500]-9,746-51,294*0.5800-513\})$$
$$= \underline{30.00}\%$$

Law-3065:

If both (**d**= D/A= Dividend Portion or Payout= 30.00%), (**t**= T/B= Tax Rate= 30.00%), (**f**= F/S= Fixed Portion= 58.00%), (**V**= Variable Cost= 9,746 millions USD), (**H**= Home Taken Dividend= 1,659 millions USD), (**s**= [S/S']-1= Sales Growth= 5.00%), (**S'**= Sales of Past Year= 48,851 millions USD), (**S**= Sales or Revenues= 51,294 millions USD), and (**I**= Interest Expense= 513 millions USD), are known, then it's (**y**= Yielding Tax Rate of Dividend planned) is:

$$\begin{aligned}\mathbf{y} &= 1\text{-}\mathbf{H}/(\mathbf{d}[1\text{-}\mathbf{t}]\{\mathbf{S'}[1+\mathbf{s}]\text{-}\mathbf{V}\text{-}\mathbf{Sf}\text{-}\mathbf{I}\} \\ &= 1\text{-}1{,}659/(0.3000[1\text{-}0.3000]\{48{,}851 \\ &\qquad [1+0.0500]\text{-}9{,}746\text{-}51{,}294 \\ &\qquad *0.5800\text{-}513\} \\ &= 30.00\%\end{aligned}$$

Law-3066:

If both (**d**= D/A= Dividend Portion or Payout= 30.00%), (**y**= Y/D= Yielding Tax Rate of Dividend= 30.00%), (**f**= F/S= Fixed Portion= 58.00%), (**V**= Variable Cost= 9,746 millions USD), (**H**= Home Taken Dividend= 1,659 millions USD), (**s**= [S/S']-1= Sales Growth= 5.00%), (**S'**= Sales of Past Year= 48,851 millions USD), (**S**= Sales or Revenues= 51,294 millions USD), and (**I**= Interest Expense= 513 millions USD), are known, then it's (**t**= Tax Rate planned) is:

$$\begin{aligned}
\mathbf{t} &= 1-\mathbf{H}/(\mathbf{d}[1-\mathbf{y}]\{\mathbf{S'}[1+\mathbf{s}]-\mathbf{V}-\mathbf{Sf}-\mathbf{I}\} \\
&= 1-1,659/(0.3000[1-0.3000]\{48,851 \\
&\qquad [1+0.0500]-9,746-51,294 \\
&\qquad *0.5800-513\} \\
&= \underline{30.00\%}
\end{aligned}$$

Law-3067:

If both (**d**= D/A= Dividend Portion or Payout= 30.00%), (**y**= Y/D= Yielding Tax Rate of Dividend= 30.00%), (**f**= F/S= Fixed Portion= 58.00%), (**V**= Variable Cost= 9,746 millions USD), (**H**= Home Taken Dividend= 1,659 millions USD), (**s**= [S/S']-1= Sales Growth= 5.00%), (**t**= T/B= Tax Rate= 30.00%), (**$**= Sales or Revenues= 51,294 millions USD), and (**I**= Interest Expense= 513 millions USD), are known, then it's (**$'**= Sales Past) must be:

$$\$' = (V+\$f+I+H/\{d[1-y][1-t]\})/[1+s]$$
$$= (9{,}746+51{,}294*0.5800+513+1{,}659 \ / \{0.3000[1-0.3000][1-0.3000]\}) \ / [1+0.0500]$$
$$= \underline{48{,}851} \text{ millions USD}$$

Law-3068:

If both (**d**= D/A= Dividend Portion or Payout= 30.00%), (**y**= Y/D= Yielding Tax Rate of Dividend= 30.00%), (**f**= F/S= Fixed Portion= 58.00%), (**V**= Variable Cost= 9,746 millions USD), (**H**= Home Taken Dividend= 1,659 millions USD), (**$'**= Sales of Past Year= 48,851 millions USD), (**t**= T/B= Tax Rate= 30.00%), (**$**= Sales or Revenues= 51,294 millions USD), and (**I**= Interest Expense= 513 millions USD), are known, then it's (**s**= Sales Growth planned) is:

$$s = (V+\$f+I+H/\{d[1-y][1-t]\})/\$'-1$$
$$= (9,746+51,294*0.5800+513+1,659$$
$$/\{0.3000[1-0.3000][1-0.3000]\})$$
$$/48,851-1$$
$$= 5.00\%$$

Law-3069:

If both (**d**= D/A= Dividend Portion or Payout= 30.00%), (**y**= Y/D= Yielding Tax Rate of Dividend= 30.00%), (**f**= F/S= Fixed Portion= 58.00%), (**s**= [S/S']-1= Sales Growth= 5.00%), (**H**= Home Taken Dividend= 1,659 millions USD), (**S'**= Sales of Past Year= 48,851 millions USD), (**t**= T/B= Tax Rate= 30.00%), (**S**= Sales or Revenues= 51,294 millions USD), and (**I**= Interest Expense= 513 millions USD), are known, then it's (**V**= Variable Cost planned) is:

$$V = S'[1+s] - Sf - I - H/\{d[1-y][1-t]\}$$
$$= 48,851[1+0.0500] - 51,294*0.5800$$
$$\quad -513 - 1,659/\{0.3000[1-0.3000]$$
$$\quad [1-0.3000]\}$$
$$= 9,746 \text{ millions USD}$$

Law-3070:

If both (**d**= D/A= Dividend Portion or Payout= 30.00%), (**y**= Y/D= Yielding Tax Rate of Dividend= 30.00%), (**f**= F/S= Fixed Portion= 58.00%), (**s**= [S/S']-1= Sales Growth= 5.00%), (**H**= Home Taken Dividend= 1,659 millions USD), (**S'**= Sales of Past Year= 48,851 millions USD), (**t**= T/B= Tax Rate= 30.00%), (**V**= Variable Cost= 9,746 millions USD), and (**I**= Interest Expense= 513 millions USD), are known, then it's (**S**= Sales or Revenues planned) is:

$$S = (S'[1+s] - V - I - H/\{d[1-y][1-t]\})/f$$
$$= 48,851[1+0.0500] - 9,746 - 513 - 1,659$$
$$\quad /\{0.3000[1-0.3000][1-0.3000]\}$$
$$\quad /0.5800$$
$$= 51,294 \text{ millions USD}$$

Law-3071:

If both (**d**= D/A= Dividend Portion or Payout= 30.00%), (**y**= Y/D= Yielding Tax Rate of Dividend= 30.00%), (**$**= Sales or Revenues= 51,294 millions USD), (**s**= [S/S']-1= Sales Growth= 5.00%), (**H**= Home Taken Dividend= 1,659 millions USD), (**$'**= Sales of Past Year= 48,851 millions USD), (**t**= T/B= Tax Rate= 30.00%), (**V**= Variable Cost= 9,746 millions USD), and (**I**= Interest Expense= 513 millions USD), are known, then it's (**f**= Fixed Portion planned) is:

$$f = (\$'[1+s]-V-I-H/\{d[1-y][1-t]\})/\$$$
$$= 48,851[1+0.0500]-9,746-513-1,659$$
$$/\{0.3000[1-0.3000][1-0.3000]\}$$
$$/51,294$$
$$= \underline{58.00\%}$$

Law-3072:

If both (**d**= D/A= Dividend Portion or Payout= 30.00%), (**y**= Y/D= Yielding Tax Rate of Dividend= 30.00%), (**S**= Sales or Revenues= 51,294 millions USD), (**s**= [S/S']-1= Sales Growth= 5.00%), (**H**= Home Taken Dividend= 1,659 millions USD), (**S'**= Sales of Past Year= 48,851 millions USD), (**t**= T/B= Tax Rate= 30.00%), (**V**= Variable Cost= 9,746 millions USD), and (**f**= F/S= Fixed Portion= 58.00%), are known, then it's (**I**= Interest Expense planned) is:

$$I = S'[1+s]-V-Sf-H/\{d[1-y][1-t]\}$$
$$= 48,851[1+0.0500]-9,746-51,294$$
$$*0.5800-1,659/\{0.3000$$
$$[1-0.3000][1-0.3000]\}$$
$$= \underline{513} \text{ millions USD}$$

Law-3073:

If both (**d**= D/A= Dividend Portion or Payout= 30.00%), (**y**= Y/D= Yielding Tax Rate of Dividend= 30.00%), (**$**= Sales or Revenues= 51,294 millions USD), (**s**= [S/S']-1= Sales Growth= 5.00%), (**i**= I/S= Interest Portion= 1.00%), (**$'**= Sales of Past Year= 48,851 millions USD), (**t**= T/B= Tax Rate= 30.00%), (**V**= Variable Cost= 9,746 millions USD), and (**f**= F/S= Fixed Portion= 58.00%), are known, then it's (**H**= Home Taken Dividend planned) is:

$$H = d[1-y][1-t]\{\$'[1+s]-V-\$f-\$i\}$$
$$= d[1-y][1-t]\{\$'[1+s]-V-\$[f+i]\}$$
$$= 0.3000[1-0.3000][1-0.3000]\{48,851[1+0.0500]-9,746-51,294[0.5800+0.0100]\}$$
$$= 1,659 \text{ millions USD}$$

Law-3074:

If both (**H**= Home Taken Dividend= 1,659 millions USD), (**y**= Y/D= Yielding Tax Rate of Dividend= 30.00%), (**$**= Sales or Revenues= 51,294 millions USD), (**s**= [S/S']-1= Sales Growth= 5.00%), (**i**= I/S= Interest Portion= 1.00%), (**$'**= Sales of Past Year= 48,851 millions USD), (**t**= T/B= Tax Rate= 30.00%), (**V**= Variable Cost= 9,746 millions USD), and (**f**= F/S= Fixed Portion= 58.00%), are known, then it's (**d**= Dividend Portion or Payout planned) is:

$$\begin{aligned}
\mathbf{d} &= \mathbf{H}/([1-\mathbf{y}][1-\mathbf{t}]\{\mathbf{\$'}[1+\mathbf{s}]-\mathbf{V}-\mathbf{\$f}-\mathbf{\$i}\}) \\
&= \mathbf{H}/([1-\mathbf{y}][1-\mathbf{t}]\{\mathbf{\$'}[1+\mathbf{s}]-\mathbf{V}-\mathbf{\$}[\mathbf{f}+\mathbf{i}]\}) \\
&= 1,659/([1-0.3000][1-0.3000]\{48,851 \\
&\quad [1+0.0500]-9,746-51,294 \\
&\quad [0.5800+0.0100]\}) \\
&= 30.00\%
\end{aligned}$$

Law-3075:

If both (**H**= Home Taken Dividend= 1,659 millions USD), (**d**= D/A= Dividend Portion or Payout= 30.00%), (**$**= Sales or Revenues= 51,294 millions USD), (**s**= [S/S']-1= Sales Growth= 5.00%), (**i**= I/S= Interest Portion= 1.00%), (**$'**= Sales of Past Year= 48,851 millions USD), (**t**= T/B= Tax Rate= 30.00%), (**V**= Variable Cost= 9,746 millions USD), and (**f**= F/S= Fixed Portion= 58.00%), are known, then it's (**y**= Yielding Tax Rate of Dividend planned) is:

$$y = 1 - \mathbf{H}/(\mathbf{d}[1-\mathbf{t}]\{\mathbf{\$'}[1+\mathbf{s}] - \mathbf{V} - \mathbf{\$f} - \mathbf{\$i}\})$$
$$= 1 - \mathbf{H}/(\mathbf{d}[1-\mathbf{t}]\{\mathbf{\$'}[1+\mathbf{s}] - \mathbf{V} - \mathbf{\$}[\mathbf{f}+\mathbf{i}]\})$$
$$= 1 - 1{,}659/(0.3000[1-0.3000]\{48{,}851[1+0.0500] - 9{,}746 - 51{,}294[0.5800+0.0100]\})$$
$$= \underline{30.00\%}$$

Law-3076:

If both (**H**= Home Taken Dividend= 1,659 millions USD), (**d**= D/A= Dividend Portion or Payout= 30.00%), (**S**= Sales or Revenues= 51,294 millions USD), (**s**= [S/S']-1= Sales Growth= 5.00%), (**i**= I/S= Interest Portion= 1.00%), (**S'**= Sales of Past Year= 48,851 millions USD), (**y**= Y/D= Yielding Tax Rate of Dividend= 30.00%), (**V**= Variable Cost= 9,746 millions USD), and (**f**= F/S= Fixed Portion= 58.00%), are known, then it's (**t**= Tax Rate planned) is:

$$\begin{aligned}\mathbf{t} &= 1\text{-}\mathbf{H}/(\mathbf{d}[1\text{-}\mathbf{y}]\{\mathbf{S'}[1+\mathbf{s}]\text{-}\mathbf{V}\text{-}\mathbf{Sf}\text{-}\mathbf{Si}\}) \\ &= 1\text{-}\mathbf{H}/(\mathbf{d}[1\text{-}\mathbf{y}]\{\mathbf{S'}[1+\mathbf{s}]\text{-}\mathbf{V}\text{-}\mathbf{S}[\mathbf{f}+\mathbf{i}]\}) \\ &= 1\text{-}1{,}659/(0.3000[1\text{-}0.3000]\{48{,}851 \\ &\qquad [1+0.0500]\text{-}9{,}746\text{-}51{,}294 \\ &\qquad [0.5800+0.0100]\}) \\ &= 30.00\%\end{aligned}$$

Law-3077:

If both (**H**= Home Taken Dividend= 1,659 millions USD), (**d**= D/A= Dividend Portion or Payout= 30.00%), (**$**= Sales or Revenues= 51,294 millions USD), (**s**= [S/S']-1= Sales Growth= 5.00%), (**i**= I/S= Interest Portion= 1.00%), (**t**= T/B= Tax Rate= 30.00%), (**y**= Y/D= Yielding Tax Rate of Dividend= 30.00%), (**V**= Variable Cost= 9,746 millions USD), and (**f**= F/S= Fixed Portion= 58.00%), are known, then it's (**$'**= Sales Past) must be:

$'= (**V**+**$**[**f**+**i**]+**H**/{**d**[1-**y**][1-**t**]})/[1+**s**]
   = (9,746+51,294[0.5800+0.0100]
        +1,659/{0.3000 [1-0.3000]
        [1-0.3000]})/[1+0.0500]
   = 48,851 millions USD

Law-3078:

If both (**H**= Home Taken Dividend= 1,659 millions USD), (**d**= D/A= Dividend Portion or Payout= 30.00%), (**$**= Sales or Revenues= 51,294 millions USD), (**$'**= Sales of Past Year= 48,851 millions USD), (**i**= I/S= Interest Portion= 1.00%), (**t**= T/B= Tax Rate= 30.00%), (**y**= Y/D= Yielding Tax Rate of Dividend= 30.00%), (**V**= Variable Cost= 9,746 millions USD), and (**f**= F/S= Fixed Portion= 58.00%), are known, then it's (**s**= Sales Growth planned) is:

$$s = (V + \$[f+i] + H/\{d[1-y][1-t]\})/\$' - 1$$
$$= (9,746 + 51,294[0.5800 + 0.0100]$$
$$+ 1,659/\{0.3000[1-0.3000]$$
$$[1-0.3000]\})/48,851 - 1$$
$$= \underline{5.00\%}$$

Law-3079:
If both (**H**= Home Taken Dividend= 1,659 millions USD), (**d**= D/A= Dividend Portion or Payout= 30.00%), (**$**= Sales or Revenues= 51,294 millions USD), (**$'**= Sales of Past Year= 48,851 millions USD), (**i**= I/S= Interest Portion= 1.00%), (**t**= T/B= Tax Rate= 30.00%), (**y**= Y/D= Yielding Tax Rate of Dividend= 30.00%), (**s**= [S/S']-1= Sales Growth= 5.00%), and (**f**= F/S= Fixed Portion= 58.00%), are known, then it's (**V**= Variable Cost planned) is:

$$V = \$'[1+s] - \$[f+i] - H/\{d[1-y][1-t]\}$$
$$= 48,851[1+0.0500] - 51,294[0.5800$$
$$+0.0100] - 1,659/\{0.3000$$
$$[1-0.3000][1-0.3000]\})$$
$$= \underline{9,746} \text{ millions USD}$$

Law-3080:
If both (**H**= Home Taken Dividend= 1,659 millions USD), (**d**= D/A= Dividend Portion or Payout= 30.00%), (**V**= Variable Cost= 9,746 millions USD), (**$'**= Sales of Past Year= 48,851 millions USD), (**i**= I/S= Interest Portion= 1.00%), (**t**= T/B= Tax Rate= 30.00%), (**y**= Y/D= Yielding Tax Rate of Dividend= 30.00%), (**s**= [S/S']-1= Sales Growth= 5.00%), and (**f**= F/S= Fixed Portion= 58.00%), are known, then it's (**$**= Sales or Revenues planned) is:

$$\$ = (\$'[1+s] - V - H/\{d[1-y][1-t]\})/[f+i]$$
$$= 48,851[1+0.0500] - 9,746 - 1,659$$
$$/\{0.3000[1-0.3000]$$
$$[1-0.3000]\})/[0.5800+0.0100]$$
$$= \underline{51,294} \text{ millions USD}$$

Law-3081:

If both (**H**= Home Taken Dividend= 1,659 millions USD), (**d**= D/A= Dividend Portion or Payout= 30.00%), (**V**= Variable Cost= 9,746 millions USD), (**$'**= Sales of Past Year= 48,851 millions USD), (**i**= I/S= Interest Portion= 1.00%), (**t**= T/B= Tax Rate= 30.00%), (**y**= Y/D= Yielding Tax Rate of Dividend= 30.00%), (**s**= [S/S']-1= Sales Growth= 5.00%), and (**$**= Sales or Revenues= 51,294 millions USD), are known, then it's (**f**= Fixed Portion planned) is:

$$f = (\$'[1+s]-V-H/\{d[1-y][1-t]\})/\$-i$$
$$= 48,851[1+0.0500]-9,746-1,659$$
$$/\{0.3000[1-0.3000][1-0.3000]\})$$
$$/51,294-0.0100$$
$$= \underline{58.00\%}$$

Law-3082:

If both (**H**= Home Taken Dividend= 1,659 millions USD), (**d**= D/A= Dividend Portion or Payout= 30.00%), (**V**= Variable Cost= 9,746 millions USD), (**$'**= Sales of Past Year= 48,851 millions USD), (**f**= F/S= Fixed Portion= 58.00%), (**t**= T/B= Tax Rate= 30.00%), (**y**= Y/D= Yielding Tax Rate of Dividend= 30.00%), (**s**= [S/S']-1= Sales Growth= 5.00%), and (**$**= Sales or Revenues= 51,294 millions USD), are known, then it's (**i**= Interest Portion planned) is:

$$i = (\$'[1+s]-V-H/\{d[1-y][1-t]\})/\$-f$$
$$= 48,851[1+0.0500]-9,746-1,659$$
$$/\{0.3000[1-0.3000]$$
$$[1-0.3000]\})/51,294-0.5800$$
$$= \underline{1.00\%}$$

Law-3083:

If both (**i**= I/S= Interest Portion= 1.00%), (**d**= D/A= Dividend Portion or Payout= 30.00%), (**V**= Variable Cost= 9,746 millions USD), (**$'**= Sales of Past Year= 48,851 millions USD), (**f**= F/S= Fixed Portion= 58.00%), (**t**= T/B= Tax Rate= 30.00%), (**y**= Y/D= Yielding Tax Rate of Dividend= 30.00%), (**s**= [S/S']-1= Sales Growth= 5.00%), and (**$**= Sales or Revenues= 51,294 millions USD), are known, then it's (**H**= Home Taken Dividend planned) is:

$$H = d[1-y][1-t]\{\$'[1+s]-V-\$'i[1+s]-\$f\}$$
$$= d[1-y][1-t]\{\$'[1+s][1-i]-V-\$f\}$$
$$= 0.3000[1-0.3000][1-0.3000]\{48,851$$
$$[1+0.0500][1-0.0100]-9,746$$
$$-51,294*0.5800\}$$
$$= \underline{1,659} \text{ millions USD}$$

Finance Construction-4, *Tim Asikin, Steve Asikin, Indra Senihardja*

<u>Law-3084</u>:

If both (i= I/S= Interest Portion= <u>1.00</u>%), (H= Home Taken Dividend= <u>1,659</u> millions USD), (V= Variable Cost= <u>9,746</u> millions USD), (S'= Sales of Past Year= <u>48,851</u> millions USD), (f= F/S= Fixed Portion= <u>58.00</u>%), (t= T/B= Tax Rate= <u>30.00</u>%), (y= Y/D= Yielding Tax Rate of Dividend= <u>30.00</u>%), (s= [S/S']-1= Sales Growth= <u>5.00</u>%), and (S= Sales or Revenues= <u>51,294</u> millions USD), are known, then it's (d= Dividend Portion or Payout planned) is:

d = H/([1-y][1-t]{S'[1+s]-V-$S'i$[1+s]-Sf})
   = H/([1-y][1-t]{S'[1+s][1-i]-V-Sf})
   = 1,659[1-0.3000][1-0.3000]{48,851
       [1+0.0500][1-0.0100]-9,746
       -51,294*0.5800}
   = <u>30.00</u>%

Law-3085:

If both (**i**= I/S= Interest Portion= 1.00%), (**H**= Home Taken Dividend= 1,659 millions USD), (**V**= Variable Cost= 9,746 millions USD), (**S'**= Sales of Past Year= 48,851 millions USD), (**f**= F/S= Fixed Portion= 58.00%), (**t**= T/B= Tax Rate= 30.00%), (**d**= D/A= Dividend Portion or Payout= 30.00%), (**s**= [S/S']-1= Sales Growth= 5.00%), and (**$**= Sales or Revenues= 51,294 millions USD), are known, then it's (**y**= Yielding Tax Rate of Dividend planned) is:

$$y = 1 - H/(d[1-t]\{S'[1+s] - V - S'i[1+s] - \$f\})$$
$$= 1 - H/(d[1-t]\{S'[1+s][1-i] - V - \$f\})$$
$$= 1 - 1,659/(0.3000[1-0.3000]\{48,851[1+0.0500][1-0.0100] - 9,746 - 51,294*0.5800\})$$
$$= \underline{30.00\%}$$

Law-3086:

If both (**i**= I/S= Interest Portion= 1.00%), (**H**= Home Taken Dividend= 1,659 millions USD), (**V**= Variable Cost= 9,746 millions USD), (**$'**= Sales of Past Year= 48,851 millions USD), (**f**= F/S= Fixed Portion= 58.00%), (**y**= Y/D= Yielding Tax Rate of Dividend= 30.00%), (**d**= D/A= Dividend Portion or Payout= 30.00%), (**s**= [S/S']-1= Sales Growth= 5.00%), and (**$**= Sales or Revenues= 51,294 millions USD), are known, then it's (**t**= Tax Rate planned) is:

$$\begin{aligned}
\mathbf{t} &= 1-\mathbf{H}/(\mathbf{d}[1-\mathbf{y}]\{\mathbf{\$'}[1+\mathbf{s}]-\mathbf{V}-\mathbf{\$'i}[1+\mathbf{s}]-\mathbf{\$f}\}) \\
&= 1-\mathbf{H}/(\mathbf{d}[1-\mathbf{y}]\{\mathbf{\$'}[1+\mathbf{s}][1-\mathbf{i}]-\mathbf{V}-\mathbf{\$f}\}) \\
&= 1-1,659/(0.3000[1-0.3000]\{48,851 \\
&\qquad [1+0.0500][1-0.0100]-9,746 \\
&\qquad -51,294*0.5800\} \\
&= 30.00\%
\end{aligned}$$

Law-3087:

If both (**i**= I/S= Interest Portion= 1.00%), (**H**= Home Taken Dividend= 1,659 millions USD), (**V**= Variable Cost= 9,746 millions USD), (**t**= T/B= Tax Rate= 30.00%), (**f**= F/S= Fixed Portion= 58.00%), (**y**= Y/D= Yielding Tax Rate of Dividend= 30.00%), (**d**= D/A= Dividend Portion or Payout= 30.00%), (**s**= [S/S']-1= Sales Growth= 5.00%), and (**$**= Sales or Revenues= 51,294 millions USD), are known, then it's (**$'**= Sales Past) must be:

$'= (**V**+**$f**+**H**/{**d**[1-**y**][1-**t**]})/{[1+**s**][1-**i**]}
  = (9,746+51,294*0.5800+1,659
       /(0.3000[1-0.3000]
       [1-0.3000]})/{[1+0.0500]
       [1-0.0100]
  = 48,851 millions USD

Law-3088:

If both (**i**= I/S= Interest Portion= 1.00%), (**H**= Home Taken Dividend= 1,659 millions USD), (**V**= Variable Cost= 9,746 millions USD), (**t**= T/B= Tax Rate= 30.00%), (**f**= F/S= Fixed Portion= 58.00%), (**y**= Y/D= Yielding Tax Rate of Dividend= 30.00%), (**d**= D/A= Dividend Portion or Payout= 30.00%), (**$'**= Sales of Past Year= 48,851 millions USD), and (**$**= Sales or Revenues= 51,294 millions USD), are known, then it's (**s**= Sales Growth planned) is:

$s = (V+\$f+H/\{d[1-y][1-t]\})/\{\$'[1-i]\}-1$
$= (9,746+51,294*0.5800+1,659$
$/(0.3000[1-0.3000]$
$[1-0.3000]\})/\{48,851$
$[1-0.0100]-1$
$= 5.00\%$

Law-3089:

If both (s= [S/S']-1= Sales Growth= 5.00%), (H= Home Taken Dividend= 1,659 millions USD), (V= Variable Cost= 9,746 millions USD), (t= T/B= Tax Rate= 30.00%), (f= F/S= Fixed Portion= 58.00%), (y= Y/D= Yielding Tax Rate of Dividend= 30.00%), (d= D/A= Dividend Portion or Payout= 30.00%), (S'= Sales of Past Year= 48,851 millions USD), and (S= Sales or Revenues= 51,294 millions USD), are known, then it's (i= Interest Portion planned) is:

i = 1-(V+Sf+H/{d[1-y][1-t]})/{S'[1+s]}
   = 1-(9,746+51,294*0.5800+1,659
       /(0.3000[1-0.3000]
       [1-0.3000]})/{48,851
       [1+0.0500]
   = 1.00%

Law-3090:
If both (**s**= [S/S']-1= Sales Growth= 5.00%), (**H**= Home Taken Dividend= 1,659 millions USD), (**i**= I/S= Interest Portion= 1.00%), (**t**= T/B= Tax Rate= 30.00%), (**f**= F/S=Fixed Portion= 58.00%), (**y**= Y/D= Yielding Tax Rate of Dividend= 30.00%), (**d**= D/A= Dividend Portion or Payout= 30.00%), (**S'**= Sales of Past Year= 48,851 millions USD), and (**S**= Sales or Revenues= 51,294 millions USD), are known, then it's (**V**= Variable Cost planned) is:

$$V = S'[1+s][1-i] - Sf - H/\{d[1-y][1-t]\}/\{S'[1+s]\}$$
$$= 48,851[1+0.0500][1-0.0100] - 51,294$$
$$*0.5800 - 1.659/\{0.3000$$
$$[1-0.3000][1-0.3000]\}]$$
$$= 9,746 \text{ millions USD}$$

Law-3091:
If both (**s**= [S/S']-1= Sales Growth= 5.00%), (**H**= Home Taken Dividend= 1,659 millions USD), (**i**= I/S= Interest Portion= 1.00%), (**t**= T/B= Tax Rate= 30.00%), (**f**= F/S= Fixed Portion= 58.00%), (**y**= Y/D= Yielding Tax Rate of Dividend= 30.00%), (**d**= D/A= Dividend Portion or Payout= 30.00%), (**S'**= Sales of Past Year= 48,851 millions USD), and (**V**= Variable Cost= 9,746 millions USD), are known, then it's (**S**= Sales or Revenues planned) is:

$$S = (S'[1+s][1-i] - V - H\{d[1-y][1-t]\})/f$$
$$= 48,851[1+0.0500][1-0.0100] - 9,746$$
$$-1,659/\{0.3000[1-0.3000]$$
$$[1-0.3000]\}/0.5800$$
$$= 51,294 \text{ millions USD}$$

Law-3092:

If both (s= [S/S']-1= Sales Growth= 5.00%), (H= Home Taken Dividend= 1,659 millions USD), (i= I/S= Interest Portion= 1.00%), (t= T/B= Tax Rate= 30.00%), (S= Sales or Revenues= 51,294 millions USD), (y= Y/D= Yielding Tax Rate of Dividend= 30.00%), (d= D/A= Dividend Portion or Payout= 30.00%), (S'= Sales of Past Year= 48,851 millions USD), and (V= Variable Cost= 9,746 millions USD), are known, then it's (f= Fixed Portion planned) is:

$$f = (S'[1+s][1-i]-V-H/\{d[1-y][1-t]\})/S$$
$$= 48,851[1+0.0500][1-0.0100]-9,746$$
$$-1,659/\{0.3000[1-0.3000]$$
$$[1-0.3000]\}/51,294$$
$$= 58.00\%$$

Law-3093:

If both (s= [S/S']-1= Sales Growth= 5.00%), (i= I/S= Interest Portion= 1.00%), (t= T/B= Tax Rate= 30.00%), (y= Y/D= Yielding Tax Rate of Dividend= 30.00%), (d= D/A= Dividend Portion or Payout= 30.00%), (f= F/S= Fixed Portion= 58.00%), (S'= Sales of Past Year= 48,851 millions USD), and (V= Variable Cost= 9,746 millions USD), are known, then it's (H= Home Taken Dividend planned) is:

$$H = d[1-y][1-t]\{S'[1+s]-V-S'f[1+s]-S'i[1+s]\}$$
$$= d[1-y][1-t]\{S'[1+s][1-f-i]-V\}$$
$$= 0.3000[1-0.3000][1-0.3000]\{48,851$$
$$[1+0.0500][1-0.5800-0.0100]$$
$$-9,746\}$$
$$= 1,659 \text{ millions USD}$$

Finance Construction-4, *Tim Asikin, Steve Asikin, Indra Senihardja*

Law-3094:

If both (s= [S/S']-1= Sales Growth= 5.00%), (i= I/S= Interest Portion= 1.00%), (t= T/B= Tax Rate= 30.00%), (y= Y/D= Yielding Tax Rate of Dividend= 30.00%), (H= Home Taken Dividend= 1,659 millions USD), (f= F/S= Fixed Portion= 58.00%), (S'= Sales of Past Year= 48,851 millions USD), and (V= Variable Cost= 9,746 millions USD), are known, then it's (d= Dividend Portion or Payout planned) is:

d= H/([1-y][1-t]{S'[1+s]-V-$S'f$[1+s]-$S'i$[1+s]})
　　= H/([1-y][1-t]{S'[1+s][1-f-i]-V})
　　= 1,659/([1-0.3000][1-0.3000]{48,851
　　　　[1+0.0500][1-0.5800-0.0100]
　　　　-9,746})
　= 30.00%

Law-3095:

If both (s= [S/S']-1= Sales Growth= 5.00%), (i= I/S= Interest Portion= 1.00%), (t= T/B= Tax Rate= 30.00%), (d= D/A= Dividend Portion or Payout= 30.00%), (H= Home Taken Dividend= 1,659 millions USD), (f= F/S= Fixed Portion= 58.00%), (S'= Sales of Past Year= 48,851 millions USD), and (V= Variable Cost= 9,746 millions USD), are known, then it's (y= Yielding Tax Rate of Dividend planned) is:

$$y = 1 - H/(d[1-t]\{S'[1+s] - V - S'f[1+s] - S'i[1+s]\})$$
$$= 1 - H/(d[1-t]\{S'[1+s][1-f-i] - V\})$$
$$= 1 - 1{,}659/(0.3000[1-0.3000]\{48{,}851[1+0.0500][1-0.5800-0.0100] - 9{,}746\})$$
$$= \underline{30.00\%}$$

Law-3096:
If both (s= [S/S']-1= Sales Growth= 5.00%), (i= I/S= Interest Portion= 1.00%), (y= Y/D= Yielding Tax Rate of Dividend= 30.00%), (d= D/A= Dividend Portion or Payout= 30.00%), (H= Home Taken Dividend= 1,659 millions USD), (f= F/S= Fixed Portion= 58.00%), (S'= Sales of Past Year= 48,851 millions USD), and (V= Variable Cost= 9,746 millions USD), are known, then it's (t= Tax Rate planned) is:

t= 1-H/(d[1-y]{S'[1+s]-V-$S'f$[1+s]-$S'i$[1+s]})
   = 1-H/(d[1-y]{S'[1+s][1-f-i]-V})
   = 1-1,659/(0.3000[1-0.3000]{48,851
         [1+0.0500][1-0.5800-0.0100]
         -9,746})
   = 30.00%

Law-3097:
If both (s= [S/S']-1= Sales Growth= 5.00%), (i= I/S= Interest Portion= 1.00%), (y= Y/D= Yielding Tax Rate of Dividend= 30.00%), (d= D/A= Dividend Portion or Payout= 30.00%), (H= Home Taken Dividend= 1,659 millions USD), (f= F/S= Fixed Portion= 58.00%), (t= T/B= Tax Rate= 30.00%), and (V= Variable Cost= 9,746 millions USD), are known, then it's (S'= Sales Past) must be:

S'= (V+H/{d[1-y][1-t]})/{[1+s][1-f-i]}
   = (9,746+1,659/{0.3000[1-0.3000]
         [1-0.3000]}){[1+0.0500]
         [1-0.5800-0.0100]}
   = 48,851 millions USD

Law-3098:

If both ($'= Sales of Past Year= 48,851 millions USD), (i= I/S= Interest Portion= 1.00%), (y= Y/D= Yielding Tax Rate of Dividend= 30.00%), (d= D/A= Dividend Portion or Payout= 30.00%), (H= Home Taken Dividend= 1,659 millions USD), (f= F/S= Fixed Portion= 58.00%), (t= T/B= Tax Rate= 30.00%), and (V= Variable Cost= 9,746 millions USD), are known, then it's (s= Sales Growth planned) is:

$$s = (V+H/\{d[1-y][1-t]\})/\{\$'[1-f-i]\}-1$$
$$= (9,746+1,659/\{0.3000[1-0.3000]$$
$$[1-0.3000]\})\{48,851[1-0.5800$$
$$-0.0100]\}-1$$
$$= 5.00\%$$

Law-3099:

If both ($'= Sales of Past Year= 48,851 millions USD), (i= I/S= Interest Portion= 1.00%), (y= Y/D= Yielding Tax Rate of Dividend= 30.00%), (d= D/A= Dividend Portion or Payout= 30.00%), (H= Home Taken Dividend= 1,659 millions USD), (s= [S/S']-1= Sales Growth= 5.00%), (t= T/B= Tax Rate= 30.00%), and (V= Variable Cost= 9,746 millions USD), are known, then it's (f= Fixed Portion planned) is:

$$f = 1-i-(V+H/\{d[1-y][1-t]\})/\{\$'[1+s]\}$$
$$= 1-0.0100-(9,746+1,659/\{0.3000$$
$$[1-0.3000][1-0.3000]\})\{48,851$$
$$[1+0.0500]\}$$
$$= 58.00\%$$

Law-3100:
If both ($**S'**$= Sales of Past Year= 48,851 millions USD), (**f**= F/S= Fixed Portion= 58.00%), (**y**= Y/D= Yielding Tax Rate of Dividend= 30.00%), (**d**= D/A= Dividend Portion or Payout= 30.00%), (**H**= Home Taken Dividend= 1,659 millions USD), (**s**= [S/S']-1= Sales Growth= 5.00%), (**t**= T/B= Tax Rate= 30.00%), and (**V**= Variable Cost= 9,746 millions USD), are known, then it's (**i**= Interest Portion planned) is:

$$\mathbf{i} = 1-\mathbf{f}-(\mathbf{V}+\mathbf{H}/\{\mathbf{d}[1-\mathbf{y}][1-\mathbf{t}]\})/\{\mathbf{S'}[1+\mathbf{s}]\}$$
$$= 1-0.5800-(9{,}746+1{,}659/\{0.3000[1-0.3000][1-0.3000]\})/\{48{,}851[1+0.0500]\}$$
$$= \underline{1.00\%}$$

Law-3101:
If both ($**S'**$= Sales of Past Year= 48,851 millions USD), (**f**= F/S= Fixed Portion= 58.00%), (**y**= Y/D= Yielding Tax Rate of Dividend= 30.00%), (**d**= D/A= Dividend Portion or Payout= 30.00%), (**H**= Home Taken Dividend= 1,659 millions USD), (**s**= [S/S']-1= Sales Growth= 5.00%), (**t**= T/B= Tax Rate= 30.00%), and (**i**= I/S= Interest Portion= 1.00%), are known, then it's (**V**= Variable Cost planned) is:

$$\mathbf{V} = \mathbf{S'}[1+\mathbf{s}][1-\mathbf{f}-\mathbf{i}]-\mathbf{H}/\{\mathbf{d}[1-\mathbf{y}][1-\mathbf{t}]\}$$
$$= 48{,}851[1+0.0500][1-0.5800-0.0100]-1{,}659/\{0.3000[1-0.3000][1-0.3000]\}$$
$$= \underline{9{,}746} \text{ millions USD}$$

Law-3102:

If both ($\$'$= Sales of Past Year= 48,851 millions USD), (F= Fixed Cost= 29,750 millions USD), (y= Y/D= Yielding Tax Rate of Dividend= 30.00%), (d= D/A= Dividend Portion or Payout= 30.00%), (v= V/S= Variable Portion= 19.00%), (s= [S/S']-1= Sales Growth= 5.00%), (t= T/B= Tax Rate= 30.00%), (I= Interest Expense= 513 millions USD), and ($\$$= Sales or Revenues= 51,294 millions USD), are known, then it's (H= Home Taken Dividend planned) is:

$H = d[1-y][1-t]\{\$'[1+s]-\$v-F-I\}$
 $= 0.3000[1-0.3000][1-0.3000]\{48,851$
   $[1+0.0500]-51,294*0.1900$
   $-29,750-513\}$
 $= \underline{1,659}$ millions USD

Law-3103:

If both ($\$'$= Sales of Past Year= 48,851 millions USD), (F= Fixed Cost= 29,750 millions USD), (y= Y/D= Yielding Tax Rate of Dividend= 30.00%), (H= Home Taken Dividend= 1,659 millions USD), (v= V/S= Variable Portion= 19.00%), (s= [S/S']-1= Sales Growth= 5.00%), (t= T/B= Tax Rate= 30.00%), (I= Interest Expense= 513 millions USD), and ($\$$= Sales or Revenues= 51,294 millions USD), are known, then it's (d= Dividend Portion or Payout planned) is:

$d = H/([1-y][1-t]\{\$'[1+s]-\$v-F-I\})$
 $= 1,659/([1-0.3000][1-0.3000]\{48,851$
   $[1+0.0500]-51,294*0.1900$
   $-29,750-513\})$
 $= \underline{30.00\%}$

Law-3104:

If both ($\$'$= Sales of Past Year= 48,851 millions USD), (F= Fixed Cost= 29,750 millions USD), (d= D/A= Dividend Portion or Payout= 30.00%), (H= Home Taken Dividend= 1,659 millions USD), (v= V/S= Variable Portion= 19.00%), (s= [S/S']-1= Sales Growth= 5.00%), (t= T/B= Tax Rate= 30.00%), (I= Interest Expense= 513 millions USD), and ($\$$= Sales or Revenues= 51,294 millions USD), are known, then it's (y= Yielding Tax Rate of Dividend planned) is:

$$y = 1 - H/(d[1-t]\{\$'[1+s] - \$v - F - I\})$$
$$= 1 - 1,659/(0.3000[1-0.3000]\{48,851[1+0.0500] - 51,294*0.1900 - 29,750 - 513\})$$
$$= 30.00\%$$

Law-3105:

If both ($\$'$= Sales of Past Year= 48,851 millions USD), (F= Fixed Cost= 29,750 millions USD), (d= D/A= Dividend Portion or Payout= 30.00%), (H= Home Taken Dividend= 1,659 millions USD), (v= V/S= Variable Portion= 19.00%), (s= [S/S']-1= Sales Growth= 5.00%), (y= Y/D= Yielding Tax Rate of Dividend= 30.00%), (I= Interest Expense= 513 millions USD), and ($\$$= Sales or Revenues= 51,294 millions USD), are known, then it's (t= Tax Rate planned) is:

$$t = 1-H/(d[1-y]\{\$'[1+s]-\$v-F-I\})$$
$$= 1-1,659/(0.3000[1-0.3000]\{48,851[1+0.0500]-51,294*0.1900-29,750-513\})$$
$$= \underline{30.00\%}$$

Law-3106:

If both (**t**= T/B= Tax Rate= 30.00%), (**F**= Fixed Cost= 29,750 millions USD), (**d**= D/A= Dividend Portion or Payout= 30.00%), (**H**= Home Taken Dividend= 1,659 millions USD), (**v**= V/S= Variable Portion= 19.00%), (**s**= [S/S']-1= Sales Growth= 5.00%), (**y**= Y/D= Yielding Tax Rate of Dividend= 30.00%), (**I**= Interest Expense= 513 millions USD), and (**S**= Sales or Revenues= 51,294 millions USD), are known, then it's (**S'**= Sales Past) must be:

$$S' = (Sv+F+I+H/\{d[1-y][1-t]\})/[1+s]$$
$$= (51,294*0.1900+29,750+513+1,659$$
$$/(0.3000[1-0.3000][1-0.3000]\}$$
$$/ [1+0.0500]$$
$$= \underline{48,851} \text{ millions USD}$$

Law-3107:

If both (**t**= T/B= Tax Rate= <u>30.00</u>%), (**F**= Fixed Cost= <u>29,750</u> millions USD), (**d**= D/A= Dividend Portion or Payout= 30.00%), (**H**= Home Taken Dividend= <u>1,659</u> millions USD), (**v**= V/S= Variable Portion= <u>19.00</u>%), (**$'**= Sales of Past Year= <u>48,851</u> millions USD), (**y**= Y/D= Yielding Tax Rate of Dividend= <u>30.00</u>%), (**I**= Interest Expense= <u>513</u> millions USD), and (**$**= Sales or Revenues= <u>51,294</u> millions USD), are known, then it's (**s**= Sales Growth planned) is:

$$s = (\$v+F+I+H/\{d[1-y][1-t]\})/\$'-1$$
$$= (51,294*0.1900+29,750+513+1,659$$
$$/\{0.3000[1-0.3000]$$
$$[1-0.3000]\})/48,851-1$$
$$= \underline{5.00}\%$$

Law-3108:

If both (**t**= T/B= Tax Rate= 30.00%), (**F**= Fixed Cost= 29,750 millions USD), (**d**= D/A= Dividend Portion or Payout= 30.00%), (**H**= Home Taken Dividend= 1,659 millions USD), (**v**= V/S= Variable Portion= 19.00%), (**$'**= Sales of Past Year= 48,851 millions USD), (**y**= Y/D= Yielding Tax Rate of Dividend= 30.00%), (**I**= Interest Expense= 513 millions USD), and (**s**= [S/S']-1= Sales Growth= 5.00%), are known, then it's (**$**= Sales or Revenues planned) is:

$$\$ = (\$'[1+s]-F-I-H/\{d[1-y][1-t]\})/v$$
$$= (48,851[1+0.0500]-29,750-513$$
$$-1,659/\{0.3000[1-0.3000]$$
$$[1-0.3000]\})/0.1900$$
$$= \underline{51,294} \text{ millions USD}$$

Law-3109:

If both (**t**= T/B= Tax Rate= 30.00%), (**F**= Fixed Cost= 29,750 millions USD), (**d**= D/A= Dividend Portion or Payout= 30.00%), (**H**= Home Taken Dividend= 1,659 millions USD), (**$**= Sales or Revenues= 51,294 millions USD), (**$'**= Sales of Past Year= 48,851 millions USD), (**y**= Y/D= Yielding Tax Rate of Dividend= 30.00%), (**I**= Interest Expense= 513 millions USD), and (**s**= [S/S']-1= Sales Growth= 5.00%), are known, then it's (**v**= Variable Portion planned) is:

$$\begin{aligned}
\mathbf{v} &= (\$'[1+\mathbf{s}]\text{-}\mathbf{F}\text{-}\mathbf{I}\text{-}\mathbf{H}/\{\mathbf{d}[1\text{-}\mathbf{y}][1\text{-}\mathbf{t}]\})/\$ \\
&= (48,851[1+0.0500]\text{-}29,750\text{-}513 \\
&\qquad \text{-}1,659/\{0.3000[1\text{-}0.3000] \\
&\qquad [1\text{-}0.3000]\})/51,294 \\
&= 19.00\%
\end{aligned}$$

Law-3110:
If both (**t**= T/B= Tax Rate= 30.00%), (**v**= V/S= Variable Portion= 19.00%), (**d**= D/A= Dividend Portion or Payout= 30.00%), (**H**= Home Taken Dividend= 1,659 millions USD), (**S**= Sales or Revenues= 51,294 millions USD), (**S'**= Sales of Past Year= 48,851 millions USD), (**y**= Y/D= Yielding Tax Rate of Dividend= 30.00%), (**I**= Interest Expense= 513 millions USD), and (**s**= [S/S']-1= Sales Growth= 5.00%), are known, then it's (**F**= Fixed Cost planned) is:

$$F = (S'[1+s] - Sv - I - H/\{d[1-y][1-t]\}$$
$$= (48,851[1+0.0500] - 51,294*0.1900$$
$$\quad -513 - 1,659/\{0.3000[1-0.3000]$$
$$\quad [1-0.3000]\}$$
$$= 29,750 \text{ millions USD}$$

Finance Construction-4, *Tim Asikin, Steve Asikin, Indra Senihardja*

<u>Law-3111</u>:

If both (t= T/B= Tax Rate= <u>30.00</u>%), (v= V/S= Variable Portion= <u>19.00</u>%), (d= D/A= Dividend Portion or Payout= 30.00%), (H= Home Taken Dividend= <u>1,659</u> millions USD), ($\$$= Sales or Revenues= <u>51,294</u> millions USD), ($\$'$= Sales of Past Year= <u>48,851</u> millions USD), (y= Y/D= Yielding Tax Rate of Dividend= <u>30.00</u>%), (F= Fixed Cost= <u>29,750</u> millions USD), and (s= [S/S']-1= Sales Growth= <u>5.00</u>%), are known, then it's (I= Interest Expense planned) is:

$I = (\$'[1+s] - \$v - F - H/\{d[1-y][1-t]\}$
$= (48,851[1+0.0500] - 51,294*0.1900$
$\qquad -29,750 - 1,659/\{0.3000$
$\qquad [1-0.3000][1-0.3000]\}$
$= \underline{513}$ millions USD

Finance Construction-4, *Tim Asikin, Steve Asikin, Indra Senihardja*

Law-3112:

If both (**t**= T/B= Tax Rate= 30.00%), (**v**= V/S= Variable Portion= 19.00%), (**d**= D/A= Dividend Portion or Payout= 30.00%), (**i**= I/S= Interest Portion= 1.00%), (**S**= Sales or Revenues= 51,294 millions USD), (**S'**= Sales of Past Year= 48,851 millions USD), (**y**= Y/D= Yielding Tax Rate of Dividend= 30.00%), (**F**= Fixed Cost= 29,750 millions USD), and (**s**= [S/S']-1= Sales Growth= 5.00%), are known, then it's (**H**= Home Taken Dividend planned) is:

$$H = d[1-y][1-t]\{S'[1+s]-Sv-F-Si\}$$
$$= d[1-y][1-t]\{S'[1+s]-S[v+i]-F\}$$
$$= 0.3000[1-0.3000][1-0.3000]\{48,851[1+0.0500]-51,294[0.1900+0.0100]-29,750\}$$
$$= \underline{1,659} \text{ millions USD}$$

Law-3113:

If both (**t**= T/B= Tax Rate= 30.00%), (**v**= V/S= Variable Portion= 19.00%), (**H**= Home Taken Dividend= 1,659 millions USD), (**i**= I/S= Interest Portion= 1.00%), (**S**= Sales or Revenues= 51,294 millions USD), (**S'**= Sales of Past Year= 48,851 millions USD), (**y**= Y/D= Yielding Tax Rate of Dividend= 30.00%), (**F**= Fixed Cost= 29,750 millions USD), and (**s**= [S/S']-1= Sales Growth= 5.00%), are known, then it's (**d**= Dividend Portion or Payout planned) is:

$$d = H/([1-y][1-t]\{S'[1+s]-Sv-F-Si\})$$
$$= H/([1-y][1-t]\{S'[1+s]-S[v+i]-F\})$$
$$= 1,659/([1-0.3000][1-0.3000]\{48,851[1+0.0500]-51,294[0.1900+0.0100]-29,750\})$$
$$= \underline{30.00\%}$$

Law-3114:

If both (**t**= T/B= Tax Rate= 30.00%), (**v**= V/S= Variable Portion= 19.00%), (**H**= Home Taken Dividend= 1,659 millions USD), (**i**= I/S= Interest Portion= 1.00%), (**$**= Sales or Revenues= 51,294 millions USD), (**$'**= Sales of Past Year= 48,851 millions USD), (**d**= D/A= Dividend Portion or Payout= 30.00%), (**F**= Fixed Cost= 29,750 millions USD), and (**s**= [S/S']-1= Sales Growth= 5.00%), are known, then it's (**y**= Yielding Tax Rate of Dividend planned) is:

$$\begin{aligned}
\mathbf{y} &= 1-\mathbf{H}/(\mathbf{d}[1-\mathbf{t}]\{\mathbf{\$'}[1+\mathbf{s}]-\mathbf{\$v}-\mathbf{F}-\mathbf{\$i}\}) \\
&= 1-\mathbf{H}/(\mathbf{d}[1-\mathbf{t}]\{\mathbf{\$'}[1+\mathbf{s}]-\mathbf{\$}[\mathbf{v}+\mathbf{i}]-\mathbf{F}\}) \\
&= 1-1,659/(0.3000[1-0.3000]\{48,851 \\
&\qquad [1+0.0500]-51,294[0.1900 \\
&\qquad +0.0100]-29,750\}) \\
&= 30.00\%
\end{aligned}$$

Law-3115:

If both (y= Y/D= Yielding Tax Rate of Dividend= 30.00%), (v= V/S= Variable Portion= 19.00%), (H= Home Taken Dividend= 1,659 millions USD), (i= I/S= Interest Portion= 1.00%), (S= Sales or Revenues= 51,294 millions USD), (S'= Sales of Past Year= 48,851 millions USD), (d= D/A= Dividend Portion or Payout= 30.00%), (F= Fixed Cost= 29,750 millions USD), and (s= [S/S']-1= Sales Growth= 5.00%), are known, then it's (t= Tax Rate planned) is:

$$\begin{aligned}
t &= 1-H/(d[1-y]\{S'[1+s]-Sv-F-Si\}) \\
&= 1-H/(d[1-y]\{S'[1+s]-S[v+i]-F\}) \\
&= 1-1,659/(0.3000[1-0.3000]\{48,851 \\
&\qquad [1+0.0500]-51,294[0.1900 \\
&\qquad +0.0100]-29,750\}) \\
&= 30.00\%
\end{aligned}$$

Law-3116:

If both (**y**= Y/D= Yielding Tax Rate of Dividend= 30.00%), (**v**= V/S= Variable Portion= 19.00%), (**H**= Home Taken Dividend= 1,659 millions USD), (**i**= I/S= Interest Portion= 1.00%), (**$**= Sales or Revenues= 51,294 millions USD), (**t**= T/B= Tax Rate= 30.00%), (**d**= D/A= Dividend Portion or Payout= 30.00%), (**F**= Fixed Cost= 29,750 millions USD), and (**s**= [S/S']-1= Sales Growth= 5.00%), are known, then it's (**$'**= Sales Past) must be:

$'= ($[**v**+**i**]+**F**+**H**/{**d**[1-**y**][1-**t**]})/[1+**s**]
 = (51,294[0.1900+0.0100]+29,750
   +1,659/{0.3000[1-0.3000]
   [1-0.3000]})/[1+0.0500]
 = 48,851 millions USD

Law-3117:

If both (y= Y/D= Yielding Tax Rate of Dividend= 30.00%), (v= V/S= Variable Portion= 19.00%), (H= Home Taken Dividend= 1,659 millions USD), (i= I/S= Interest Portion= 1.00%), (S= Sales or Revenues= 51,294 millions USD), (t= T/B= Tax Rate= 30.00%), (d= D/A= Dividend Portion or Payout= 30.00%), (F= Fixed Cost= 29,750 millions USD), and (S'= Sales of Past Year= 48,851 millions USD), are known, then it's (s= Sales Growth planned) is:

$$s = (S[v+i]+F+H/\{d[1-y][1-t]\})/S'-1$$
$$= (51,294[0.1900+0.0100]+29,750$$
$$+1,659/\{0.3000[1-0.3000]$$
$$[1-0.3000]\})/48,851-1$$
$$= 5.00\%$$

Law-3118:

If both (**y**= Y/D= Yielding Tax Rate of Dividend= 30.00%), (**v**= V/S= Variable Portion= 19.00%), (**H**= Home Taken Dividend= 1,659 millions USD), (**i**= I/S= Interest Portion= 1.00%), (**s**= [S/S']-1= Sales Growth= 5.00%), (**t**= T/B= Tax Rate= 30.00%), (**d**= D/A= Dividend Portion or Payout= 30.00%), (**F**= Fixed Cost= 29,750 millions USD), and (**$'**= Sales of Past Year= 48,851 millions USD), are known, then it's (**$**= Sales or Revenues planned) is:

$= $'[1+s]-F-H/{d[1-y][1-t]})/[v+i]
= 48,851[1+0.0500]-29,750-1,659
/{0.3000[1-0.3000]
[1-0.3000]})/[0.1900+0.0100]
= 51,294 millions USD

Law-3119:

If both (**y**= Y/D= Yielding Tax Rate of Dividend= 30.00%), (**$**= Sales or Revenues= 51,294 millions USD), (**H**= Home Taken Dividend= 1,659 millions USD), (**i**= I/S= Interest Portion= 1.00%), (**s**= [S/S']-1= Sales Growth= 5.00%), (**t**= T/B= Tax Rate= 30.00%), (**d**= D/A= Dividend Portion or Payout= 30.00%), (**F**= Fixed Cost= 29,750 millions USD), and (**$'**= Sales of Past Year= 48,851 millions USD), are known, then it's (**v**= Variable Portion planned) is:

$$v = \$'[1+s] - F - H / \{d[1-y][1-t]\}) / \$ - i$$
$$= 48,851[1+0.0500] - 29,750 - 1,659$$
$$/\{0.3000[1-0.3000]$$
$$[1-0.3000]\})/51,294 - 0.0100$$
$$= \underline{19.00}\%$$

Law-3120:

If both (y= Y/D= Yielding Tax Rate of Dividend= 30.00%), ($\$$= Sales or Revenues= 51,294 millions USD), (H= Home Taken Dividend= 1,659 millions USD), (v= V/S= Variable Portion= 19.00%), (s= [S/S']-1= Sales Growth= 5.00%), (t= T/B= Tax Rate= 30.00%), (d= D/A= Dividend Portion or Payout= 30.00%), (F= Fixed Cost= 29,750 millions USD), and ($\$'$= Sales of Past Year= 48,851 millions USD), are known, then it's (i= Interest Portion planned) is:

$$i = \$'[1+s]-F-H/\{d[1-y][1-t]\})/\$-v$$
$$= 48,851[1+0.0500]-29,750-1,659$$
$$/\{0.3000[1-0.3000]$$
$$[1-0.3000]\})/51,294-0.1900$$
$$= 1.00\%$$

Law-3121:

If both (y= Y/D= Yielding Tax Rate of Dividend= 30.00%), (S= Sales or Revenues= 51,294 millions USD), (H= Home Taken Dividend= 1,659 millions USD), (v= V/S= Variable Portion= 19.00%), (s= [S/S']-1= Sales Growth= 5.00%), (t= T/B= Tax Rate= 30.00%), (d= D/A= Dividend Portion or Payout= 30.00%), (i= I/S= Interest Portion= 1.00%), and (S'= Sales of Past Year= 48,851 millions USD), are known, then it's (F= Fixed Cost planned) is:

$$F = S'[1+s] - S[v+i] - H/\{d[1-y][1-t]\}$$
$$= 48,851[1+0.0500] - 51,294[0.1900 + 0.0100] - 1,659/\{0.3000[1-0.3000][1-0.3000]\}$$
$$= 29,750 \text{ millions USD}$$

Law-3122:

If both (y= Y/D= Yielding Tax Rate of Dividend= 30.00%), ($\$$= Sales or Revenues= 51,294 millions USD), (F= Fixed Cost= 29,750 millions USD), (v= V/S= Variable Portion= 19.00%), (s= [S/S']-1= Sales Growth= 5.00%), (t= T/B= Tax Rate= 30.00%), (d= D/A= Dividend Portion or Payout= 30.00%), (i= I/S= Interest Portion= 1.00%), and ($\$$'= Sales of Past Year= 48,851 millions USD), are known, then it's (H= Home Taken Dividend planned) is:

$$H = d[1-y][1-t]\{\$'[1+s]-\$v-F-\$'i[1+s]\}$$
$$= d[1-y][1-t]\{\$'[1+s][1-i]-\$v-F\}$$
$$= 0.3000[1-0.3000][1-0.3000]\{48,851$$
$$[1+0.0500][1-0.0100]-51,294$$
$$*0.1900-29,750\}$$
$$= \underline{1,659} \text{ millions USD}$$

Finance Construction-4, *Tim Asikin, Steve Asikin, Indra Senihardja*

<u>Law-3123</u>:

If both (**y**= Y/D= Yielding Tax Rate of Dividend= 30.00%), (**$**= Sales or Revenues= 51,294 millions USD), (**F**= Fixed Cost= 29,750 millions USD), (**v**= V/S= Variable Portion= 19.00%), (**s**= [S/S']-1= Sales Growth= 5.00%), (**t**= T/B= Tax Rate= 30.00%), (**H**= Home Taken Dividend= 1,659 millions USD), (**i**= I/S= Interest Portion= 1.00%), and (**$'**= Sales of Past Year= 48,851 millions USD), are known, then it's (**d**= Dividend Portion or Payout planned) is:

$$\begin{aligned}\mathbf{d} &= \mathbf{H}/([1\text{-}\mathbf{y}][1\text{-}\mathbf{t}]\{\mathbf{\$'}[1\text{+}\mathbf{s}]\text{-}\mathbf{\$v}\text{-}\mathbf{F}\text{-}\mathbf{\$'i}[1\text{+}\mathbf{s}]\}) \\ &= \mathbf{H}/([1\text{-}\mathbf{y}][1\text{-}\mathbf{t}]\{\mathbf{\$'}[1\text{+}\mathbf{s}][1\text{-}\mathbf{i}]\text{-}\mathbf{\$v}\text{-}\mathbf{F}\}) \\ &= 1{,}659/([1\text{-}0.3000][1\text{-}0.3000]\{48{,}851 \\ &\qquad [1\text{+}0.0500][1\text{-}0.0100]\text{-}51{,}294 \\ &\qquad *0.1900\text{-}29{,}750\}) \\ &= \underline{30.00\%}\end{aligned}$$

Law-3124:

If both (**d**= D/A= Dividend Portion or Payout= 30.00%), (**$**= Sales or Revenues= 51,294 millions USD), (**F**= Fixed Cost= 29,750 millions USD), (**v**= V/S= Variable Portion= 19.00%), (**s**= [S/S']-1= Sales Growth= 5.00%), (**t**= T/B= Tax Rate= 30.00%), (**H**= Home Taken Dividend= 1,659 millions USD), (**i**= I/S= Interest Portion= 1.00%), and (**$'**= Sales of Past Year= 48,851 millions USD), are known, then it's (**y**= Yielding Tax Rate of Dividend planned) is:

$$y = 1-H/(d[1-t]\{\$'[1+s]-\$v-F-\$'i[[1+s]]\})$$
$$= 1-H/(d[1-t]\{\$'[1+s][1-i]-\$v-F\})$$
$$= 1-1{,}659/(0.3000[1-0.3000]\{48{,}851[1+0.0500][1-0.0100]-51{,}294*0.1900-29{,}750\})$$
$$= 30.00\%$$

Law-3125:

If both (d= D/A= Dividend Portion or Payout= 30.00%), (S= Sales or Revenues= 51,294 millions USD), (y= Y/D= Yielding Tax Rate of Dividend= 30.00%), (F= Fixed Cost= 29,750 millions USD), (v= V/S= Variable Portion= 19.00%), (s= [S/S']-1= Sales Growth= 5.00%), (H= Home Taken Dividend= 1,659 millions USD), (i= I/S= Interest Portion= 1.00%), and (S'= Sales of Past Year= 48,851 millions USD), are known, then it's (t= Tax Rate planned) is:

$$t = 1-H/(d[1-y]\{S'[1+s]-Sv-F-S'i[1+s]\})$$
$$= 1-H/(d[1-y]\{S'[1+s][1-i]-Sv-F\})$$
$$= 1-1{,}659/(0.3000[1-0.3000]\{48{,}851[1+0.0500][1-0.0100]-51{,}294*0.1900-29{,}750\})$$
$$= 30.00\%$$

Law-3126:

If both (**d**= D/A=Dividend Portion or Payout= 30.00%), (**$**= Sales or Revenues= 51,294 millions USD), (**y**= Y/D= Yielding Tax Rate of Dividend= 30.00%), (**F**= Fixed Cost= 29,750 millions USD), (**v**= V/S= Variable Portion= 19.00%), (**s**= [S/S']-1= Sales Growth= 5.00%), (**H**= Home Taken Dividend= 1,659 millions USD), (**i**= I/S= Interest Portion= 1.00%), and (**t**= T/B= Tax Rate= 30.00%), are known, then it's (**$'**= Sales Past) must be:

$'= ($v+F+H/{d[1-y][1-t]})/{[1+s][1-i]}
　= (51,294*0.1900+29,750+1,659
　　　　/{0.3000 [1-0.3000]
　　　　[1-0.3000]})/{[1+0.0500]
　　　　[1-0.0100]}
　= 48,851 millions USD

Law-3127:

If both (**d**= D/A=Dividend Portion or Payout= 30.00%), (**$**= Sales or Revenues= 51,294 millions USD), (**y**= Y/D= Yielding Tax Rate of Dividend= 30.00%), (**F**= Fixed Cost= 29,750 millions USD), (**v**= V/S= Variable Portion= 19.00%), (**$'**= Sales of Past Year= 48,851 millions USD), (**H**= Home Taken Dividend= 1,659 millions USD), (**i**= I/S= Interest Portion= 1.00%), and (**t**= T/B= Tax Rate= 30.00%), are known, then it's (**s**= Sales Growth planned) is:

$$s = (\$v+F+H/\{d[1-y][1-t]\})/\{\$'[1-i]\}-1$$
$$= (51{,}294*0.1900+29{,}750+1{,}659$$
$$/\{0.3000[1-0.3000]$$
$$[1-0.3000]\})/\{48{,}851$$
$$[1-0.0100]\}-1$$
$$= \underline{5.00}\%$$

Finance Construction-4, *Tim Asikin, Steve Asikin, Indra Senihardja*

Law-3128:

If both (d= D/A= Dividend Portion or Payout= 30.00%), (S= Sales or Revenues= 51,294 millions USD), (y= Y/D= Yielding Tax Rate of Dividend= 30.00%), (F= Fixed Cost= 29,750 millions USD), (v= V/S= Variable Portion= 19.00%), (S'= Sales of Past Year= 48,851 millions USD), (H= Home Taken Dividend= 1,659 millions USD), (s= [S/S']-1= Sales Growth= 5.00%), and (t= T/B= Tax Rate= 30.00%), are known, then it's (i= Interest Portion planned) is:

i = 1-(Sv+F+H/{d[1-y][1-t]})/{S'[1+s]}
   = 1-(51,294*0.1900+29,750+1,659
         /{0.3000[1-0.3000][1-0.3000]})
         /{48,851[1+0.0500]}
   = 1.00%

Finance Construction-4, *Tim Asikin, Steve Asikin, Indra Senihardja*

Law-3129:

If both (**d**= D/A= Dividend Portion or Payout= 30.00%), (**i**= I/S= Interest Portion= 1.00%), (**y**= Y/D= Yielding Tax Rate of Dividend= 30.00%), (**F**= Fixed Cost= 29,750 millions USD), (**v**= V/S= Variable Portion= 19.00%), (**$'**= Sales of Past Year= 48,851 millions USD), (**H**= Home Taken Dividend= 1,659 millions USD), (**s**= [S/S']-1= Sales Growth= 5.00%), and (**t**= T/B= Tax Rate= 30.00%), are known, then it's (**$**= Sales or Revenues planned) is:

$= ($'[1+s][1-i]-F-H/{d[1-y][1-t]})/v
 = (48,851[1+0.0500][1-0.0100]-29,750
   -1,659/{0.3000[1-0.3000]
   [1-0.3000]})/0.1900
 = 51,294 millions USD

Law-3130:

If both (**d**= D/A=Dividend Portion or Payout= 30.00%), (**i**= I/S= Interest Portion= 1.00%), (**y**= Y/D= Yielding Tax Rate of Dividend= 30.00%), (**F**= Fixed Cost= 29,750 millions USD), (**S**= Sales or Revenues= 51,294 millions USD), (**S'**= Sales of Past Year= 48,851 millions USD), (**H**= Home Taken Dividend= 1,659 millions USD), (**s**= [S/S']-1= Sales Growth= 5.00%), and (**t**= T/B= Tax Rate= 30.00%), are known, then it's (**v**= Variable Portion planned) is:

$$v = (S'[1+s][1-i]-F-H/\{d[1-y][1-t]\})/S$$
$$= (48,851[1+0.0500][1-0.0100]-29,750$$
$$-1,659/\{0.3000[1-0.3000]$$
$$[1-0.3000]\})/51,294$$
$$= \underline{19.00\%}$$

Law-3131:

If both (**d**= D/A= Dividend Portion or Payout= 30.00%), (**i**= I/S= Interest Portion= 1.00%), (**y**= Y/D= Yielding Tax Rate of Dividend= 30.00%), (**v**= V/S= Variable Portion= 19.00%), (**$**= Sales or Revenues= 51,294 millions USD), (**$'**= Sales of Past Year= 48,851 millions USD), (**H**= Home Taken Dividend= 1,659 millions USD), (**s**= [S/S']-1= Sales Growth= 5.00%), and (**t**= T/B= Tax Rate= 30.00%), are known, then it's (**F**= Fixed Cost planned) is:

$$F = \$'[1+s][1-i] - \$v - H/\{d[1-y][1-t]\}$$
$$= 48{,}851[1+0.0500][1-0.0100] - 51{,}294 * 0.1900 - 1{,}659/\{0.3000 [1-0.3000][1-0.3000]\}$$
$$= 29{,}759 \text{ millions USD}$$

Finance Construction-4, *Tim Asikin, Steve Asikin, Indra Senihardja*

Law-3132:

If both (**d**= D/A= Dividend Portion or Payout= 30.00%), (**I**= Interest Expense= 513 millions USD), (**y**= Y/D= Yielding Tax Rate of Dividend= 30.00%), (**v**= V/S= Variable Portion= 19.00%), (**$**= Sales or Revenues= 51,294 millions USD), (**$'**= Sales of Past Year= 48,851 millions USD), (**f**= F/S= Fixed Portion= 58.00%), (**s**= [S/S']-1= Sales Growth= 5.00%), and (**t**= T/B= Tax Rate= 30.00%), are known, then it's (**H**= Human Dividend Portion planned) is:

$$H = d[1-y][1-t]\{\$'[1+s]-\$v-\$f-I\}$$
$$= d[1-y][1-t]\{\$'[1+s]-\$[v+f]-I\}$$
$$= 0.3000[1-0.3000][1-0.3000]\{48,851[1+0.0500]-51,294[0.1900+0.5800]-513\}$$
$$= 1,659 \text{ millions USD}$$

Law-3133:

If both (**H**= Home Taken Dividend= 1,659 millions USD), (**I**= Interest Expense= 513 millions USD), (**y**= Y/D= Yielding Tax Rate of Dividend= 30.00%), (**v**= V/S= Variable Portion= 19.00%), (**S**= Sales or Revenues= 51,294 millions USD), (**S'**= Sales of Past Year= 48,851 millions USD), (**f**= F/S= Fixed Portion= 58.00%), (**s**= [S/S']-1= Sales Growth= 5.00%), and (**t**= T/B= Tax Rate= 30.00%), are known, then it's (**d**= Dividend Portion or Payout planned) is:

$$\begin{aligned}
\mathbf{d} &= \mathbf{H}/([1-\mathbf{y}][1-\mathbf{t}]\{\mathbf{S'}[1+\mathbf{s}]-\mathbf{S}\mathbf{v}-\mathbf{S}\mathbf{f}-\mathbf{I}\}) \\
&= \mathbf{H}/([1-\mathbf{y}][1-\mathbf{t}]\{\mathbf{S'}[1+\mathbf{s}]-\mathbf{S}[\mathbf{v}+\mathbf{f}]-\mathbf{I}\}) \\
&= 1{,}659/([1-0.3000][1-0.3000]\{48{,}851[1+0.0500]-51{,}294[0.1900+0.5800]-513\}) \\
&= 30.00\%
\end{aligned}$$

Finance Construction-4, *Tim Asikin, Steve Asikin, Indra Senihardja*

Law-3134:

If both (**H**= Home Taken Dividend= 1,659 millions USD), (**I**= Interest Expense= 513 millions USD), (**v**= V/S= Variable Portion= 19.00%), (**d**= D/A= Dividend Portion or Payout= 30.00%), (**S**= Sales or Revenues= 51,294 millions USD), (**S'**= Sales of Past Year= 48,851 millions USD), (**f**= F/S= Fixed Portion= 58.00%), (**s**= [S/S']-1= Sales Growth= 5.00%), and (**t**= T/B= Tax Rate= 30.00%), are known, then it's (**y**= Yielding Tax Rate of Dividend planned) is:

$$y = 1-H/(d[1-t]\{S'[1+s]-Sv-Sf-I\})$$
$$= 1-H/(d[1-t]\{S'[1+s]-S[v+f]-I\})$$
$$= 1-1,659/(0.3000[1-0.3000][1-0.3000]$$
$$\{48,851[1+0.0500]-51,294$$
$$[0.1900+0.5800]-513\})$$
$$= 30.00\%$$

Law-3135:

If both (**H**= Home Taken Dividend= 1,659 millions USD), (**I**= Interest Expense= 513 millions USD), (**v**= V/S= Variable Portion= 19.00%), (**d**= D/A= Dividend Portion or Payout= 30.00%), (**S**= Sales or Revenues= 51,294 millions USD), (**S'**= Sales of Past Year= 48,851 millions USD), (**f**= F/S= Fixed Portion= 58.00%), (**s**= [S/S']-1= Sales Growth= 5.00%), and (**y**= Y/D= Yielding Tax Rate of Dividend= 30.00%), are known, then it's (**t**= Tax Rate planned) is:

$$t = 1-\mathbf{H}/(\mathbf{d}[1-\mathbf{y}]\{\mathbf{S'}[1+\mathbf{s}]-\mathbf{S}\mathbf{v}-\mathbf{S}\mathbf{f}-\mathbf{I}\})$$
$$= 1-\mathbf{H}/(\mathbf{d}[1-\mathbf{y}]\{\mathbf{S'}[1+\mathbf{s}]-\mathbf{S}[\mathbf{v}+\mathbf{f}]-\mathbf{I}\})$$
$$= 1-1,659/(0.3000[1-0.3000][1-0.3000]$$
$$\{48,851[1+0.0500]-51,294$$
$$[0.1900+0.5800]-513\})$$
$$= 30.00\%$$

Finance Construction-4, *Tim Asikin, Steve Asikin, Indra Senihardja*

Law-3136:
If both (**H**= Home Taken Dividend= 1,659 millions USD), (**I**= Interest Expense= 513 millions USD), (**v**= V/S= Variable Portion= 19.00%), (**d**= D/A= Dividend Portion or Payout= 30.00%), (**$**= Sales or Revenues= 51,294 millions USD), (**t**= T/B= Tax Rate= 30.00%), (**f**= F/S= Fixed Portion= 58.00%), (**s**= [S/S']-1= Sales Growth= 5.00%), and (**y**= Y/D= Yielding Tax Rate of Dividend= 30.00%), are known, then it's (**$'**= Sales Past) must be:

$'= ($[v+f]+I+H/{d[1-y][1-t]})/[1+s]
 = (51,294[0.1900+0.5800]+513+1,659
        /{0.3000[1-0.3000][1-0.3000]})
        /[1+0.0500]
 = 48,851 millions USD

Law-3137:
If both (**H**= Home Taken Dividend= 1,659 millions USD), (**I**= Interest Expense= 513 millions USD), (**v**= V/S= Variable Portion= 19.00%), (**d**= D/A= Dividend Portion or Payout= 30.00%), (**$**= Sales or Revenues= 51,294 millions USD), (**t**= T/B= Tax Rate= 30.00%), (**f**= F/S= Fixed Portion= 58.00%), (**$'**= Sales of Past Year= 48,851 millions USD), and (**y**= Y/D= Yielding Tax Rate of Dividend= 30.00%), are known, then it's (**s**= Sales Growth planned) is:

s= ($[v+f]+I+H/{d[1-y][1-t]})/$'-1
 = (51,294[0.1900+0.5800]+513+1,659
        /{[1-0.3000][1-0.3000]})
        /48,851-1
 = 5.00%

Law-3138:

If both (**H**= Home Taken Dividend= 1,659 millions USD), (**I**= Interest Expense= 513 millions USD), (**v**= V/S= Variable Portion= 19.00%), (**d**= D/A= Dividend Portion or Payout= 30.00%), (**s**= [S/S']-1= Sales Growth= 5.00%), (**t**= T/B= Tax Rate= 30.00%), (**f**= F/S= Fixed Portion= 58.00%), (**S'**= Sales of Past Year= 48,851 millions USD), and (**y**= Y/D= Yielding Tax Rate of Dividend= 30.00%), are known, then it's (**S**= Sales or Revenues planned) is:

$$S = (S'[1+s]-I-H/\{d[1-y][1-t]\})/[v+f]$$
$$= 48,851[1+0.0500]-513-1,659/\{0.3000[1-0.3000][1-0.3000]\})/[0.1900+0.5800]$$
$$= 51.294 \text{ millions USD}$$

Law-3139:

If both (**H**= Home Taken Dividend= 1,659 millions USD), (**I**= Interest Expense= 513 millions USD), (**$**= Sales or Revenues= 51,294 millions USD), (**s**= [S/S']-1= Sales Growth= 5.00%), (**d**= D/A= Dividend Portion or Payout= 30.00%), (**t**= T/B= Tax Rate= 30.00%), (**f**= F/S= Fixed Portion= 58.00%), (**$'**= Sales of Past Year= 48,851 millions USD), and (**y**= Y/D= Yielding Tax Rate of Dividend= 30.00%), are known, then it's (**v**= Variable Portion planned) is:

$$v = (\$'[1+s]-I-H/\{d[1-y][1-t]\})/\$-f$$

$$= 48,851[1+0.0500]-513-1,659/(0.3000$$
$$[1-0.3000][1-0.3000]\})/51,294$$
$$-0.5800$$

$$= \underline{19.00\%}$$

Law-3140:

If both (**H**= Home Taken Dividend= <u>1,659</u> millions USD), (**I**= Interest Expense= <u>513</u> millions USD), (**$**= Sales or Revenues= <u>51,294</u> millions USD), (**s**= [S/S']-1= Sales Growth= <u>5.00</u>%), (**d**= D/A= Dividend Portion or Payout= 30.00%), (**t**= T/B= Tax Rate= <u>30.00</u>%), (**v**= V/S= Variable Portion= <u>19.00</u>%), (**$'**= Sales of Past Year= <u>48,851</u> millions USD), and (**y**= Y/D= Yielding Tax Rate of Dividend= <u>30.00</u>%), are known, then it's (**f**= Fixed Portion planned) is:

$$f = (\$'[1+s] - I - H/\{d[1-y][1-t]\})/\$ - v$$
$$= 48,851[1+0.0500] - 513 - 1,659/(0.3000$$
$$[1-0.3000][1-0.3000]\})$$
$$/51,294 - 0.1900$$
$$= \underline{58.00}\%$$

Law-3141:

If both (**H**= Home Taken Dividend= 1,659 millions USD), (**f**= F/S= Fixed Portion= 58.00%), (**S**= Sales or Revenues= 51,294 millions USD), (**s**= [S/S']-1= Sales Growth= 5.00%), (**d**= D/A= Dividend Portion or Payout= 30.00%), (**t**= T/B= Tax Rate= 30.00%), (**v**= V/S= Variable Portion= 19.00%), (**S'**= Sales of Past Year= 48,851 millions USD), and (**y**= Y/D= Yielding Tax Rate of Dividend= 30.00%), are known, then it's (**I**= Interest Expense planned) is:

$$I = S'[1+s] - S[v+f] - H/\{d[1-y][1-t]\}$$
$$= 48,851[1+0.0500] - 51,294[0.1900 + 0.5800] - 1,659/\{0.3000 [1-0.3000][1-0.3000]\}$$
$$= 513 \text{ millions USD}$$

Law-3142:

If both (**i**= I/S= Interest Portion= 1.00%), (**f**= F/S= Fixed Portion= 58.00%), (**$**= Sales or Revenues= 51,294 millions USD), (**s**= [S/S']-1= Sales Growth= 5.00%), (**d**= D/A= Dividend Portion or Payout= 30.00%), (**t**= T/B= Tax Rate= 30.00%), (**v**= V/S= Variable Portion= 19.00%), (**$'**= Sales of Past Year= 48,851 millions USD), and (**y**= Y/D= Yielding Tax Rate of Dividend= 30.00%), are known, then it's (**H**= Home Taken Expense planned) is:

$$H = d[1-y][1-t]\{\$'[1+s]-\$v-\$f-\$i\}$$
$$= d[1-y][1-t]\{\$'[1+s]-\$[v+f+i]\}$$
$$= 0.3000[1-0.3000][1-0.3000]\{48,851[1+0.0500]-51,294[0.1900+0.5800+0.0100]\}$$
$$= 1,659 \text{ millions USD}$$

Law-3143:

If both (**i**= I/S= Interest Portion= 1.00%), (**f**= F/S= Fixed Portion= 58.00%), (**$**= Sales or Revenues= 51,294 millions USD), (**s**= [S/S']-1= Sales Growth= 5.00%), (**t**= T/B= Tax Rate= 30.00%), (**υ**= V/S= Variable Portion= 19.00%), (**H**= Home Taken Dividend= 1,659 millions USD),(**$'**= Sales of Past Year= 48,851 millions USD), and (**y**= Y/D= Yielding Tax Rate of Dividend= 30.00%), are known, then it's (**d**= Dividend Portion or Payout planned) is:

$$d = H/([1-y][1-t]\{\$'[1+s]-\$υ-\$f-\$i\})$$
$$= H/([1-y][1-t]\{\$'[1+s]-\$[υ+f+i]\})$$
$$= 1{,}659/([1-0.3000][1-0.3000]\{48{,}851[1+0.0500]-51{,}294[0.1900+0.5800+0.0100]\})$$
$$= 30.00\%$$

Law-3144:

If both (**i**= I/S= Interest Portion= 1.00%), (**f**= F/S= Fixed Portion= 58.00%), (**$**= Sales or Revenues= 51,294 millions USD), (**s**= [S/S']-1= Sales Growth= 5.00%), (**t**= T/B= Tax Rate= 30.00%), (**v**= V/S= Variable Portion= 19.00%), (**H**= Home Taken Dividend= 1,659 millions USD),(**$'**= Sales of Past Year= 48,851 millions USD), and (**d**= D/A= Dividend Portion or Payout= 30.00%), are known, then it's (**y**= Yielding Tax Rate of Dividend planned) is:

$$y = H/(d[1-t]\{\$'[1+s]-\$v-\$f-\$i\})$$
$$= 1-H/(d[1-t]\{\$'[1+s]-\$[v+f+i]\})$$
$$= 1-1,659/(0.3000[1-0.3000]\{48,851[1+0.0500]-51,294[0.1900+0.5800+0.0100]\})$$
$$= \underline{30.00\%}$$

Finance Construction-4, *Tim Asikin, Steve Asikin, Indra Senihardja*

Law-3145:

If both (**i**= I/S= Interest Portion= 1.00%), (**f**= F/S= Fixed Portion= 58.00%), (**$**= Sales or Revenues= 51,294 millions USD), (**s**= [S/S']-1= Sales Growth= 5.00%), (**y**= Y/D= Yielding Tax Rate of Dividend= 30.00%), (**v**= V/S= Variable Portion= 19.00%), (**H**= Home Taken Dividend= 1,659 millions USD),(**$'**= Sales of Past Year= 48,851 millions USD), and (**d**= D/A= Dividend Portion or Payout= 30.00%), are known, then it's (**t**= Tax Rate planned) is:

t = 1-**H**/(**d**[1-**y**]{**$'**[1+**s**]-**$v**-**$f**-**$i**})
  = 1-**H**/(**d**[1-**y**]{**$'**[1+**s**]-**$**[**v**+**f**+**i**]})
  = 1-1,659/(0.3000[1-0.3000]{48,851[1+0.0500]-51,294[0.1900+0.5800+0.0100]})
  = 30.00%

Law-3146:
If both (i= I/S= Interest Portion= 1.00%), (f= F/S= Fixed Portion= 58.00%), ($\$$= Sales or Revenues= 51,294 millions USD), (s= [S/S']-1= Sales Growth= 5.00%), (y= Y/D= Yielding Tax Rate of Dividend= 30.00%), (v= V/S= Variable Portion= 19.00%), (H= Home Taken Dividend= 1,659 millions USD), (t= T/B= Tax Rate= 30.00%), and (d= D/A= Dividend Portion or Payout= 30.00%), are known, then it's ($\$'$= Sales Past) must be:

$\$'$ = ($\$[v+f+i]+H/\{d[1-y][1-t]\}$)/[1+$s$]
  = (51,294[0.1900+0.5800+0.0100]
     +1,659/{0.3000][1-0.3000
     [1-0.3000]/[1+0.0500]
  = 48,851 millions USD

Law-3147:
If both (i= I/S= Interest Portion= 1.00%), (f= F/S= Fixed Portion= 58.00%), ($\$$= Sales or Revenues= 51,294 millions USD), ($\$'$= Sales of Past Year= 48,851 millions USD), (y= Y/D= Yielding Tax Rate of Dividend= 30.00%), (v= V/S= Variable Portion= 19.00%), (H= Home Taken Dividend= 1,659 millions USD), (t= T/B= Tax Rate= 30.00%), and (d= D/A= Dividend Portion or Payout= 30.00%), are known, then it's (s= Sales Growth planned) is:

s = ($\$[v+f+i]+H/\{d[1-y][1-t]\}$)/$\$'$-1
  = (51,294[0.1900+0.5800+0.0100]
     +1,659/{0.3000][1-0.3000
     [1-0.3000]/48,851-1
  = 5.00%

Law-3148:

If both (i= I/S= Interest Portion= 1.00%), (f= F/S= Fixed Portion= 58.00%), (s= [S/S']-1= Sales Growth= 5.00%), ($\$'$= Sales of Past Year= 48,851 millions USD), (y= Y/D= Yielding Tax Rate of Dividend= 30.00%), (v= V/S= Variable Portion= 19.00%), (H= Home Taken Dividend= 1,659 millions USD), (t= T/B= Tax Rate= 30.00%), and (d= D/A= Dividend Portion or Payout= 30.00%), are known, then it's ($\$$= Sales or Revenues planned) is:

$\$ = (\$'[1+s]-H/\{d[1-y][1-t]\})/[v+f+i]$
$= (48,851[1+0.0500]-1,659/\{0.3000$
$[1-0.3000][1-0.3000]\})$
$/[0.1900+0.5800+0.0100]$
$= \underline{51,294}$ millions USD

Law-3149:

If both (i= I/S= Interest Portion= 1.00%), (f= F/S= Fixed Portion= 58.00%), (s= [S/S']-1= Sales Growth= 5.00%), ($\$'$= Sales of Past Year= 48,851 millions USD), (y= Y/D= Yielding Tax Rate of Dividend= 30.00%), ($\$$= Sales or Revenues= 51,294 millions USD), (H= Home Taken Dividend= 1,659 millions USD), (t= T/B= Tax Rate= 30.00%), and (d= D/A= Dividend Portion or Payout= 30.00%), are known, then it's (v= Variable Portion planned) is:

$v = (\$'[1+s]-H/\{d[1-y][1-t]\})/\$-f-i$
$= (48,851[1+0.0500]-1,659/\{0.3000$
$[1-0.3000][1-0.3000]\})$
$/51,294-0.5800-0.0100$
$= \underline{19.00}\%$

Law-3150:

If both (i= I/S= Interest Portion= 1.00%), (v= V/S= Variable Portion= 19.00%), (s= [S/S']-1= Sales Growth= 5.00%), (S'= Sales of Past Year= 48,851 millions USD), (y= Y/D= Yielding Tax Rate of Dividend= 30.00%), (S= Sales or Revenues= 51,294 millions USD), (H= Home Taken Dividend= 1,659 millions USD), (t= T/B= Tax Rate= 30.00%), and (d= D/A= Dividend Portion or Payout= 30.00%), are known, then it's (f= Fixed Portion planned) is:

f = (S'[1+s]-H/{d[1-y][1-t]})/S-v-i
 = (48,851[1+0.0500]-1,659/{0.3000
      [1-0.3000][1-0.3000]})
      /51,294-0.1900-0.0100
 = 58.00%

Law-3151:

If both (f= F/S= Fixed Portion= 58.00%), (v= V/S= Variable Portion= 19.00%), (s= [S/S']-1= Sales Growth= 5.00%), (S'= Sales of Past Year= 48,851 millions USD), (y= Y/D= Yielding Tax Rate of Dividend= 30.00%), (S= Sales or Revenues= 51,294 millions USD), (H= Home Taken Dividend= 1,659 millions USD), (t= T/B= Tax Rate= 30.00%), and (d= D/A= Dividend Portion or Payout= 30.00%), are known, then it's (i= Interest Portion planned) is:

i = (S'[1+s]-H/{d[1-y][1-t]})/S-v-f
 = (48,851[1+0.0500]-1,659/{0.3000
      [1-0.3000][1-0.3000]})
      /51,294-0.1900-0.5800
 = 1.00%

Law-3152:

If both (**f**= F/S=Fixed Portion= 58.00%), (**v**= V/S= Variable Portion= 19.00%), (**s**= [S/S']-1= Sales Growth= 5.00%), (**$'**= Sales of Past Year= 48,851 millions USD), (**y**= Y/D= Yielding Tax Rate of Dividend= 30.00%), (**$**= Sales or Revenues= 51,294 millions USD), (**i**= I/S= Interest Portion= 1.00%), (**t**= T/B= Tax Rate= 30.00%), and (**d**= D/A= Dividend Portion or Payout= 30.00%), are known, then it's (**H**= Home Taken Dividend planned) is:

$$\begin{aligned}
\mathbf{H} &= \mathbf{d}[1-\mathbf{y}][1-\mathbf{t}]\{\$'[1+\mathbf{s}]-\$\mathbf{v}-\$\mathbf{f}-\$'\mathbf{i}[1+\mathbf{s}]\} \\
&= \mathbf{d}[1-\mathbf{y}][1-\mathbf{t}]\{\$'[1+\mathbf{s}][1-\mathbf{i}]-\$[\mathbf{v}+\mathbf{f}]\} \\
&= 0.3000[1-0.3000][1-0.3000]\{48,851 \\
&\qquad [1+0.0500][1-0.0100]-51,294 \\
&\qquad [0.1900+0.5800]\} \\
&= \underline{1,659} \text{ millions USD}
\end{aligned}$$

Law-3153:

If both (**f**= F/S= Fixed Portion= 58.00%), (**v**= V/S= Variable Portion= 19.00%), (**s**= [S/S']-1= Sales Growth= 5.00%), (**$'**= Sales of Past Year= 48,851 millions USD), (**y**= Y/D= Yielding Tax Rate of Dividend= 30.00%), (**$**= Sales or Revenues= 51,294 millions USD), (**i**= I/S= Interest Portion= 1.00%), (**t**= T/B= Tax Rate= 30.00%), and (**H**= Home Taken Dividend= 1,659 millions USD), are known, then it's (**d**= Dividend Portion or Payout planned) is:

$$d = H/([1-y][1-t]\{\$'[1+s]-\$v-\$f-\$'i[1+s]\})$$
$$= H/([1-y][1-t]\{\$'[1+s][1-i]-\$[v+f]\})$$
$$= 1{,}659/([1-0.3000][1-0.3000]\{48{,}851[1+0.0500][1-0.0100]-51{,}294[0.1900+0.5800]\})$$
$$= 30.00\%$$

Finance Construction-4, *Tim Asikin, Steve Asikin, Indra Senihardja*

Law-3154:

If both (**f**= F/S= Fixed Portion= 58.00%), (**v**= V/S= Variable Portion= 19.00%), (**s**= [S/S']-1= Sales Growth= 5.00%), (**$'**= Sales of Past Year= 48,851 millions USD), (**d**= D/A= Dividend Portion or Payout= 30.00%), (**$**= Sales or Revenues= 51,294 millions USD), (**i**= I/S= Interest Portion= 1.00%), (**t**= T/B= Tax Rate= 30.00%), and (**H**= Home Taken Dividend= 1,659 millions USD),are known, then it's (**y**= Yielding Tax Rate of Dividend planned) is:

$$y = 1 - H/(d[1-t]\{\$'[1+s] - \$v - \$f - \$'i[1+s]\})$$
$$= 1 - H/(d[1-t]\{\$'[1+s][1-i] - \$[v+f]\})$$
$$= 1 - 1{,}659/(0.3000[1-0.3000]\{48{,}851$$
$$[1+0.0500][1-0.0100] - 51{,}294$$
$$[0.1900+0.5800]\}$$
$$= 30.00\%$$

Law-3155:

If both (**f**= F/S= Fixed Portion= 58.00%), (**v**= V/S= Variable Portion= 19.00%), (**s**= [S/S']-1= Sales Growth= 5.00%), (**$'**= Sales of Past Year= 48,851 millions USD), (**d**= D/A= Dividend Portion or Payout= 30.00%), (**$**= Sales or Revenues= 51,294 millions USD), (**i**= I/S= Interest Portion= 1.00%), (**y**= Y/D= Yielding Tax Rate of Dividend= 30.00%), and (**H**= Home Taken Dividend= 1,659 millions USD), are known, then it's (**y**= Yielding Tax Rate of Dividend planned) is:

$$t = 1 - H/(d[1-y]\{\$'[1+s][1-i]-\$[v+f]\})$$
$$= 1-1,659/(\ 0.3000[1-0.3000]\{48,851$$
$$[1+0.0500][1-0.0100]-51,294$$
$$[0.1900+0.5800]\})$$
$$= \underline{30.00\%}$$

Law-3156:

If both (**f**= F/S= Fixed Portion= 58.00%), (**v**= V/S= Variable Portion= 19.00%), (**s**= [S/S']-1= Sales Growth= 5.00%), (**d**= D/A= Dividend Portion or Payout= 30.00%), (**t**= T/B= Tax Rate= 30.00%), (**$**= Sales or Revenues= 51,294 millions USD), (**i**= I/S= Interest Portion= 1.00%), (**y**= Y/D= Yielding Tax Rate of Dividend= 30.00%), and (**H**= Home Taken Dividend= 1,659 millions USD), are known, then it's (**$'**= Sales Past) must be:

$'= ($[v+f]+H/{d[1-y][1-t]})/{[1+s][1-i]}
  = (51,294[0.1900+0.5800]+1,659
      /{0.3000[1-0.3000][1-0.3000]})
      /{[1+0.0500][1-0.0100]}
  = 48,851 millions USD

Law-3157:

If both (**f**= F/S= Fixed Portion= 58.00%), (**v**= V/S= Variable Portion= 19.00%), (**$'**= Sales of Past Year= 48,851 millions USD), (**d**= D/A= Dividend Portion or Payout= 30.00%), (**t**= T/B= Tax Rate= 30.00%), (**$**= Sales or Revenues= 51,294 millions USD), (**i**= I/S= Interest Portion= 1.00%), (**y**= Y/D= Yielding Tax Rate of Dividend= 30.00%), and (**H**= Home Taken Dividend= 1,659 millions USD), are known, then it's (**s**= Sales Growth planned) is:

s= ($[v+f]+H/{d[1-y][1-t]})/{$'[1-i]}-1
  = (51,294[0.1900+0.5800]+1,659
      /{0.3000[1-0.3000][1-0.3000]})
      /{48,851[1-0.0100]}-1
  = 5.00%

Law-3158:

If both (**f**= F/S= Fixed Portion= 58.00%), (**v**= V/S= Variable Portion= 19.00%), (**$'**= Sales of Past Year= 48,851 millions USD), (**d**= D/A= Dividend Portion or Payout= 30.00%), (**t**= T/B= Tax Rate= 30.00%), (**$**= Sales or Revenues= 51,294 millions USD), (**s**= [S/S']-1= Sales Growth= 5.00%), (**y**= Y/D= Yielding Tax Rate of Dividend= 30.00%), and (**H**= Home Taken Dividend= 1,659 millions USD), are known, then it's (**i**= Interest Portion planned) is:

$$i = 1-(\$[v+f]+H/\{d[1-y][1-t]\})/\{\$'[1+s]\}$$
$$= 1-(51,294[0.1900+0.5800]+1,659$$
$$/\{0.3000[1-0.3000][1-0.3000]\})$$
$$/\{48,851[1+0.0500]\}$$
$$= 1.00\%$$

Law-3159:

If both (**f**= F/S= Fixed Portion= 58.00%), (**v**= V/S= Variable Portion= 19.00%), (**$'**= Sales of Past Year= 48,851 millions USD), (**d**= D/A= Dividend Portion or Payout= 30.00%), (**t**= T/B= Tax Rate= 30.00%), (**i**= I/S= Interest Portion= 1.00%), (**s**= [S/S']-1= Sales Growth= 5.00%), (**y**= Y/D= Yielding Tax Rate of Dividend= 30.00%), and (**H**= Home Taken Dividend= 1,659 millions USD), are known, then it's (**$**= Sales or Revenues planned) is:

$$\$ = (\$'[1+s][1-i]-H/\{d[1-y][1-t]\})/[v+f]$$
$$= (48,851[1+0.0500][1-0.0100]-1,659$$
$$/\{0.3000[1-0.3000][1-0.3000]\})$$
$$/ [0.1900+0.5800]$$
$$= 51,294 \text{ millions USD}$$

Law-3160:

If both (**f**= F/S= Fixed Portion= 58.00%), (**$**= Sales or Revenues= 51,294 millions USD), (**$'**= Sales of Past Year= 48,851 millions USD), (**d**= D/A= Dividend Portion or Payout= 30.00%), (**t**= T/B= Tax Rate= 30.00%), (**i**= I/S= Interest Portion= 1.00%), (**s**= [S/S']-1= Sales Growth= 5.00%), (**y**= Y/D= Yielding Tax Rate of Dividend= 30.00%), and (**H**= Home Taken Dividend= 1,659 millions USD), are known, then it's (**v**= Variable Portion planned) is:

$$v = (\$'[1+s][1-i]-H/\{d[1-y][1-t]\})/\$-f$$
$$= (48,851[1+0.0500][1-0.0100]-1,659$$
$$/\{0.3000[1-0.3000][1-0.3000]\})$$
$$/51,294-0.5800$$
$$= 19.00\%$$

Law-3161:

If both (**v**= V/S= Variable Portion= 19.00%), (**$**= Sales or Revenues= 51,294 millions USD), (**$'**= Sales of Past Year= 48,851 millions USD), (**d**= D/A= Dividend Portion or Payout= 30.00%), (**t**= T/B= Tax Rate= 30.00%), (**i**= I/S= Interest Portion= 1.00%), (**s**= [S/S']-1= Sales Growth= 5.00%), (**y**= Y/D= Yielding Tax Rate of Dividend= 30.00%), and (**H**= Home Taken Dividend= 1,659 millions USD), are known, then it's (**f**= Fixed Portion planned) is:

$$f = (\$'[1+s][1-i]-H/\{d[1-y][1-t]\})/\$-v$$
$$= (48,851[1+0.0500][1-0.0100]-1,659$$
$$/\{0.3000[1-0.3000][1-0.3000]\})$$
$$/51,294-0.1900$$
$$= 58.00\%$$

Law-3162:

If both (v= V/S= Variable Portion= 19.00%), (S= Sales or Revenues= 51,294 millions USD), (S'= Sales of Past Year= 48,851 millions USD), (f= F/S= Fixed Portion= 58.00%), (d= D/A= Dividend Portion or Payout= 30.00%), (t= T/B= Tax Rate= 30.00%), (I= Interest Expense= 513 millions USD), (s= [S/S']-1= Sales Growth= 5.00%), and (y= Y/D= Yielding Tax Rate of Dividend= 30.00%), and are known, then it's (H= Home Taken Dividend planned) is:

$$H = d[1-y][1-t]\{S'[1+s]-Sv-S'f[1+s]-I\}$$
$$= d[1-y][1-t]\{S'[1+s][1-f]-Sv-I\}$$
$$= 0.3000[1-0.3000][1-0.3000]\{48,851[1+0.0500][1-0.5800]-51,294*0.1900-513\}$$
$$= \underline{1,659} \text{ millions USD}$$

Finance Construction-4, *Tim Asikin, Steve Asikin, Indra Senihardja*

Law-3163:

If both (**v**= V/S= Variable Portion= 19.00%), (**$**= Sales or Revenues= 51,294 millions USD), (**$'**= Sales of Past Year= 48,851 millions USD), (**f**= F/S= Fixed Portion= 58.00%), (**H**= Home Taken Dividend= 1,659 millions USD), (**t**= T/B= Tax Rate= 30.00%), (**I**= Interest Expense= 513 millions USD), (**s**= [S/S']-1= Sales Growth= 5.00%), and (**y**= Y/D= Yielding Tax Rate of Dividend= 30.00%), are known, then it's (**d**= Dividend Portion or Payout planned) is:

$$d = H/([1-y][1-t]\{\$'[1+s]-\$v-\$'f[1+s]-I\})$$
$$= H/([1-y][1-t]\{\$'[1+s][1-f]-\$v-I\})$$
$$= 1,659/([1-0.3000][1-0.3000]\{48,851[1+0.0500][1-0.5800]-51,294*0.1900-513\})$$
$$= 30.00\%$$

Law-3164:

If both (**v**= V/S= Variable Portion= 19.00%), (**S**= Sales or Revenues= 51,294 millions USD), (**S'**= Sales of Past Year= 48,851 millions USD), (**f**= F/S= Fixed Portion= 58.00%), (**H**= Home Taken Dividend= 1,659 millions USD), (**t**= T/B= Tax Rate= 30.00%), (**I**= Interest Expense= 513 millions USD), (**s**= [S/S']-1= Sales Growth= 5.00%), and (**d**= D/A= Dividend Portion or Payout= 30.00%), are known, then it's (**y**= Yielding Tax Rate of Dividend planned) is:

$$y = 1 - H/(d[1-t]\{S'[1+s] - Sv - S'f[1+s] - I\})$$
$$= 1 - H/(d[1-t]\{S'[1+s][1-f] - Sv - I\})$$
$$= 1 - 1{,}659/(0.3000[1-0.3000]\{48{,}851[1+0.0500][1-0.5800] - 51{,}294 * 0.1900 - 513\})$$
$$= \underline{30.00\%}$$

Finance Construction-4, *Tim Asikin, Steve Asikin, Indra Senihardja*

Law-3165:

If both (v= V/S= Variable Portion= 19.00%), (S= Sales or Revenues= 51,294 millions USD), (S'= Sales of Past Year= 48,851 millions USD), (f= F/S=Fixed Portion= 58.00%), (H= Home Taken Dividend= 1,659 millions USD), (y= Y/D= Yielding Tax Rate of Dividend= 30.00%), (I= Interest Expense= 513 millions USD), (s= [S/S']-1= Sales Growth= 5.00%), and (d= D/A= Dividend Portion or Payout= 30.00%), are known, then it's (t= Tax Rate planned) is:

t = 1-H/(d[1-y]{S'[1+s]-Sv-$S'f$[1+s]-I})
  = 1-H/(d[1-y]{S'[1+s][1-f]-Sv-I})
  = 1-1,659/(0.3000[1-0.3000]{48,851[1+0.0500][1-0.5800]-51,294*0.1900-513})
  = 30.00%

Law-3166:

If both (**v**= V/S= Variable Portion= 19.00%), (**S**= Sales or Revenues= 51,294 millions USD), (**f**= F/S= Fixed Portion= 58.00%), (**H**= Home Taken Dividend= 1,659 millions USD), (**y**= Y/D= Yielding Tax Rate of Dividend= 30.00%), (**t**= T/B= Tax Rate= 30.00%), (**I**= Interest Expense= 513 millions USD), (**s**= [S/S']-1= Sales Growth= 5.00%), and (**d**= D/A= Dividend Portion or Payout= 30.00%), are known, then it's (**S'**= Sales Past) must be:

$$S' = (Sv + I + H/\{d[1-y][1-t]\})/\{[1+s][1-f]\})$$
$$= (51,294*0.1900 + 513 + 1,659/\{0.3000[1-0.3000][1-0.3000]\})/\{[1+0.0500][1-0.5800]\}$$
$$= \underline{48,851} \text{ millions USD}$$

Law-3167:

If both (**v**= V/S= Variable Portion= 19.00%), (**$**= Sales or Revenues= 51,294 millions USD), (**f**= F/S= Fixed Portion= 58.00%), (**H**= Home Taken Dividend= 1,659 millions USD), (**y**= Y/D= Yielding Tax Rate of Dividend= 30.00%), (**t**= T/B= Tax Rate= 30.00%), (**I**= Interest Expense= 513 millions USD), (**$'**= Sales of Past Year= 48,851 millions USD), and (**d**= D/A= Dividend Portion or Payout= 30.00%), are known, then it's (**s**= Sales Growth planned) is:

$$s = (\$v + I + H / \{d[1-y][1-t]\}) / \{\$'[1-f]\} - 1$$
$$= (51{,}294 * 0.1900 + 513 + 1{,}659 / \{0.3000 [1-0.3000][1-0.3000]\}) / \{48{,}851 [1-0.5800]\} - 1$$
$$= 5.00\%$$

Law-3168:

If both (**v**= V/S= Variable Portion= 19.00%), (**$**= Sales or Revenues= 51,294 millions USD), (**s**= [S/S']-1= Sales Growth= 5.00%), (**H**= Home Taken Dividend= 1,659 millions USD), (**y**= Y/D= Yielding Tax Rate of Dividend= 30.00%), (**t**= T/B= Tax Rate= 30.00%), (**I**= Interest Expense= 513 millions USD), (**$'**= Sales of Past Year= 48,851 millions USD), and (**d**= D/A= Dividend Portion or Payout= 30.00%), are known, then it's (**f**= Fixed Portion planned) is:

$$f = 1-(\$v+I+H/\{d[1-y][1-t]\})/\{\$'[1+s]\}$$
$$= 1-(51,294*0.1900+513+1,659$$
$$/\{0.3000[1-0.3000][1-0.3000]\})$$
$$/\{48,851[1+0.0500]\}$$
$$= 58.00\%$$

Finance Construction-4, *Tim Asikin, Steve Asikin, Indra Senihardja*

Law-3169:

If both (v= V/S= Variable Portion= 19.00%), (f= F/S= Fixed Portion= 58.00%), (s= [S/S']-1= Sales Growth= 5.00%), (H= Home Taken Dividend= 1,659 millions USD), (y= Y/D= Yielding Tax Rate of Dividend= 30.00%), (t= T/B= Tax Rate= 30.00%), (I= Interest Expense= 513 millions USD), (S'= Sales of Past Year= 48,851 millions USD), and (d= D/A= Dividend Portion or Payout= 30.00%), are known, then it's (S= Sales or Revenues planned) is:

$$S = (S'[1+s][1-f]-I-H/\{d[1-y][1-t]\})/v$$
$$= (48,851[1+0.0500][1-0.5800]-513$$
$$-1,659/\{0.3000[1-0.3000]$$
$$[1-0.3000]\})/0.1900$$
$$= 51,294 \text{ millions USD}$$

Law-3170:

If both ($\$$= Sales or Revenues= 51,294 millions USD), (f= F/S= Fixed Portion= 58.00%), (s= [S/S']-1= Sales Growth= 5.00%), (H= Home Taken Dividend= 1,659 millions USD), (y= Y/D= Yielding Tax Rate of Dividend= 30.00%), (t= T/B= Tax Rate= 30.00%), (I= Interest Expense= 513 millions USD), ($\$'$= Sales of Past Year= 48,851 millions USD), and (d= D/A= Dividend Portion or Payout= 30.00%), are known, then it's (v= Variable Portion planned) is:

$$v = (\$'[1+s][1-f]-I-H/\{d[1-y][1-t]\})/\$$$
$$= (48,851[1+0.0500][1-0.5800]-513$$
$$-1,659/\{0.3000[1-0.3000]$$
$$[1-0.3000]\})/51,294$$
$$= 19.00\%$$

Law-3171:

If both ($= Sales or Revenues= 51,294 millions USD), (f= F/S= Fixed Portion= 58.00%), (s= [S/S']-1= Sales Growth= 5.00%), (H= Home Taken Dividend= 1,659 millions USD), (y= Y/D= Yielding Tax Rate of Dividend= 30.00%), (t= T/B= Tax Rate= 30.00%), (v= V/S= Variable Portion= 19.00%), ($'= Sales of Past Year= 48,851 millions USD), and (d= D/A= Dividend Portion or Payout= 30.00%), are known, then it's (I= Interest Expense planned) is:

$$I = \$'[1+s][1-f] - \$v - H/\{d[1-y][1-t]\}$$
$$= 48,851[1+0.0500][1-0.5800] - 51,294$$
$$*0.1900 - 1,659/\{0.3000$$
$$[1-0.3000][1-0.3000]\}$$
$$= 513 \text{ millions USD}$$

Finance Construction-4, *Tim Asikin, Steve Asikin, Indra Senihardja*

Law-3172:

If both ($\$$= Sales or Revenues= 51,294 millions USD), (f= F/S= Fixed Portion= 58.00%), (s= [S/S']-1= Sales Growth= 5.00%), (i= I/S= Interest Portion= 1.00%), (y= Y/D= Yielding Tax Rate of Dividend= 30.00%), (t= T/B= Tax Rate= 30.00%), (v= V/S= Variable Portion= 19.00%), ($\$'$= Sales of Past Year= 48,851 millions USD), and (d= D/A= Dividend Portion or Payout= 30.00%), are known, then it's (H= Home Taken Dividend planned) is:

$H = d[1-y][1-t]\{\$'[1+s]-\$v-\$'f[1+s]-\$i\}$
$= d[1-y][1-t]\{\$'[1+s][1-f]-\$[v+i]\}$
$= 0.3000[1-0.3000][1-0.3000]\{48,851$
$[1+0.0500][1-0.5800]-51,294$
$[0.1900+0.0100]\}$
$= \underline{1,659}$ millions USD

Law-3173:

If both (S= Sales or Revenues= 51,294 millions USD), (f= F/S= Fixed Portion= 58.00%), (s= [S/S']-1= Sales Growth= 5.00%), (i= I/S= Interest Portion= 1.00%), (y= Y/D= Yielding Tax Rate of Dividend= 30.00%), (t= T/B= Tax Rate= 30.00%), (v= V/S= Variable Portion= 19.00%), (S'= Sales of Past Year= 48,851 millions USD), and (H= Home Taken Dividend= 1,659 millions USD), are known, then it's (d= Dividend Portion or Payout planned) is:

$$d = H/([1-y][1-t]\{S'[1+s]-Sv-S'f[1+s]-Si\})$$
$$= H/([1-y][1-t]\{S'[1+s][1-f]-S[v+i]\})$$
$$= 1{,}659/([1-0.3000][1-0.3000]\{48{,}851[1+0.0500][1-0.5800]-51{,}294[0.1900+0.0100]\})$$
$$= 30.00\%$$

Law-3174:

If both ($= Sales or Revenues= 51,294 millions USD), (f= F/S= Fixed Portion= 58.00%), (s= [S/S']-1= Sales Growth= 5.00%), (i= I/S= Interest Portion= 1.00%), (d= D/A= Dividend Portion or Payout= 30.00%), (t= T/B= Tax Rate= 30.00%), (v= V/S= Variable Portion= 19.00%), ($'= Sales of Past Year= 48,851 millions USD), and (H= Home Taken Dividend= 1,659 millions USD), are known, then it's (y= Yielding Tax Rate of Dividend planned) is:

$$y = 1-H/(d[1-t]\{\$'[1+s]-\$v-\$'f[1+s]-\$i\})$$
$$= 1-H/(d[1-t]\{\$'[1+s][1-f]-\$[v+i]\})$$
$$= 1-1,659/(0.3000[1-0.3000]\{48,851[1+0.0500][1-0.5800]-51,294[0.1900+0.0100]\})$$
$$= \underline{30.00\%}$$

Law-3175:

If both (**$** = Sales or Revenues = 51,294 millions USD), (**f** = F/S = Fixed Portion = 58.00%), (**s** = [S/S']-1 = Sales Growth = 5.00%), (**i** = I/S = Interest Portion = 1.00%), (**d** = D/A = Dividend Portion or Payout = 30.00%), (**y** = Y/D = Yielding Tax Rate of Dividend = 30.00%), (**v** = V/S = Variable Portion = 19.00%), (**$'** = Sales of Past Year = 48,851 millions USD), and (**H** = Home Taken Dividend = 1,659 millions USD), are known, then it's (**t** = Tax Rate planned) is:

$$t = 1-H/(d[1-y]\{\$'[1+s]-\$v-\$'f[1+s]-\$i\})$$
$$= 1-H/(d[1-y]\{\$'[1+s][1-f]-\$[v+i]\})$$
$$= 1-1,659/(0.3000[1-0.3000]\{48,851[1+0.0500][1-0.5800]-51,294[0.1900+0.0100]\})$$
$$= 30.00\%$$

Finance Construction-4, *Tim Asikin, Steve Asikin, Indra Senihardja*

<u>Law-3176</u>:

If both ($= Sales or Revenues= 51,294 millions USD), (**f**= F/S= Fixed Portion= 58.00%), (**s**= [S/S']-1= Sales Growth= 5.00%), (**i**= I/S= Interest Portion= 1.00%), (**d**= D/A= Dividend Portion or Payout= 30.00%), (**y**= Y/D= Yielding Tax Rate of Dividend= 30.00%), (**v**= V/S= Variable Portion= 19.00%), (**t**= T/B= Tax Rate= 30.00%), and (**H**= Home Taken Dividend= 1,659 millions USD), are known, then it's ($'= Sales Past) must be:

$'= ($[**v**+**i**]+**H**/{**d**[1-**y**][1-**t**]})/{[1+**s**][1-**f**]}
 = (51.294[0.1900+0.0100]+1,659/{0.3000 [1-0.3000][1-0.3000]})
  /{[1+0.0500][1-0.5800]}
 = 48,851 millions USD

Law-3177:

If both ($= Sales or Revenues= 51,294 millions USD), (f= F/S= Fixed Portion= 58.00%), ($'= Sales of Past Year= 48,851 millions USD), (i= I/S= Interest Portion= 1.00%), (d= D/A= Dividend Portion or Payout= 30.00%), (y= Y/D= Yielding Tax Rate of Dividend= 30.00%), (v= V/S= Variable Portion= 19.00%), (t= T/B= Tax Rate= 30.00%), and (H= Home Taken Dividend= 1,659 millions USD), are known, then it's (s= Sales Growth planned) is:

s = ($[v+i]+H/\{d[1-y][1-t]\})/\{$'[1-f]\}-1
= (51,294[0.1900+0.0100]+1,659
   /{0.3000[1-0.3000][1-0.3000]})
   /{48,851[1-0.5800]}-1
= 5.00%

Law-3178:

If both ($= Sales or Revenues= 51,294 millions USD), (s= [S/S']-1= Sales Growth= 5.00%), ($'= Sales of Past Year= 48,851 millions USD), (i= I/S= Interest Portion= 1.00%), (d= D/A= Dividend Portion or Payout= 30.00%), (y= Y/D= Yielding Tax Rate of Dividend= 30.00%), (v= V/S= Variable Portion= 19.00%), (t= T/B= Tax Rate= 30.00%), and (H= Home Taken Dividend= 1,659 millions USD), are known, then it's (f= Fixed Portion planned) is:

f = 1-($[v+i]+H/\{d[1-y][1-t]\})/\{$'[1+s]\}
= 1-(51,294[0.1900+0.0100]+1,659
   /{0.3000[1-0.3000][1-0.3000]})
   /{48,851[1+0.0500]}
= 58.00%

Law-3179:

If both (**f**= F/S=Fixed Portion= 58.00%), (**s**= [S/S']-1= Sales Growth= 5.00%), (**$'**= Sales of Past Year= 48,851 millions USD), (**i**= I/S= Interest Portion= 1.00%), (**d**= D/A= Dividend Portion or Payout= 30.00%), (**y**= Y/D= Yielding Tax Rate of Dividend= 30.00%), (**v**= V/S= Variable Portion= 19.00%), (**t**= T/B= Tax Rate= 30.00%), and (**H**= Home Taken Dividend= 1,659 millions USD), are known, then it's (**$**= Sales or Revenues planned) is:

$= ($'[1+s][1-f]-H/{d[1-y][1-t]})/[v+i]
 = (48,851[1+0.0500][1-0.5800]
   -1,659/{0.3000[1-0.3000]
   [1-0.3000]})/[0.1900+0.0100]
 = 51,294 millions USD

Law-3180:

If both (**f**= F/S= Fixed Portion= 58.00%), (**s**= [S/S']-1= Sales Growth= 5.00%), (**$'**= Sales of Past Year= 48,851 millions USD), (**i**= I/S= Interest Portion= 1.00%), (**d**= D/A= Dividend Portion or Payout= 30.00%), (**y**= Y/D= Yielding Tax Rate of Dividend= 30.00%), (**$**= Sales or Revenues= 51,294 millions USD), (**t**= T/B= Tax Rate= 30.00%), and (**H**= Home Taken Dividend= 1,659 millions USD), are known, then it's (**v**= Variable Portion planned) is:

v= ($'[1+s][1-f]-H/{d[1-y][1-t]})/$-i
 = (48,851[1+0.0500][1-0.5800]
   -1,659/{0.3000[1-0.3000]
   [1-0.3000]})/51,294-0.0100
 = 19.00%

Law-3181:

If both (**f**= F/S= Fixed Portion= 58.00%), (**s**= [S/S']-1= Sales Growth= 5.00%), (**$'**= Sales of Past Year= 48,851 millions USD), (**v**= V/S= Variable Portion= 19.00%), (**d**= D/A= Dividend Portion or Payout= 30.00%), (**y**= Y/D= Yielding Tax Rate of Dividend= 30.00%), (**$**= Sales or Revenues= 51,294 millions USD), (**t**= T/B= Tax Rate= 30.00%), and (**H**= Home Taken Dividend= 1,659 millions USD), are known, then it's (**i**= Interest Portion planned) is:

$$\begin{aligned}\mathbf{i} &= (\mathbf{\$'}[1+\mathbf{s}][1-\mathbf{f}]-\mathbf{H}/\{\mathbf{d}[1-\mathbf{y}][1-\mathbf{t}]\})/\mathbf{\$}-\mathbf{v}\\ &= (48{,}851[1+0.0500][1-0.5800]\\ &\quad -1{,}659/\{0.3000[1-0.3000]\\ &\quad [1-0.3000]\})/51{,}294-0.1900\\ &= 1.00\%\end{aligned}$$

Law-3182:

If both (**f**= F/S= Fixed Portion= 58.00%), (**s**= [S/S']-1= Sales Growth= 5.00%), (**$'**= Sales of Past Year= 48,851 millions USD), (**v**= V/S= Variable Portion= 19.00%), (**d**= D/A=Dividend Portion or Payout= 30.00%), (**y**= Y/D= Yielding Tax Rate of Dividend= 30.00%), (**$**= Sales or Revenues= 51,294 millions USD), (**t**= T/B= Tax Rate= 30.00%), and (**i**= I/S= Interest Portion= 1.00%), are known, then it's (**H**= Home Taken Dividend planned) is:

$$\begin{aligned} \mathbf{H} &= \mathbf{d}[1\text{-}\mathbf{y}][1\text{-}\mathbf{t}]\{\$'[1+\mathbf{s}]\text{-}\$\mathbf{v}\text{-}\$'\mathbf{f}[1+\mathbf{s}]\text{-}\$'\mathbf{i}[1+\mathbf{s}]\} \\ &= \mathbf{d}[1\text{-}\mathbf{y}][1\text{-}\mathbf{t}]\{\$'[1+\mathbf{s}][1\text{-}\mathbf{f}\text{-}\mathbf{i}]\text{-}\$\mathbf{v}\} \\ &= 0.3000[1\text{-}0.3000][1\text{-}0.3000]\{48,851 \\ &\qquad [1+0.0500][1\text{-}0.5800\text{-}0.0100] \\ &\qquad \text{-}51,294*0.1900\} \\ &= \underline{1,659} \text{ millions USD} \end{aligned}$$

Law-3183:

If both (**f**= F/S= Fixed Portion= 58.00%), (**s**= [S/S']-1= Sales Growth= 5.00%), (**$'**= Sales of Past Year= 48,851 millions USD), (**v**= V/S= Variable Portion= 19.00%), (**H**= Home Taken Dividend= 1,659 millions USD), (**y**= Y/D= Yielding Tax Rate of Dividend= 30.00%), (**$**= Sales or Revenues= 51,294 millions USD), (**t**= T/B= Tax Rate= 30.00%), and (**i**= I/S= Interest Portion= 1.00%), are known, then it's (**d**= Dividend Portion or Payout planned) is:

$$d = H/([1-y][1-t]\{\$'[1+s]-\$v-\$'f[1+s]-\$'i[1+s]\})$$
$$= H/([1-y][1-t]\{\$'[1+s][1-f-i]-\$v\})$$
$$= 1,659/([1-0.3000][1-0.3000]\{48,851$$
$$[1+0.0500][1-0.5800-0.0100]$$
$$-51,294*0.1900\})$$
$$= 30.00\%$$

Finance Construction-4, *Tim Asikin, Steve Asikin, Indra Senihardja*

Law-3184:

If both (**f**= F/S= Fixed Portion= 58.00%), (**s**= [S/S']-1= Sales Growth= 5.00%), (**S'**= Sales of Past Year= 48,851 millions USD), (**v**= V/S= Variable Portion= 19.00%), (**H**= Home Taken Dividend= 1,659 millions USD), (**d**= D/A= Dividend Portion or Payout= 30.00%), (**S**= Sales or Revenues= 51,294 millions USD), (**t**= T/B= Tax Rate= 30.00%), and (**i**= I/S= Interest Portion= 1.00%), are known, then it's (**y**= Yielding Tax Rate of Dividend planned) is:

$$y = 1 - H/(d[1-t]\{S'[1+s] - Sv - S'f[1+s] - S'i[1+s]\})$$
$$= 1 - H/(d[1-t]\{S'[1+s][1-f-i] - Sv\})$$
$$= 1 - 1{,}659/(0.3000[1-0.3000]\{48{,}851[1+0.0500][1-0.5800-0.0100] - 51{,}294*0.1900\})$$
$$= 30.00\%$$

Finance Construction-4, *Tim Asikin, Steve Asikin, Indra Senihardja*

Law-3185:

If both (**f**= F/S=Fixed Portion= 58.00%), (**s**= [S/S']-1= Sales Growth= 5.00%), (**$'**= Sales of Past Year= 48,851 millions USD), (**v**= V/S= Variable Portion= 19.00%), (**H**= Home Taken Dividend= 1,659 millions USD), (**d**= D/A= Dividend Portion or Payout= 30.00%), (**$**= Sales or Revenues= 51,294 millions USD), (**y**= Y/D= Yielding Tax Rate of Dividend= 30.00%), and (**i**= I/S= Interest Portion= 1.00%), are known, then it's (**t**= Tax Rate planned) is:

$$\begin{aligned}
\mathbf{t} &= 1-\mathbf{H}/(\mathbf{d}[1-\mathbf{y}]\{\mathbf{\$'}[1+\mathbf{s}]-\mathbf{\$v}-\mathbf{\$'f}[1+\mathbf{s}]-\mathbf{\$'i}[1+\mathbf{s}]\}) \\
&= 1-\mathbf{H}/(\mathbf{d}[1-\mathbf{y}]\{\mathbf{\$'}[1+\mathbf{s}][1-\mathbf{f}-\mathbf{i}]-\mathbf{\$v}\}) \\
&= 1-1{,}659/(0.3000[1-0.3000]\{48{,}851 \\
&\qquad [1+0.0500][1-0.5800-0.0100] \\
&\qquad -51{,}294*0.1900\}) \\
&= 30.00\%
\end{aligned}$$

Law-3186:

If both (f= F/S=Fixed Portion= 58.00%), (s= [S/S']-1= Sales Growth= 5.00%), (t= T/B= Tax Rate= 30.00%), (v= V/S= Variable Portion= 19.00%), (H= Home Taken Dividend= 1,659 millions USD), (d= D/A= Dividend Portion or Payout= 30.00%), ($\$$= Sales or Revenues= 51,294 millions USD), (y= Y/D= Yielding Tax Rate of Dividend= 30.00%), and (i= I/S= Interest Portion= 1.00%), are known, then it's ($\$'$= Sales Past) must be:

$$\$' = (\$v + H/\{d[1-y][1-t]\})/\{[1+s][1-f-i]\}$$
$$= (51{,}294 * 0.1900 + 1{,}659/\{0.3000[1-0.3000][1-0.3000]\})/\{[1+0.0500][1-0.5800-0.0100]\}$$
$$= 48{,}851 \text{ millions USD}$$

Finance Construction-4, *Tim Asikin, Steve Asikin, Indra Senihardja*

Law-3187:

If both (**f**= F/S= Fixed Portion= 58.00%), ($'= Sales of Past Year= 48,851 millions USD), (**t**= T/B= Tax Rate= 30.00%), (**v**= V/S= Variable Portion= 19.00%), (**H**= Home Taken Dividend= 1,659 millions USD), (**d**= D/A= Dividend Portion or Payout= 30.00%), ($= Sales or Revenues= 51,294 millions USD), (**y**= Y/D= Yielding Tax Rate of Dividend= 30.00%), and (**i**= I/S= Interest Portion= 1.00%), are known, then it's (**s**= Sales Growth planned) is:

s = ($\$v$+**H**/{**d**[1-**y**][1-**t**]})/{$\$'$[1-**f**-**i**]}-1
 = (51,294*0.1900+1,659/{0.3000
    [1-0.3000][1-0.3000]})/{48,851
    [1-0.5800-0.0100]}-1
 = 5.00%

Law-3188:

If both (s= [S/S']-1= Sales Growth= 5.00%), ($\$'$= Sales of Past Year= 48,851 millions USD), (t= T/B= Tax Rate= 30.00%), (v= V/S= Variable Portion= 19.00%), (H= Home Taken Dividend= 1,659 millions USD), (d= D/A= Dividend Portion or Payout= 30.00%), ($\$$= Sales or Revenues= 51,294 millions USD), (y= Y/D= Yielding Tax Rate of Dividend= 30.00%), and (i= I/S= Interest Portion= 1.00%), are known, then it's (f= Fixed Portion planned) is:

f = 1-i-($\$v$+$H$/{$d$[1-$y$][1-$t$]})/{$\$'$[1+$s$]}
 = 1-0.0100-(51,294*0.1900+1,659
  /{0.3000[1-0.3000]
   [1-0.3000]})/{48,851
   [1+0.0500]}
 = 58.00%

Finance Construction-4, *Tim Asikin, Steve Asikin, Indra Senihardja*

Law-3189:

If both (s= [S/S']-1= Sales Growth= 5.00%), (S'= Sales of Past Year= 48,851 millions USD), (t= T/B= Tax Rate= 30.00%), (v= V/S= Variable Portion= 19.00%), (H= Home Taken Dividend= 1,659 millions USD), (d= D/A= Dividend Portion or Payout= 30.00%), (S= Sales or Revenues= 51,294 millions USD), (y= Y/D= Yielding Tax Rate of Dividend= 30.00%), and (f= F/S= Fixed Portion= 58.00%), are known, then it's (i= Interest Portion planned) is:

i= 1-f-(Sv+H/{d[1-y][1-t]})/{S'[1+s]}
    = 1-0.5800-(51,294*0.1900+1,659
        /{0.3000[1-0.3000]
        [1-0.3000]})/{48,851
        [1+0.0500]}
    = 1.00%

Finance Construction-4, *Tim Asikin, Steve Asikin, Indra Senihardja*

Law-3190:
If both (s= [S/S']-1= Sales Growth= 5.00%), ($\$'$= Sales of Past Year= 48,851 millions USD), (t= T/B= Tax Rate= 30.00%), (v= V/S= Variable Portion= 19.00%), (H= Home Taken Dividend= 1,659 millions USD), (d= D/A= Dividend Portion or Payout= 30.00%), (i= I/S= Interest Portion= 1.00%), (y= Y/D= Yielding Tax Rate of Dividend= 30.00%), and (f= F/S= Fixed Portion= 58.00%), are known, then it's ($\$$= Sales or Revenues planned) is:

$\$ = (\$'[1+s][1-f-i]-H/\{d[1-y][1-t]\})/v$
= (48,851[1-0.5800-0.0100]-1,659
  /{0.3000[1-0.3000]
  [1-0.3000]})/0.1900
= 51,294 millions USD

Law-3191:
If both (s= [S/S']-1= Sales Growth= 5.00%), ($\$'$= Sales of Past Year= 48,851 millions USD), (t= T/B= Tax Rate= 30.00%), ($\$$= Sales or Revenues= 51,294 millions USD), (H= Home Taken Dividend= 1,659 millions USD), (d= D/A= Dividend Portion or Payout= 30.00%), (i= I/S= Interest Portion= 1.00%), (y= Y/D= Yielding Tax Rate of Dividend= 30.00%), and (f= F/S= Fixed Portion= 58.00%), are known, then it's (v= Variable Portion planned) is:

$v = (\$'[1+s][1-f-i]-H/\{d[1-y][1-t]\})/\$$
= (48,851[1-0.5800-0.0100]-1,659
  /{0.3000[1-0.3000]
  [1-0.3000]})/51,294
= 19.00%

Law-3192:

If both (s= [S/S']-1= Sales Growth= 5.00%), (S'= Sales of Past Year= 48,851 millions USD), (t= T/B= Tax Rate= 30.00%), (v= V/S= Variable Portion= 19.00%), (d= D/A= Dividend Portion or Payout= 30.00%), (I= Interest Expense= 513 millions USD), (y= Y/D= Yielding Tax Rate of Dividend= 30.00%), and (F= Fixed Cost= 29,750 millions USD), are known, then it's (H= Home Taken Dividend planned) is:

$$H = d[1-y][1-t]\{S'[1+s]-S'v[1+s]-F-I\}$$
$$= d[1-y][1-t]\{S'[1+s][1-v]-F-I\}$$
$$= 0.3000[1-0.3000][1-0.3000]\{48,851$$
$$[1-0.1900]-29,750-513\}$$
$$= 1,659 \text{ millions USD}$$

Law-3193:

If both (s= [S/S']-1= Sales Growth= 5.00%), (S'= Sales of Past Year= 48,851 millions USD), (t= T/B= Tax Rate= 30.00%), (v= V/S= Variable Portion= 19.00%), (H= Home Taken Dividend= 1,659 millions USD), (I= Interest Expense= 513 millions USD), (y= Y/D= Yielding Tax Rate of Dividend= 30.00%), and (F= Fixed Cost= 29,750 millions USD), are known, then it's (d= Dividend Portion or Payout planned) is:

$$d = H/([1-y][1-t]\{S'[1+s]-S'v[1+s]-F-I\})$$
$$= H/([1-y][1-t]\{S'[1+s][1-v]-F-I\})$$
$$= 1,659/([1-0.3000][1-0.3000]\{48,851$$
$$[1-0.1900]-29,750-513\})$$
$$= 30.00\%$$

Law-3194:

If both (s= [S/S']-1= Sales Growth= 5.00%), ($\$'$= Sales of Past Year= 48,851 millions USD), (t= T/B= Tax Rate= 30.00%), (v= V/S= Variable Portion= 19.00%), (H= Home Taken Dividend= 1,659 millions USD), (I= Interest Expense= 513 millions USD), (d= D/A= Dividend Portion or Payout= 30.00%), and (F= Fixed Cost= 29,750 millions USD), are known, then it's (y= Yielding Tax Rate of Dividend planned) is:

y= 1-H/(d[1-t]{$\$'$[1+$s$]-$\$'v$[1+$s$]-$F$-$I$})
= 1-H/(d[1-t]{$\$'$[1+$s$][1-$v$]-$F$-$I$})
= 1-1,659/(0.3000[1-0.3000]{48,851
  [1-0.1900]-29,750-513})
= 30.00%

Law-3195:

If both (s= [S/S']-1= Sales Growth= 5.00%), ($\$'$= Sales of Past Year= 48,851 millions USD), (y= Y/D= Yielding Tax Rate of Dividend= 30.00%), (v= V/S= Variable Portion= 19.00%), (H= Home Taken Dividend= 1,659 millions USD), (I= Interest Expense= 513 millions USD), (d= D/A= Dividend Portion or Payout= 30.00%), and (F= Fixed Cost= 29,750 millions USD), are known, then it's (t= Tax Rate planned) is:

t= 1-H/(d[1-y]{$\$'$[1+$s$]-$\$'v$[1+$s$]-$F$-$I$})
= 1-H/(d[1-y]{$\$'$[1+$s$][1-$v$]-$F$-$I$})
= 1-1,659/(0.3000[1-0.3000]{48,851
  [1-0.1900]-29,750-513})
= 30.00%

Law-3196:

If both (**s**= [S/S']-1= Sales Growth= 5.00%), (**t**= T/B= Tax Rate= 30.00%), (**y**= Y/D= Yielding Tax Rate of Dividend= 30.00%), (**v**= V/S= Variable Portion= 19.00%), (**H**= Home Taken Dividend= 1,659 millions USD), (**I**= Interest Expense= 513 millions USD), (**d**= D/A= Dividend Portion or Payout= 30.00%), and (**F**= Fixed Cost= 29,750 millions USD), are known, then it's (**S'**= Sales Past) must be:

$$S' = (F+I+H/\{d[1-y][1-t]\})/\{[1+s][1-v]\}$$
$$= (29{,}750+513+1{,}659/\{0.3000[1-0.3000][1-0.3000]\})/\{[1+0.0500][1-0.1900]\}$$
$$= 48{,}851 \text{ millions USD}$$

Law-3197:

If both (**S'**= Sales of Past Year= 48,851 millions USD), (**t**= T/B= Tax Rate= 30.00%), (**y**= Y/D= Yielding Tax Rate of Dividend= 30.00%), (**v**= V/S= Variable Portion= 19.00%), (**H**= Home Taken Dividend= 1,659 millions USD), (**I**= Interest Expense= 513 millions USD), (**d**= D/A= Dividend Portion or Payout= 30.00%), and (**F**= Fixed Cost= 29,750 millions USD), are known, then it's (**s**= Sales Growth planned) is:

$$s = (F+I+H/\{d[1-y][1-t]\})/\{S'[1-v]\}-1$$
$$= (29{,}750+513+1{,}659/\{0.3000[1-0.3000][1-0.3000]\})/\{48{,}851[1-0.1900]\}-1$$
$$= 5.00\%$$

Law-3198:
If both ($'= Sales of Past Year= 48,851 millions USD), (t= T/B= Tax Rate= 30.00%), (y= Y/D= Yielding Tax Rate of Dividend= 30.00%), (s= [S/S']-1= Sales Growth= 5.00%), (H= Home Taken Dividend= 1,659 millions USD), (I= Interest Expense= 513 millions USD), (d= D/A= Dividend Portion or Payout= 30.00%), and (F= Fixed Cost= 29,750 millions USD), are known, then it's (v= Variable Portion planned) is:

v= 1-(F+I+H/{d[1-y][1-t]})/{$'[1+s]}
  = 1-(29,750+513+1,659/{0.3000
        [1-0.3000][1-0.3000]})/{48,851
        [1+0.0500]}
  = 19.00%

Law-3199:

If both ($'= Sales of Past Year= 48,851 millions USD), (t= T/B= Tax Rate= 30.00%), (y= Y/D= Yielding Tax Rate of Dividend= 30.00%), (s= [S/S']-1= Sales Growth= 5.00%), (H= Home Taken Dividend= 1,659 millions USD), (I= Interest Expense= 513 millions USD), (d= D/A= Dividend Portion or Payout= 30.00%), and (v= V/S= Variable Portion= 19.00%), are known, then it's (F= Fixed Cost planned) is:

$$F = \$'[1+s][1-v] - I - H/\{d[1-y][1-t]\}$$
$$= 48,851[1+0.0500][1-0.1900] - 513$$
$$\quad -1,659/\{0.3000[1-0.3000]$$
$$\quad [1-0.3000]\}$$
$$= 29,750 \text{ millions USD}$$

Law-3200:

If both ($'= Sales of Past Year= 48,851 millions USD), (t= T/B= Tax Rate= 30.00%), (y= Y/D= Yielding Tax Rate of Dividend= 30.00%), (s= [S/S']-1= Sales Growth= 5.00%), (H= Home Taken Dividend= 1,659 millions USD), (F= Fixed Cost= 29,750 millions USD), (d= D/A= Dividend Portion or Payout= 30.00%), and (v= V/S= Variable Portion= 19.00%), are known, then it's (I= Interest Expense planned) is:

$$I = \$'[1+s][1-v] - F - H/\{d[1-y][1-t]\}$$
$$= 48,851[1+0.0500][1-0.1900] - 29,750$$
$$\quad -1,659/\{0.3000[1-0.3000]$$
$$\quad [1-0.3000]\}$$
$$= 513 \text{ millions USD}$$

Law-3201:

If both ($\$'$= Sales of Past Year= 48,851 millions USD), (**t**= T/B= Tax Rate= 30.00%), (**y**= Y/D= Yielding Tax Rate of Dividend= 30.00%), (**s**= [S/S']-1= Sales Growth= 5.00%), (**i**= I/S= Interest Portion= 1.00%), (**F**= Fixed Cost= 29,750 millions USD), ($\$$= Sales or Revenues= 51,294 millions USD), (**d**= D/A= Dividend Portion or Payout= 30.00%), and (**v**= V/S= Variable Portion= 19.00%), are known, then it's (**H**= Home Taken Dividend planned) is:

$$\begin{aligned}
\mathbf{H} &= \mathbf{d}[1-\mathbf{y}][1-\mathbf{t}]\{\$'[1+\mathbf{s}]-\$'\mathbf{v}[1+\mathbf{s}]-\mathbf{F}-\$\mathbf{i}\} \\
&= \mathbf{d}[1-\mathbf{y}][1-\mathbf{t}]\{\$'[1+\mathbf{s}][1-\mathbf{v}]-\mathbf{F}-\$\mathbf{i}\} \\
&= 0.3000[1-0.3000][1-0.3000]\{48,851 \\
&\qquad [1+0.0500][1-0.1900]-29,750 \\
&\qquad -51,294*0.0100\} \\
&= \underline{1,659} \text{ millions USD}
\end{aligned}$$

Law-3202:

If both ($**'**= Sales of Past Year= 48,851 millions USD), (**t**= T/B= Tax Rate= 30.00%), (**y**= Y/D= Yielding Tax Rate of Dividend= 30.00%), (**s**= [S/S']-1= Sales Growth= 5.00%), (**i**= I/S= Interest Portion= 1.00%), (**F**= Fixed Cost= 29,750 millions USD), ($= Sales or Revenues= 51,294 millions USD), (**H**= Home Taken Dividend= 1,659 millions USD), and (**v**= V/S= Variable Portion= 19.00%), are known, then it's (**d**= Dividend Portion or Payout planned) is:

$$d = H/([1-y][1-t]\{\$'[1+s]-\$'v[1+s]-F-\$i\})$$
$$= H/([1-y][1-t]\{\$'[1+s][1-v]-F-\$i\})$$
$$= 1,659/([1-0.3000][1-0.3000]\{48,851\\ [1+0.0500][1-0.1900]-29,750\\ -51,294*0.0100\})$$
$$= 30.00\%$$

Law-3203:

If both ($'= Sales of Past Year= 48,851 millions USD), (t= T/B= Tax Rate= 30.00%), (d= D/A= Dividend Portion or Payout= 30.00%), (s= [S/S']-1= Sales Growth= 5.00%), (i= I/S= Interest Portion= 1.00%), (F= Fixed Cost= 29,750 millions USD), ($= Sales or Revenues= 51,294 millions USD), (H= Home Taken Dividend= 1,659 millions USD), and (v= V/S= Variable Portion= 19.00%), are known, then it's (y= Yielding Tax Rate of Dividend planned) is:

$$y = 1 - H/(d[1-t]\{\$'[1+s] - \$'v[1+s] - F - \$i\})$$
$$= 1 - H/(d[1-t]\{\$'[1+s][1-v] - F - \$i\})$$
$$= 1 - 1{,}659/(0.3000[1-0.3000]\{48{,}851[1+0.0500][1-0.1900] - 29{,}750 - 51{,}294*0.0100\})$$
$$= \underline{30.00\%}$$

Law-3204:

If both ($\$'$= Sales of Past Year= <u>48,851</u> millions USD), (y= Y/D= Yielding Tax Rate of Dividend= <u>30.00</u>%), (d= D/A= Dividend Portion or Payout= 30.00%), (s= [S/S']-1= Sales Growth= <u>5.00</u>%), (i= I/S= Interest Portion= <u>1.00</u>%), (F= Fixed Cost= <u>29,750</u> millions USD), ($\$$= Sales or Revenues= <u>51,294</u> millions USD), (H= Home Taken Dividend= <u>1,659</u> millions USD), and (v= V/S= Variable Portion= <u>19.00</u>%), are known, then it's (t= Yielding Tax Rate of Dividend planned) is:

t= 1-**H**/(**d**[1-**y**]{$\$'$[1+**s**][1-**v**]-**F**-**\$i**})
 =1-1,659/(0.3000[1-0.3000]{48,851
 [1+0.0500][1-0.1900]-29,750
 -51,294*0.0100})
 = <u>30.00</u>%

Law-3205:

If both (**t**= T/B= Tax Rate= 30.00%), (**y**= Y/D= Yielding Tax Rate of Dividend= 30.00%), (**d**= D/A= Dividend Portion or Payout= 30.00%), (**s**= [S/S']-1= Sales Growth= 5.00%), (**i**= I/S= Interest Portion= 1.00%), (**F**= Fixed Cost= 29,750 millions USD), (**$**= Sales or Revenues= 51,294 millions USD), (**H**= Home Taken Dividend= 1,659 millions USD), and (**v**= V/S= Variable Portion= 19.00%), are known, then it's (**$'**= Sales Past) must be:

$$\$' = (F+\$i+H/\{d[1-y][1-t]\})/\{[1+s][1-v]\}$$
$$= (29,750+51,294*0.0100+1,659$$
$$/(0.3000[1-0.3000][1-0.3000]\})$$
$$/\{[1+0.0500][1-0.1900]\}$$
$$= \underline{48,851} \text{ millions USD}$$

Finance Construction-4, *Tim Asikin, Steve Asikin, Indra Senihardja*

Law-3206:

If both (**t**= T/B= Tax Rate= 30.00%), (**y**= Y/D= Yielding Tax Rate of Dividend= 30.00%), (**d**= D/A= Dividend Portion or Payout= 30.00%), (**S'**= Sales of Past Year= 48,851 millions USD), (**i**= I/S= Interest Portion= 1.00%), (**F**= Fixed Cost= 29,750 millions USD), (**S**= Sales or Revenues= 51,294 millions USD), (**H**= Home Taken Dividend= 1,659 millions USD), and (**v**= V/S= Variable Portion= 19.00%), are known, then it's (**s**= Sales Growth planned) is:

$$s = (F+Si+H/\{d[1-y][1-t]\})/\{S'[1-v]\}-1$$
$$= (29,750+51,294*0.0100+1,659$$
$$/(0.3000[1-0.3000][1-0.3000]\})$$
$$/\{48,851[1-0.1900]\}-1$$
$$= 5.00\%$$

Law-3207:

If both (**t**= T/B= Tax Rate= 30.00%), (**y**= Y/D= Yielding Tax Rate of Dividend= 30.00%), (**d**= D/A= Dividend Portion or Payout= 30.00%), (**$'**= Sales of Past Year= 48,851 millions USD), (**i**= I/S= Interest Portion= 1.00%), (**F**= Fixed Cost= 29,750 millions USD), (**$**= Sales or Revenues= 51,294 millions USD), (**H**= Home Taken Dividend= 1,659 millions USD), and (**s**= [S/S']-1= Sales Growth= 5.00%), are known, then it's (**v**= Variable Portion planned) is:

$$v = 1-(F+\$i+H/\{d[1-y][1-t]\})/\{\$'[1+s]\}$$
$$= 1-(29,750+51,294*0.0100+1,659$$
$$/(0.3000[1-0.3000][1-0.3000]\})$$
$$/\{48,851[1+0.0500]\}$$
$$= 19.00\%$$

Law-3208:

If both (**t**= T/B= Tax Rate= 30.00%), (**y**= Y/D= Yielding Tax Rate of Dividend= 30.00%), (**d**= D/A= Dividend Portion or Payout= 30.00%), (**$'**= Sales of Past Year= 48,851 millions USD), (**i**= I/S= Interest Portion= 1.00%), (**v**= V/S= Variable Portion= 19.00%), (**$**= Sales or Revenues= 51,294 millions USD), (**H**= Home Taken Dividend= 1,659 millions USD), and (**s**= [S/S']-1= Sales Growth= 5.00%), are known, then it's (**F**= Fixed Cost planned) is:

$$F = \$'[1+s][1-v]-\$i-H/\{d[1-y][1-t]\}$$
$$= 48,851[1+0.0500][1-0.1900]-51,294$$
$$*0.0100-1,659/\{0.3000$$
$$[1-0.3000][1-0.3000]\})$$
$$= 29,750 \text{ millions USD}$$

Finance Construction-4, *Tim Asikin, Steve Asikin, Indra Senihardja*

Law-3209:

If both (**t**= T/B= Tax Rate= 30.00%), (**y**= Y/D= Yielding Tax Rate of Dividend= 30.00%), (**d**= D/A= Dividend Portion or Payout= 30.00%), (**$'**= Sales of Past Year= 48,851 millions USD), (**i**= I/S= Interest Portion= 1.00%), (**v**= V/S= Variable Portion= 19.00%), (**F**= Fixed Cost= 29,750 millions USD), (**H**= Home Taken Dividend= 1,659 millions USD), and (**s**= [S/S']-1= Sales Growth= 5.00%), are known, then it's (**$**= Sales or Revenues planned) is:

$= ($'[1+s][1-v]-F-H/{d[1-y][1-t]})/i
    = (48,851[1+0.0500][1-0.1900]-29,750
       -1,659/{0.3000[1-0.3000]
       [1-0.3000]})/0.0100
    = 51,294 millions USD

Law-3210:

If both (**t**= T/B= Tax Rate= 30.00%), (**y**= Y/D= Yielding Tax Rate of Dividend= 30.00%), (**d**= D/A= Dividend Portion or Payout= 30.00%), (**$'**= Sales of Past Year= 48,851 millions USD), (**$**= Sales or Revenues= 51,294 millions USD), (**v**= V/S= Variable Portion= 19.00%), (**F**= Fixed Cost= 29,750 millions USD), (**H**= Home Taken Dividend= 1,659 millions USD), and (**s**= [S/S']-1= Sales Growth= 5.00%), are known, then it's (**i**= Interest Portion planned) is:

$i = (\$'[1+s][1-v]-F-H/\{d[1-y][1-t]\})/\$$
$= (48,851[1+0.0500][1-0.1900]-29,750$
$\quad -1,659/\{0.3000[1-0.3000]$
$\quad [1-0.3000]\})/51,294$
$= \underline{1.00\%}$

Finance Construction-4, *Tim Asikin, Steve Asikin, Indra Senihardja*

Law-3211:

If both (t= T/B= Tax Rate= 30.00%), (y= Y/D= Yielding Tax Rate of Dividend= 30.00%), (d= D/A= Dividend Portion or Payout= 30.00%), (S'= Sales of Past Year= 48,851 millions USD), (S= Sales or Revenues= 51,294 millions USD), (v= V/S= Variable Portion= 19.00%), (F= Fixed Cost= 29,750 millions USD), (i= I/S= Interest Portion= 1.00%), and (s= [S/S']-1= Sales Growth= 5.00%), are known, then it's (H= Home Taken Dividend planned) is:

$$H = d[1-y][1-t]\{S'[1+s] - S'v[1+s] - F - S'i[1+s]\}$$
$$= d[1-y][1-t]\{S'[1+s][1-v-i] - F\}$$
$$= 0.3000[1-0.3000][1-0.3000]\{48,851$$
$$[1+0.0500][1-0.1900-0.0100]$$
$$-29,750\}$$
$$= \underline{1,659} \text{ millions USD}$$

Law-3212:

If both (**t**= T/B= Tax Rate= 30.00%), (**y**= Y/D= Yielding Tax Rate of Dividend= 30.00%), (**H**= Home Taken Dividend= 1,659 millions USD), (**$'**= Sales of Past Year= 48,851 millions USD), (**$**= Sales or Revenues= 51,294 millions USD), (**v**= V/S= Variable Portion= 19.00%), (**F**= Fixed Cost= 29,750 millions USD), (**i**= I/S= Interest Portion= 1.00%), and (**s**= [S/S']-1= Sales Growth= 5.00%), are known, then it's (**d**= Dividend Portion or Payout planned) is:

$$\begin{aligned}
\mathbf{d} &= \mathbf{H}/([1-\mathbf{y}][1-\mathbf{t}]\{\$'[1+\mathbf{s}]-\$'\mathbf{v}[1+\mathbf{s}]-\mathbf{F}-\$'\mathbf{i}[1+\mathbf{s}]\}) \\
&= \mathbf{H}/([1-\mathbf{y}][1-\mathbf{t}]\{\$'[1+\mathbf{s}][1-\mathbf{v}-\mathbf{i}]-\mathbf{F}\}) \\
&= 1,659/([1-0.3000][1-0.3000]\{48,851 \\
&\qquad [1+0.0500][1-0.1900-0.0100] \\
&\qquad -29,750\}) \\
&= \underline{30.00\%}
\end{aligned}$$

Finance Construction-4, *Tim Asikin, Steve Asikin, Indra Senihardja*

Law-3213:

If both (**t**= T/B= Tax Rate= 30.00%), (**d**= D/A= Dividend Portion or Payout= 30.00%), (**H**= Home Taken Dividend= 1,659 millions USD), (**$'**= Sales of Past Year= 48,851 millions USD), (**$**= Sales or Revenues= 51,294 millions USD), (**v**= V/S= Variable Portion= 19.00%), (**F**= Fixed Cost= 29,750 millions USD), (**i**= I/S= Interest Portion= 1.00%), and (**s**= [S/S']-1= Sales Growth= 5.00%), are known, then it's (**y**= Yielding Tax Rate of Dividend planned) is:

$$y = 1-\mathbf{H}/(\mathbf{d}[1-\mathbf{t}]\{\$'[1+\mathbf{s}]-\$'\mathbf{v}[1+\mathbf{s}]-\mathbf{F}-\$'\mathbf{i}[1+\mathbf{s}]\})$$
$$= 1-\mathbf{H}/(\mathbf{d}[1-\mathbf{t}]\{\$'[1+\mathbf{s}][1-\mathbf{v}-\mathbf{i}]-\mathbf{F}\})$$
$$= 1-1{,}659/(0.3000[1-0.3000]\{48{,}851[1+0.0500][1-0.1900-0.0100]-29{,}750\})$$
$$= 30.00\%$$

Law-3214:

If both (y= Y/D= Yielding Tax Rate of Dividend= 30.00%), (d= D/A= Dividend Portion or Payout= 30.00%), (H= Home Taken Dividend= 1,659 millions USD), (S'= Sales of Past Year= 48,851 millions USD), (S= Sales or Revenues= 51,294 millions USD), (v= V/S= Variable Portion= 19.00%), (F= Fixed Cost= 29,750 millions USD), (i= I/S= Interest Portion= 1.00%), and (s= [S/S']-1= Sales Growth= 5.00%), are known, then it's (t= Tax Rate planned) is:

t= 1-H/(d[1-y]{S'[1+s][1-v-i]-F})
  = 1-1,659/(0.3000[1-0.3000]{48,851
      [1+0.0500][1-0.1900-0.0100]
      -29,750})
  = 30.00%

Law-3215:

If both (y= Y/D= Yielding Tax Rate of Dividend= 30.00%), (d= D/A= Dividend Portion or Payout= 30.00%), (H= Home Taken Dividend= 1,659 millions USD), (t= T/B= Tax Rate= 30.00%), (S= Sales or Revenues= 51,294 millions USD), (v= V/S= Variable Portion= 19.00%), (F= Fixed Cost= 29,750 millions USD), (i= I/S= Interest Portion= 1.00%), and (s= [S/S']-1= Sales Growth= 5.00%), are known, then it's (S'= Sales Past) must be:

$$S' = (F+H/\{d[1-y][1-t]\})/[\{1+s\}[1-v-i]]$$
$$= (29{,}750 + 1{,}659/\{0.3000\,[1-0.3000]\,[1-0.3000]\})/\{[1+0.0500]\,[1-0.1900-0.0100]\}$$
$$= \underline{48{,}851} \text{ millions USD}$$

Law-3216:

If both (y= Y/D= Yielding Tax Rate of Dividend= 30.00%), (d= D/A= Dividend Portion or Payout= 30.00%), (H= Home Taken Dividend= 1,659 millions USD), (t= T/B= Tax Rate= 30.00%), (S= Sales or Revenues= 51,294 millions USD), (v= V/S= Variable Portion= 19.00%), (F= Fixed Cost= 29,750 millions USD), (i= I/S= Interest Portion= 1.00%), and (S'= Sales of Past Year= 48,851 millions USD), are known, then it's (s= Sales Growth planned) is:

$$s = (F+H/\{d[1-y][1-t]\})/\{S'[1-v-i]\} - 1$$
$$= (29{,}750 + 1{,}659/\{0.3000[1-0.3000]\,[1-0.3000]\})/\{48{,}851[1-0.1900-0.0100]\} - 1$$
$$= \underline{5.00\%}$$

Law-3217:

If both (y= Y/D= Yielding Tax Rate of Dividend= 30.00%), (d= D/A= Dividend Portion or Payout= 30.00%), (H= Home Taken Dividend= 1,659 millions USD), (t= T/B= Tax Rate= 30.00%), (S= Sales or Revenues= 51,294 millions USD), (s= [S/S']-1= Sales Growth= 5.00%), (F= Fixed Cost= 29,750 millions USD), (i= I/S= Interest Portion= 1.00%), and (S'= Sales of Past Year= 48,851 millions USD), are known, then it's (v= Variable Portion planned) is:

v = 1-i-(F+H/{d[1-y][1-t]})/{S'[1+s]}
    = 1-0.0100-(29,750+1,659/{0.3000 [1-0.3000][1-0.3000]})/{48,851 [1+0.0500]}
    = 19.00%

Finance Construction-4, *Tim Asikin, Steve Asikin, Indra Senihardja*

Law-3218:

If both (y= Y/D= Yielding Tax Rate of Dividend= 30.00%), (d= D/A= Dividend Portion or Payout= 30.00%), (H= Home Taken Dividend= 1,659 millions USD), (t= T/B= Tax Rate= 30.00%), (S= Sales or Revenues= 51,294 millions USD), (s= [S/S']-1= Sales Growth= 5.00%), (F= Fixed Cost= 29,750 millions USD), (v= V/S= Variable Portion= 19.00%), and (S'= Sales of Past Year= 48,851 millions USD), are known, then it's (i= Interest Portion planned) is:

i = 1-v-(F+H/{d[1-y][1-t]})/{S'[1+s]}
 = 1-0.1900-(29,750+1,659/{0.3000
   [1-0.3000][1-0.3000]})/{48,851
   [1+0.0500]}
 = 1.00%

Law-3219:

If both (y= Y/D= Yielding Tax Rate of Dividend= 30.00%), (d= D/A= Dividend Portion or Payout= 30.00%), (H= Home Taken Dividend= 1,659 millions USD), (t= T/B= Tax Rate= 30.00%), (S= Sales or Revenues= 51,294 millions USD), (s= [S/S']-1= Sales Growth= 5.00%), (i= I/S= Interest Portion= 1.00%), (v= V/S= Variable Portion= 19.00%), and (S'= Sales of Past Year= 48,851 millions USD), are known, then it's (F= Fixed Cost planned) is:

$$F = S'[1+s][1-v-i] - H/\{d[1-y][1-t]\}$$
$$= 48,851[1+0.0500][1-0.1900-0.0100]$$
$$\quad -1,659/\{0.3000[1-0.3000]$$
$$\quad [1-0.3000]\}$$
$$= \underline{29,750} \text{ millions USD}$$

Law-3220:

If both (y= Y/D= Yielding Tax Rate of Dividend= <u>30.00</u>%), (d= Dividend Portion or Payout= <u>30.00</u>% millions USD), (f= F/S= Fixed Portion= <u>58.00</u>%), ($\$$= Sales or Revenues= <u>51,294</u> millions USD), (s= [S/S']-1= Sales Growth= <u>5.00</u>%), (t= T/B= Tax Rate= <u>30.00</u>%), (I= Interest Expense= <u>513</u> millions USD), (v= V/S= Variable Portion= <u>19.00</u>%), and ($\$'$= Sales of Past Year= <u>48,851</u> millions USD), are known, then it's (H= Home Taken Dividend planned) is:

$$H = d[1-y][1-t]\{\$'[1+s]-\$'v[1+s]-\$f-I\}$$
$$= d[1-y][1-t]\{\$'[1+s][1-v]-\$f-I\}$$
$$= 1,659/([1-0.3000][1-0.3000]\{48,851$$
$$[1+0.0500][1-0.1900]-51,294$$
$$*0.5800-513\})$$
$$= \underline{30.00}\%$$

Law-3221:

If both (y= Y/D= Yielding Tax Rate of Dividend= 30.00%), (H= Home Taken Dividend= 1,659 millions USD), (f= F/S= Fixed Portion= 58.00%), (S= Sales or Revenues= 51,294 millions USD), (s= [S/S']-1= Sales Growth= 5.00%), (t= T/B= Tax Rate= 30.00%), (I= Interest Expense= 513 millions USD), (v= V/S= Variable Portion= 19.00%), and (S'= Sales of Past Year= 48,851 millions USD), are known, then it's (d= Dividend Portion planned) is:

d = H/([1-y][1-t]{S'[1+s]-$S'v$[1+s]-Sf-I]})
 = H/([1-y][1-t]{S'[1+s][1-v]-Sf-I})
 = 1,659/([1-0.3000][1-0.3000]{48,851 [1+0.0500][1-0.1900]-51,294 *0.5800-513})
 = 30.00%

Law-3222:

If both (**d**= D/A= Dividend Portion or Payout= 30.00%), (**H**= Home Taken Dividend= 1,659 millions USD), (**f**= F/S= Fixed Portion= 58.00%), (**$**= Sales or Revenues= 51,294 millions USD), (**s**= [S/S']-1= Sales Growth= 5.00%), (**t**= T/B= Tax Rate= 30.00%), (**I**= Interest Expense= 513 millions USD), (**v**= V/S= Variable Portion= 19.00%), and (**$'**= Sales of Past Year= 48,851 millions USD), are known, then it's (**y**= Yielding Tax Rate of Dividend planned) is:

$$y = 1 - H/(d[1-t]\{\$'[1+s] - \$'v[1+s] - \$f - I\})$$
$$= 1 - H/(d[1-t]\{\$'[1+s][1-v] - \$f - I\})$$
$$= 1 - 1{,}659/(0.3000\,[1-0.3000]\{48{,}851\,[1+0.0500][1-0.1900] - 51{,}294*0.5800 - 513\})$$
$$= 30.00\%$$

Law-3223:

If both (**d**= D/A= Dividend Portion or Payout= 30.00%), (**H**= Home Taken Dividend= 1,659 millions USD), (**f**= F/S= Fixed Portion= 58.00%), (**S**= Sales or Revenues= 51,294 millions USD), (**s**= [S/S']-1= Sales Growth= 5.00%), (**y**= Y/D= Yielding Tax Rate of Dividend= 30.00%), (**I**= Interest Expense= 513 millions USD), (**v**= V/S= Variable Portion= 19.00%), and (**S'**= Sales of Past Year= 48,851 millions USD), are known, then it's (**t**= Tax Rate planned) is:

$$t = 1 - H/(d[1-y]\{S'[1+s][1-v]-Sf-I\})$$
$$= 1-1,659/(0.3000[1-0.3000]\{48,851[1+0.0500][1-0.1900]-51,294*0.5800-513\})$$
$$= \underline{30.00\%}$$

Finance Construction-4, *Tim Asikin, Steve Asikin, Indra Senihardja*

Law-3224:

If both (**d**= D/A= Dividend Portion or Payout= 30.00%), (**H**= Home Taken Dividend= 1,659 millions USD),(**f**= F/S=Fixed Portion= 58.00%), (**$**= Sales or Revenues= 51,294 millions USD), (**s**= [S/S']-1= Sales Growth= 5.00%), (**y**= Y/D= Yielding Tax Rate of Dividend= 30.00%), (**I**= Interest Expense= 513 millions USD), (**v**= V/S= Variable Portion= 19.00%), and (**t**= T/B= Tax Rate= 30.00%), are known, then it's (**$'**= Sales Past) must be:

$' = ($f+I+H/{d[1-y][1-t]})/{[1+s][1-v]}
   = (51,294*0.5800+513+1,659/{0.3000
       [1-0.3000][1-.3000]})
       /{[1+0.0500][1-0.1900]}
   = 48,851 millions USD

Finance Construction-4, *Tim Asikin, Steve Asikin, Indra Senihardja*

Law-3225:

If both (**d**= D/A= Dividend Portion or Payout= 30.00%), (**H**= Home Taken Dividend= 1,659 millions USD),(**f**= F/S= Fixed Portion= 58.00%), (**$**= Sales or Revenues= 51,294 millions USD), (**$'**= Sales of Past Year= 48,851 millions USD), (**y**= Y/D= Yielding Tax Rate of Dividend= 30.00%), (**I**= Interest Expense= 513 millions USD), (**v**= V/S= Variable Portion= 19.00%), and (**t**= T/B= Tax Rate= 30.00%), are known, then it's (**s**= Sales Growth planned) is:

$$s= (\$f+I+H/\{d[1-y][1-t]\})/\{\$'[1-v]\}-1$$
$$= (51,294*0.5800+513+1,659/\{0.3000[1-0.3000][1-.3000]\})/\{48,851[1-0.1900]\}-1$$
$$= \underline{5.00\%}$$

Finance Construction-4, *Tim Asikin, Steve Asikin, Indra Senihardja*

Law-3226:

If both (**d**= D/A= Dividend Portion or Payout= 30.00%), (**H**= Home Taken Dividend= 1,659 millions USD),(**f**= F/S= Fixed Portion= 58.00%), (**$**= Sales or Revenues= 51,294 millions USD), (**$'**= Sales of Past Year= 48,851 millions USD), (**y**= Y/D= Yielding Tax Rate of Dividend= 30.00%), (**I**= Interest Expense= 513 millions USD), (**s**= [S/S']-1= Sales Growth= 5.00%), and (**t**= T/B= Tax Rate= 30.00%), are known, then it's (**v**= Variable Portion planned) is:

$$v = 1-(\$f+I+H/\{d[1-y][1-t]\})/\{\$'[1+s]\}$$
$$= 1-(51{,}294*0.5800+513+1{,}659/\{0.3000[1-0.3000][1-0.3000]\})/\{48{,}851[1+0.0500]\}$$
$$= 19.00\%$$

Law-3227:

If both (**d**= D/A= Dividend Portion or Payout= 30.00%), (**H**= Home Taken Dividend= 1,659 millions USD), (**f**= F/S= Fixed Portion= 58.00%), (**v**= V/S= Variable Portion= 19.00%), (**$'**= Sales of Past Year= 48,851 millions USD), (**y**= Y/D= Yielding Tax Rate of Dividend= 30.00%), (**I**= Interest Expense= 513 millions USD), (**s**= [S/S']-1= Sales Growth= 5.00%), and (**t**= T/B= Tax Rate= 30.00%), are known, then it's (**$**= Sales or Revenues planned) is:

$$\$ = (\$'[1+s][1-v]-I-H/\{d[1-y][1-t]\})/f$$
$$= (48,851[1+0.0500][1-0.1900]-513$$
$$-1,659/\{0.3000[1-0.3000]$$
$$[1-0.3000]\})/0.5800$$
$$= \underline{51,294} \text{ millions USD}$$

Finance Construction-4, *Tim Asikin, Steve Asikin, Indra Senihardja*

Law-3228:

If both (**d**= D/A= Dividend Portion or Payout= 30.00%), (**H**= Home Taken Dividend= 1,659 millions USD), (**S**= Sales or Revenues= 51,294 millions USD), (**v**= V/S= Variable Portion= 19.00%), (**S'**= Sales of Past Year= 48,851 millions USD), (**y**= Y/D= Yielding Tax Rate of Dividend= 30.00%), (**I**= Interest Expense= 513 millions USD), (**s**= [S/S']-1= Sales Growth= 5.00%), and (**t**= T/B= Tax Rate= 30.00%), are known, then it's (**f**= Fixed Portion planned) is:

$$f = (S'[1+s][1-v]-I-H/\{d[1-y][1-t]\})/S$$
$$= (48{,}851[1+0.0500][1-0.1900]-513 \\ -1{,}659/\{0.3000[1-0.3000] \\ [1-0.3000]\})/51{,}294$$
$$= 58.00\%$$

Law-3229:

If both (**d**= D/A= Dividend Portion or Payout= 30.00%), (**H**= Home Taken Dividend= 1,659 millions USD), (**S**= Sales or Revenues= 51,294 millions USD), (**v**= V/S= Variable Portion= 19.00%), (**S'**= Sales of Past Year= 48,851 millions USD), (**y**= Y/D= Yielding Tax Rate of Dividend= 30.00%), (**f**= F/S= Fixed Portion= 58.00%), (**s**= [S/S']-1= Sales Growth= 5.00%), and (**t**= T/B= Tax Rate= 30.00%), are known, then it's (**I**= Interest Expense planned) is:

$$I = (S'[1+s][1-v]-Sf-H/\{d[1-y][1-t]\}$$
$$= (48,851[1+0.0500][1-0.1900]-51,294$$
$$*0.5800-1,659/\{0.3000$$
$$[1-0.3000][1-0.3000]\}$$
$$= \underline{513} \text{ millions USD}$$

Law-3230:

If both (**d**= D/A= Dividend Portion or Payout= 30.00%), (**i**= I/S= Interest Portion= 1.00%), (**$**= Sales or Revenues= 51,294 millions USD), (**v**= V/S= Variable Portion= 19.00%), (**$'**= Sales of Past Year= 48,851 millions USD), (**y**= Y/D= Yielding Tax Rate of Dividend= 30.00%), (**f**= F/S= Fixed Portion= 58.00%), (**s**= [S/S']-1= Sales Growth= 5.00%), and (**t**= T/B= Tax Rate= 30.00%), are known, then it's (**H**= Home Taken Dividend planned) is:

$$H = d[1-y][1-t]\{\$'[1+s] - \$'v[1+s] - \$f - \$i\}$$
$$= d[1-y][1-t]\{\$'[1+s][1-v] - \$[f+i]\}$$
$$= 0.3000[1-0.3000][1-0.3000]\{48,851[1+0.0500][1-0.1900] - 51,294[0.5800+0.0100]\}$$
$$= \underline{1,659} \text{ millions USD}$$

Law-3231:

If both (**H**= Home Taken Dividend= <u>1,659</u> millions USD), (**i**= I/S= Interest Portion= <u>1.00</u>%), (**$**= Sales or Revenues= <u>51,294</u> millions USD), (**v**= V/S= Variable Portion= <u>19.00</u>%), (**$'**= Sales of Past Year= <u>48,851</u> millions USD), (**y**= Y/D= Yielding Tax Rate of Dividend= <u>30.00</u>%), (**f**= F/S= Fixed Portion= <u>58.00</u>%), (**s**= [S/S']-1= Sales Growth= <u>5.00</u>%), and (**t**= T/B= Tax Rate= <u>30.00</u>%), are known, then it's (**d**= Dividend Portion planned) is:

$$d = H/([1-y][1-t]\{\$'[1+s][1-v]-\$[f+i]\})$$
$$= 1,659/([1-0.3000][1-0.3000]\{48,851$$
$$[1+0.0500][1-0.1900]-51,294$$
$$[0.5800+0.0100]\})$$
$$= \underline{30.00\%}$$

Law-3232:

If both (**H**= Home Taken Dividend= <u>1,659</u> millions USD), (**i**= I/S= Interest Portion= <u>1.00</u>%), (**$**= Sales or Revenues= <u>51,294</u> millions USD), (**v**= V/S= Variable Portion= <u>19.00</u>%), (**$'**= Sales of Past Year= <u>48,851</u> millions USD), (**d**= D/A= Dividend Portion or Payout= 30.00%), (**f**= F/S= Fixed Portion= <u>58.00</u>%), (**s**= [S/S']-1= Sales Growth= <u>5.00</u>%), and (**t**= T/B= Tax Rate= <u>30.00</u>%), are known, then it's (**y**= Yielding Tax Rate of Dividend planned) is:

$$y = 1 - H/(d[1-t]\{\$'[1+s][1-v]-\$[f+i]\})$$
$$= 1 - 1,659/(0.3000\,[1-0.3000]\{48,851$$
$$[1+0.0500][1-0.1900]-51,294$$
$$[0.5800+0.0100]\})$$
$$= \underline{30.00\%}$$

Law-3233:

If both (**H**= Home Taken Dividend= 1,659 millions USD), (**i**= I/S= Interest Portion= 1.00%), (**$**= Sales or Revenues= 51,294 millions USD), (**v**= V/S= Variable Portion= 19.00%), (**$'**= Sales of Past Year= 48,851 millions USD), (**d**= D/A= Dividend Portion or Payout= 30.00%), (**f**= F/S= Fixed Portion= 58.00%), (**s**= [S/S']-1= Sales Growth= 5.00%), and (**y**= Y/D= Yielding Tax Rate of Dividend= 30.00%), are known, then it's (**t**= Tax Rate planned) is:

$$t = 1 - H/(d[1-y]\{\$'[1+s][1-v] - \$[f+i]\})$$
$$= 1 - 1{,}659/(0.3000\,[1-0.3000]\{48{,}851\,[1+0.0500][1-0.1900] - 51{,}294\,[0.5800+0.0100]\})$$
$$= 30.00\%$$

Law-3234:

If both (**H**= Home Taken Dividend= 1,659 millions USD), (**i**= I/S= Interest Portion= 1.00%), (**$**= Sales or Revenues= 51,294 millions USD), (**v**= V/S= Variable Portion= 19.00%), (**t**= T/B= Tax Rate= 30.00%), (**d**= D/A= Dividend Portion or Payout= 30.00%), (**f**= F/S= Fixed Portion= 58.00%), (**s**= [S/S']-1= Sales Growth= 5.00%), and (**y**= Y/D= Yielding Tax Rate of Dividend= 30.00%), are known, then it's (**$'**= Sales Past) must be:

$$\$' = (\$[f+i] + H/\{d[1-y][1-t]\})/\{[1+s][1-v]\}$$
$$= (51{,}294[0.5800+0.0100] + 1{,}659/\{0.3000[1-0.3000][1-0.3000]\})/\{[1+0.0500][1-0.1900]\}$$
$$= 48{,}851 \text{ millions USD}$$

Law-3235:

If both (**H**= Home Taken Dividend= 1,659 millions USD), (**i**= I/S= Interest Portion= 1.00%), (**$**= Sales or Revenues= 51,294 millions USD), (**v**= V/S= Variable Portion= 19.00%), (**t**= T/B= Tax Rate= 30.00%), (**d**= D/A= Dividend Portion or Payout= 30.00%), (**f**= F/S= Fixed Portion= 58.00%), (**$'**= Sales of Past Year= 48,851 millions USD), and (**y**= Y/D= Yielding Tax Rate of Dividend= 30.00%), are known, then it's (**s**= Sales Growth planned) is:

$$s = (\$[f+i]+H/\{d[1-y][1-t]\})/\{\$'[1-v]\}-1$$
$$= (51,294[0.5800+0.0100]+1,659$$
$$/\{0.3000[1-0.3000][1-0.3000]\})$$
$$/\{48,851[1-0.1900]\}-1$$
$$= \underline{5.00\%}$$

Law-3236:

If both (**H**= Home Taken Dividend= 1,659 millions USD), (**i**= I/S= Interest Portion= 1.00%), (**$**= Sales or Revenues= 51,294 millions USD), (**s**= [S/S']-1= Sales Growth= 5.00%), (**t**= T/B= Tax Rate= 30.00%), (**d**= D/A= Dividend Portion or Payout= 30.00%), (**f**= F/S= Fixed Portion= 58.00%), (**$'**= Sales of Past Year= 48,851 millions USD), and (**y**= Y/D= Yielding Tax Rate of Dividend= 30.00%), are known, then it's (**v**= Variable Portion planned) is:

$$v = 1-(\$[f+i]+H/\{d[1-y][1-t]\})/\{\$'[1+s]\}$$
$$= 1-(51{,}294[0.5800+0.0100]+1{,}659$$
$$/\{0.3000\,[1-0.3000]$$
$$[1-0.3000]\})$$
$$/\{48{,}851[1+0.0500]\}$$
$$= \underline{19.00\%}$$

Law-3237:
If both (**H**= Home Taken Dividend= 1,659 millions USD), (**i**= I/S= Interest Portion= 1.00%), (**s**= [S/S']-1= Sales Growth= 5.00%), (**t**= T/B= Tax Rate= 30.00%), (**d**= D/A= Dividend Portion or Payout= 30.00%), (**f**= F/S= Fixed Portion= 58.00%),(**v**= V/S= Variable Portion= 19.00%), (**S'**= Sales of Past Year= 48,851 millions USD), and (**y**= Y/D= Yielding Tax Rate of Dividend= 30.00%), are known, then it's (**S**= Sales or Revenues planned) is:

$$S = (S'[1+s][1-v]-H/\{d[1-y][1-t]\})/[f+i]$$
$$= (48,851[1+0.0500][1-0.1900]-1,659$$
$$/\{0.3000[1-0.3000][1-0.3000]\})$$
$$/ [0.5800+0.0100]$$
$$= \underline{51,294} \text{ millions USD}$$

Law-3238:
If both (**H**= Home Taken Dividend= 1,659 millions USD), (**i**= I/S= Interest Portion= 1.00%), (**s**= [S/S']-1= Sales Growth= 5.00%), (**t**= T/B= Tax Rate= 30.00%), (**d**= D/A= Dividend Portion or Payout= 30.00%), (**S**= Sales or Revenues= 51,294 millions USD), (**v**= V/S= Variable Portion= 19.00%), (**S'**= Sales of Past Year= 48,851 millions USD), and (**y**= Y/D= Yielding Tax Rate of Dividend= 30.00%), are known, then it's (**f**= Fixed Portion planned) is:

$$f = (S'[1+s][1-v]-H/\{d[1-y][1-t]\})/S-i$$
$$= (48,851[1+0.0500][1-0.1900]-1,659$$
$$/\{0.3000[1-0.3000][1-0.3000]\})$$
$$/ 51,294-0.0100$$
$$= \underline{58.00}\%$$

Law-3239:

If both (**H**= Home Taken Dividend= 1,659 millions USD), (**f**= F/S= Fixed Portion= 58.00%), (**s**= [S/S']-1= Sales Growth= 5.00%), (**t**= T/B= Tax Rate= 30.00%), (**d**= D/A= Dividend Portion or Payout= 30.00%), (**$**= Sales or Revenues= 51,294 millions USD), (**v**= V/S= Variable Portion= 19.00%), (**$'**= Sales of Past Year= 48,851 millions USD), and (**y**= Y/D= Yielding Tax Rate of Dividend= 30.00%), are known, then it's (**i**= Interest Portion planned) is:

$$i = (\$'[1+s][1-v]-H/\{d[1-y][1-t]\})/\$-f$$
$$= (48,851[1+0.0500][1-0.1900]-1,659$$
$$/\{0.3000[1-0.3000][1-0.3000]\})$$
$$/51,294-0.5800$$
$$= 1.00\%$$

Law-3240:

If both (i= I/S= Interest Portion= 1.00%), (f= F/S= Fixed Portion= 58.00%), (s= [S/S']-1= Sales Growth= 5.00%), (t= T/B= Tax Rate= 30.00%), (d= D/A= Dividend Portion or Payout= 30.00%), (S= Sales or Revenues= 51,294 millions USD), (v= V/S= Variable Portion= 19.00%), (S'= Sales of Past Year= 48,851 millions USD), and (y= Y/D= Yielding Tax Rate of Dividend= 30.00%), are known, then it's (H= Home Taken Dividend planned) is:

H= d[1-y][1-t]{S'[1+s]-S'v[1+s]-$S f$-S'i[1+s]}
  = d[1-y][1-t]{S'[1+s][1-v-i]-$S f$}
  = 0.3000[1-0.3000][1-0.3000]{48,851
     [1+0.0500][1-0.1900-0.0100]
     -51,294*0.5800}
  = 1,659 millions USD

Law-3241:

If both (**i**= I/S= Interest Portion= 1.00%), (**f**= F/S= Fixed Portion= 58.00%), (**s**= [S/S']-1= Sales Growth= 5.00%), (**t**= T/B= Tax Rate= 30.00%), (**H**= Home Taken Dividend= 1,659 millions USD), (**$**= Sales or Revenues= 51,294 millions USD), (**v**= V/S= Variable Portion= 19.00%), (**$'**= Sales of Past Year= 48,851 millions USD), and (**y**= Y/D= Yielding Tax Rate of Dividend= 30.00%), are known, then it's (**d**= Dividend Portion or Payout planned) is:

$$\begin{aligned}
\mathbf{d} &= \mathbf{H}/([1-\mathbf{y}][1-\mathbf{t}]\{\$'[1+\mathbf{s}]-\$'\mathbf{v}[1+\mathbf{s}]-\$\mathbf{f}-\$'\mathbf{i}[1+\mathbf{s}]\}) \\
&= \mathbf{H}/([1-\mathbf{y}][1-\mathbf{t}]\{\$'[1+\mathbf{s}][1-\mathbf{v}-\mathbf{i}]-\$\mathbf{f}\}) \\
&= 1{,}659/(\,[1-0.3000][1-0.3000]\{48{,}851 \\
&\qquad\qquad [1+0.0500][1-0.1900-0.0100] \\
&\qquad\qquad -51{,}294*0.5800\}) \\
&= 30.00\%
\end{aligned}$$

Law-3242:

If both (i= I/S= Interest Portion= 1.00%), (f= F/S= Fixed Portion= 58.00%), (s= [S/S']-1= Sales Growth= 5.00%), (t= T/B= Tax Rate= 30.00%), (H= Home Taken Dividend= 1,659 millions USD), (S= Sales or Revenues= 51,294 millions USD), (v= V/S= Variable Portion= 19.00%), (S'= Sales of Past Year= 48,851 millions USD), and (d= D/A=Dividend Portion or Payout= 30.00%), are known, then it's (y= Yielding Tax Rate of Dividend planned) is:

y= 1-H/(d[1-t]{S'[1+s][1-v-i]-Sf})
  = 1-1,659/(0.3000 [1-0.3000]{48,851 [1+0.0500][1-0.1900-0.0100] -51,294*0.5800})
  = 30.00%

Law-3243:

If both (**i**= I/S= Interest Portion= 1.00%), (**f**= F/S= Fixed Portion= 58.00%), (**s**= [S/S']-1= Sales Growth= 5.00%), (**y**= Y/D= Yielding Tax Rate of Dividend= 30.00%), (**H**= Home Taken Dividend= 1,659 millions USD), (**$**= Sales or Revenues= 51,294 millions USD), (**v**= V/S= Variable Portion= 19.00%), (**$'**= Sales of Past Year= 48,851 millions USD), and (**d**= D/A= Dividend Portion or Payout= 30.00%), are known, then it's (**t**= Tax Rate planned) is:

$$t = 1-H/(d[1-y]\{\$'[1+s][1-v-i]-\$f\})$$
$$= 1-1,659/(0.3000[1-0.3000]\{48,851$$
$$[1+0.0500][1-0.1900-0.0100]$$
$$-51,294*0.5800\})$$
$$= 30.00\%$$

Law-3244:

If both (**i**= I/S= Interest Portion= 1.00%), (**f**= F/S= Fixed Portion= 58.00%), (**s**= [S/S']-1= Sales Growth= 5.00%), (**y**= Y/D= Yielding Tax Rate of Dividend= 30.00%), (**H**= Home Taken Dividend= 1,659 millions USD), (**$**= Sales or Revenues= 51,294 millions USD), (**v**= V/S= Variable Portion= 19.00%), (**t**= T/B= Tax Rate= 30.00%), and (**d**= D/A= Dividend Portion or Payout= 30.00%), are known, then it's (**$'**= Sales Past) must be:

$' = ($f+H/{d[1-y][1-t]})/{[1+s][1-v-i]}
= (51,294*0.5800+1,659/{0.3000
[1-0.3000][1-0.3000]})
/{[1+0.0500][1-0.1900
-0.0100]}
= 48,851 millions USD

Law-3245:

If both (**i**= I/S= Interest Portion= 1.00%), (**f**= F/S= Fixed Portion= 58.00%), (**$'**= Sales of Past Year= 48,851 millions USD), (**y**= Y/D= Yielding Tax Rate of Dividend= 30.00%), (**H**= Home Taken Dividend= 1,659 millions USD), (**$**= Sales or Revenues= 51,294 millions USD), (**v**= V/S= Variable Portion= 19.00%), (**t**= T/B= Tax Rate= 30.00%), and (**d**= D/A= Dividend Portion or Payout= 30.00%), are known, then it's (**s**= Sales Growth planned) is:

$$s = (\$f + H/\{d[1-y][1-t]\})/\{\$'[1-v-i]\} - 1$$
$$= (51{,}294 * 0.5800 + 1{,}659/\{0.3000\ [1-0.3000][1-0.3000]\})\ /\{48{,}851[1-0.1900-0.0100]\} - 1$$
$$= 5.00\%$$

Law-3246:

If both (i= I/S= Interest Portion= 1.00%), (f= F/S= Fixed Portion= 58.00%), ($\$'$= Sales of Past Year= 48,851 millions USD), (y= Y/D= Yielding Tax Rate of Dividend= 30.00%), (H= Home Taken Dividend= 1,659 millions USD), ($\$$= Sales or Revenues= 51,294 millions USD), (s= [S/S']-1= Sales Growth= 5.00%), (t= T/B= Tax Rate= 30.00%), and (d= D/A= Dividend Portion or Payout= 30.00%), are known, then it's (v= Variable Portion planned) is:

v= 1-i-($\$f$+$H$/{$d$[1-$y$][1-$t$]})/{$\$'$[1+$s$]}
 = 1-0.0100-(51,294*0.5800+1,659
   /{0.3000[1-0.3000][1-0.3000]})
   /{48,851[1+0.0500]}
 = 19.00%

Finance Construction-4, *Tim Asikin, Steve Asikin, Indra Senihardja*

Law-3247:

If both (**v**= V/S= Variable Portion= 19.00%), (**f**= F/S= Fixed Portion= 58.00%), (**$'**= Sales of Past Year= 48,851 millions USD), (**y**= Y/D= Yielding Tax Rate of Dividend= 30.00%), (**H**= Home Taken Dividend= 1,659 millions USD), (**$**= Sales or Revenues= 51,294 millions USD), (**s**= [S/S']-1= Sales Growth= 5.00%), (**t**= T/B= Tax Rate= 30.00%), and (**d**= D/A= Dividend Portion or Payout= 30.00%), are known, then it's (**i**= Interest Portion planned) is:

$$i = 1-v-(\$f+H/\{d[1-y][1-t]\})/\{\$'[1+s]\}$$
$$= 1-0.1900-(51,294*0.5800+1,659$$
$$/\{0.3000[1-0.3000][1-0.3000]\})$$
$$/\{48,851[1+0.0500]\}$$
$$= 1.00\%$$

Law-3248:

If both (**v**= V/S= Variable Portion= 19.00%), (**f**= F/S= Fixed Portion= 58.00%), (**$'**= Sales of Past Year= 48,851 millions USD), (**y**= Y/D= Yielding Tax Rate of Dividend= 30.00%), (**H**= Home Taken Dividend= 1,659 millions USD), (**i**= I/S= Interest Portion= 1.00%), (**s**= [S/S']-1= Sales Growth= 5.00%), (**t**= T/B= Tax Rate= 30.00%), and (**d**= D/A= Dividend Portion or Payout= 30.00%), are known, then it's (**$**= Sales or Revenues planned) is:

$= ($'[1+s][1-v-i]-H/{d[1-y][1-t]})/f
  = (48,851[1+0.0500][1-0.1900-0.0100]
    -1,659/{0.3000[1-0.3000]
    [1-0.3000]})/0.5800
  = 51,294 millions USD

Law-3249:

If both (**v**= V/S= Variable Portion= 19.00%), (**$'**= Sales of Past Year= 48,851 millions USD), (**y**= Y/D= Yielding Tax Rate of Dividend= 30.00%), (**H**= Home Taken Dividend= 1,659 millions USD), (**i**= I/S= Interest Portion= 1.00%), (**s**= [S/S']-1= Sales Growth= 5.00%), (**$**= Sales or Revenues= 51,294 millions USD), (**t**= T/B= Tax Rate= 30.00%), and (**d**= D/A= Dividend Portion or Payout= 30.00%), are known, then it's (**f**= Fixed Portion planned) is:

f= ($'[1+s][1-v-i]-H/{d[1-y][1-t]})/$
  = (48,851[1+0.0500][1-0.1900-0.0100]
    -1,659/{0.3000[1-0.3000]
    [1-0.3000]})/51,294
  = 58.00%

Law-3250:

If both (v= V/S= Variable Portion= 19.00%), ($\$'$= Sales of Past Year= 48,851 millions USD), (y= Y/D= Yielding Tax Rate of Dividend= 30.00%), (f= F/S= Fixed Portion= 58.00%), (s= [S/S']-1= Sales Growth= 5.00%), ($\$$= Sales or Revenues= 51,294 millions USD), (I= Interest Expense= 513 millions USD), (t= T/B= Tax Rate= 30.00%), and (d= D/A= Dividend Portion or Payout= 30.00%), are known, then it's (H= Home Taken Dividend planned) is:

H = d[1-y][1-t]{$\$'$[1+$s$]-$\$'v$[1+$s$]-$\$'f$[1+$s$]-$I$}
 = d[1-y][1-t]{$\$'$[1+$s$][1-$v$-$f$]-$I$}
 = 0.3000[1-0.3000][1-0.3000]{48,851
  [1+0.0500][1-0.1900-0.5800]
  -513}
 = 1,659 millions USD

Law-3251:

If both (v= V/S= Variable Portion= 19.00%), (S'= Sales of Past Year= 48,851 millions USD), (y= Y/D= Yielding Tax Rate of Dividend= 30.00%), (f= F/S= Fixed Portion= 58.00%), (s= [S/S']-1= Sales Growth= 5.00%), (S= Sales or Revenues= 51,294 millions USD), (I= Interest Expense= 513 millions USD), (t= T/B= Tax Rate= 30.00%), and (H= Home Taken Dividend= 1,659 millions USD), are known, then it's (d= Dividend Portion or Payout planned) is:

$$d = H/([1-y][1-t]\{S'[1+s][1-v-f]-I\})$$
$$= 1{,}659/([1-0.3000][1-0.3000]\{48{,}851[1+0.0500][1-0.1900-0.5800]-513\})$$
$$= 30.00\%$$

Finance Construction-4, *Tim Asikin, Steve Asikin, Indra Senihardja*

Law-3252:

If both (**v**= V/S= Variable Portion= 19.00%), (**$'**= Sales of Past Year= 48,851 millions USD), (**d**= D/A= Dividend Portion or Payout= 30.00%), (**f**= F/S= Fixed Portion= 58.00%), (**s**= [S/S']-1= Sales Growth= 5.00%), (**$**= Sales or Revenues= 51,294 millions USD), (**I**= Interest Expense= 513 millions USD), (**t**= T/B= Tax Rate= 30.00%), and (**H**= Home Taken Dividend= 1,659 millions USD), are known, then it's (**y**= Yielding Tax Rate of Dividend planned) is:

$$y= 1-H/(d[1-t]\{\$'[1+s][1-v-f]-I\})$$
$$= 1-1,659/(0.3000[1-0.3000]\{48,851[1+0.0500][1-0.1900-0.5800]-513\})$$
$$= \underline{30.00\%}$$

Law-3253:

If both (**v**= V/S= Variable Portion= 19.00%), (**$'**= Sales of Past Year= 48,851 millions USD), (**d**= D/A= Dividend Portion or Payout= 30.00%), (**f**= F/S= Fixed Portion= 58.00%), (**s**= [S/S']-1= Sales Growth= 5.00%), (**$**= Sales or Revenues= 51,294 millions USD), (**I**= Interest Expense= 513 millions USD), (**y**= Y/D= Yielding Tax Rate of Dividend= 30.00%), and (**H**= Home Taken Dividend= 1,659 millions USD), are known, then it's (**t**= Tax Rate planned) is:

$$t = 1 - H/(d[1-y]\{\$'[1+s][1-v-f]-I\})$$
$$= 1-1,659/(0.3000[1-0.3000]\{48,851[1+0.0500][1-0.1900-0.5800]-513\})$$
$$= \underline{30.00\%}$$

Law-3254:

If both (v= V/S= Variable Portion= 19.00%), (d= D/A= Dividend Portion or Payout= 30.00%), (f= F/S=Fixed Portion= 58.00%), (s= [S/S']-1= Sales Growth= 5.00%), (t= T/B= Tax Rate= 30.00%), (S= Sales or Revenues= 51,294 millions USD), (I= Interest Expense= 513 millions USD), (y= Y/D= Yielding Tax Rate of Dividend= 30.00%), and (H= Home Taken Dividend= 1,659 millions USD), are known, then it's (S'= Sales Past) must be:

$$S' = (I+H/\{d[1-y][1-t]\})/\{[1+s][1-v-f]\}$$
$$= (513+1,659/\{0.3000[1-0.3000][1-0.3000]\})/\{[1+0.0500][1-0.1900-0.5800]\}$$
$$= 48{,}851 \text{ millions USD}$$

Law-3255:

If both (v= V/S= Variable Portion= 19.00%), (d= D/A= Dividend Portion or Payout= 30.00%), (f= F/S= Fixed Portion= 58.00%), (S'= Sales of Past Year= 48,851 millions USD), (t= T/B= Tax Rate= 30.00%), (S= Sales or Revenues= 51,294 millions USD), (I= Interest Expense= 513 millions USD), (y= Y/D= Yielding Tax Rate of Dividend= 30.00%), and (H= Home Taken Dividend= 1,659 millions USD), are known, then it's (s= Sales Growth planned) is:

$$s = (I+H/\{d[1-y][1-t]\})/\{S'[1-v-f]\}-1$$
$$= (513+1,659/\{0.3000[1-0.3000][1-0.3000]\})/\{48,851[1-0.1900-0.5800]\}-1$$
$$= 5.00\%$$

Law-3256:

If both (s= [S/S']-1= Sales Growth= 5.00%), (d= D/A= Dividend Portion or Payout= 30.00%), (f= F/S= Fixed Portion= 58.00%), ($\$'$= Sales of Past Year= 48,851 millions USD), (t= T/B= Tax Rate= 30.00%), ($\$$= Sales or Revenues= 51,294 millions USD), (I= Interest Expense= 513 millions USD), (y= Y/D= Yielding Tax Rate of Dividend= 30.00%), and (H= Home Taken Dividend= 1,659 millions USD), are known, then it's (v= Variable Portion planned) is:

$$v = 1-f-(I+H/\{d[1-y][1-t]\})/\{\$'[1+s]\}$$
$$= 1-0.5800-(513+1,659/\{0.3000[1-0.3000][1-0.3000]\})/\{48,851[1+0.0500]\}$$
$$= 19.00\%$$

Law-3257:

If both (**s**= [S/S']-1= Sales Growth= 5.00%), (**d**= D/A= Dividend Portion or Payout= 30.00%), (**v**= V/S= Variable Portion= 19.00%), (**S'**= Sales of Past Year= 48,851 millions USD), (**t**= T/B= Tax Rate= 30.00%), (**S**= Sales or Revenues= 51,294 millions USD), (**I**= Interest Expense= 513 millions USD), (**y**= Y/D= Yielding Tax Rate of Dividend= 30.00%), and (**H**= Home Taken Dividend= 1,659 millions USD), are known, then it's (**f**= Fixed Portion planned) is:

f= 1-**v**-(**I**+**H**/{**d**[1-**y**][1-**t**]})/{**S'**[1+**s**]}
   = 1-0.1900-(513+1,659/{0.3000
     [1-0.3000][1-0.3000]})/{48,851
     [1+0.0500]}
   = 58.00%

Law-3258:

If both (**s**= [S/S']-1= Sales Growth= 5.00%), (**d**= D/A= Dividend Portion or Payout= 30.00%), (**v**= V/S= Variable Portion= 19.00%), (**$'**= Sales of Past Year= 48,851 millions USD), (**t**= T/B= Tax Rate= 30.00%), (**$**= Sales or Revenues= 51,294 millions USD), (**f**= F/S= Fixed Portion= 58.00%), (**y**= Y/D= Yielding Tax Rate of Dividend= 30.00%), and (**H**= Home Taken Dividend= 1,659 millions USD), are known, then it's (**I**= Interest Expense planned) is:

$$I = \$'[1+s][1-v-f] - H/\{d[1-y][1-t]\}$$
$$= (48,851[1+0.0500][1-0.1900$$
$$-0.5800] - 1,659/\{0.3000$$
$$[1-0.3000][1-0.3000]\}$$
$$= \underline{513} \text{ millions USD}$$

Law-3259:

If both (s= [S/S']-1= Sales Growth= 5.00%), (d= D/A= Dividend Portion or Payout= 30.00%), (v= V/S= Variable Portion= 19.00%), (S'= Sales of Past Year= 48,851 millions USD), (t= T/B= Tax Rate= 30.00%), (S= Sales or Revenues= 51,294 millions USD), (f= F/S= Fixed Portion= 58.00%), (y= Y/D= Yielding Tax Rate of Dividend= 30.00%), and (i= I/S= Interest Portion= 1.00%), are known, then it's (H= Home Taken dividend planned) is:

$$H = d[1-y][1-t]\{S'[1+s]-S'v[1+s]-S'f[1+s]-Si\}$$
$$= d[1-y][1-t]\{S'[1+s][1-v-f]-Si\}$$
$$= 0.3000[1-0.3000][1-0.3000]\{48,851$$
$$[1+0.0500][1-0.1900-0.5800]$$
$$-51,294*0.0100\}$$
$$= 1,659 \text{ millions USD}$$

Law-3260:

If both (s= [S/S']-1= Sales Growth= 5.00%), (H= Home Taken Dividend= 1,659 millions USD), (v= V/S= Variable Portion= 19.00%), (S'= Sales of Past Year= 48,851 millions USD), (t= T/B= Tax Rate= 30.00%), (S= Sales or Revenues= 51,294 millions USD), (f= F/S=Fixed Portion= 58.00%), (y= Y/D= Yielding Tax Rate of Dividend= 30.00%), and (i= I/S= Interest Portion= 1.00%), are known, then it's (d= Dividend Portion or Payout planned) is:

$$d = H/([1-y][1-t]\{S'[1+s][1-v-f]-Si\})$$
$$= 1,659/([1-0.3000][1-0.3000]\{48,851$$
$$[1+0.0500][1-0.1900-0.5800]$$
$$-51,294*0.0100\})$$
$$= 30.00\%$$

Law-3261:

If both (s= [S/S']-1= Sales Growth= 5.00%), (H= Home Taken Dividend= 1,659 millions USD), (v= V/S= Variable Portion= 19.00%), S'= Sales of Past Year= 48,851 millions USD), (t= T/B= Tax Rate= 30.00%), (S= Sales or Revenues= 51,294 millions USD), (f= F/S= Fixed Portion= 58.00%), (d= D/A= Dividend Portion or Payout= 30.00%), and (i= I/S= Interest Portion= 1.00%), are known, then it's (y= Yielding Tax Rate of Dividend planned) is:

$$y = 1-H/(d[1-t]\{S'[1+s][1-v-f]-Si\})$$
$$= 1-1,659/(0.3000[1-0.3000]\{48,851$$
$$[1+0.0500][1-0.1900-0.5800]$$
$$-51,294*0.0100\})$$
$$= 30.00\%$$

Law-3262:

If both (**s**= [S/S']-1= Sales Growth= 5.00%), (**H**= Home Taken Dividend= 1,659 millions USD), (**v**= V/S= Variable Portion= 19.00%), (**S'**= Sales of Past Year= 48,851 millions USD), (**y**= Y/D= Yielding Tax Rate of Dividend= 30.00%), (**S**= Sales or Revenues= 51,294 millions USD), (**f**= F/S= Fixed Portion= 58.00%), (**d**= D/A= Dividend Portion or Payout= 30.00%), and (**i**= I/S= Interest Portion= 1.00%), are known, then it's (**t**= Tax Rate planned) is:

$$\begin{aligned}
\mathbf{t} &= 1-\mathbf{H}/(\mathbf{d}[1-\mathbf{y}]\{\mathbf{S'}[1+\mathbf{s}][1-\mathbf{v}-\mathbf{f}]-\mathbf{Si}\}) \\
&= 1-1{,}659/(0.3000[1-0.3000]\{48{,}851 \\
&\qquad [1+0.0500][1-0.1900-0.5800] \\
&\qquad -51{,}294*0.0100\}) \\
&= 30.00\%
\end{aligned}$$

Law-3263:

If both (**s**= [S/S']-1= Sales Growth= 5.00%), (**H**= Home Taken Dividend= 1,659 millions USD), (**v**= V/S= Variable Portion= 19.00%), (**t**= T/B= Tax Rate= 30.00%), (**y**= Y/D= Yielding Tax Rate of Dividend= 30.00%), (**$**= Sales or Revenues= 51,294 millions USD), (**f**= F/S= Fixed Portion= 58.00%), (**d**= D/A= Dividend Portion or Payout= 30.00%), and (**i**= I/S= Interest Portion= 1.00%), are known, then it's (**$'**= Sales Past) must be:

$'= ($i+H/{d[1-y][1-t]})/{[1+s][1-v-f]}
  = (51,294*0.0100+1,659/{0.3000
        [1-0.3000][1-0.3000]})
        /{[1+0.0500][1-0.1900
        -0.5800]}
  = 48,851 millions USD

Finance Construction-4, *Tim Asikin, Steve Asikin, Indra Senihardja*

Law-3264:

If both ($\$'$= Sales of Past Year= 48,851 millions USD), (**H**= Home Taken Dividend= 1,659 millions USD), (**v**= V/S= Variable Portion= 19.00%), (**t**= T/B= Tax Rate= 30.00%), (**y**= Y/D= Yielding Tax Rate of Dividend= 30.00%), ($\$$= Sales or Revenues= 51,294 millions USD), (**f**= F/S= Fixed Portion= 58.00%), (**d**= D/A= Dividend Portion or Payout= 30.00%), and (**i**= I/S= Interest Portion= 1.00%), are known, then it's (**s**= Sales Growth planned) is:

$s = (\$i + H/\{d[1-y][1-t]\})/\{\$'[1-v-f]\} - 1$
$ = (51{,}294 * 0.0100 + 1{,}659/\{0.3000$
$[1-0.3000][1-0.3000]\})/\{48{,}851$
$\phantom{s = (51,294*0.0100+1,659/\{}[1-0.1900-0.5800]\} - 1$
$ = 5.00\%$

Law-3265:

If both ($\$'$= Sales of Past Year= 48,851 millions USD), (H= Home Taken Dividend= 1,659 millions USD), (s= [S/S']-1= Sales Growth= 5.00%), (t= T/B= Tax Rate= 30.00%), (y= Y/D= Yielding Tax Rate of Dividend= 30.00%), ($\$$= Sales or Revenues= 51,294 millions USD), (f= F/S= Fixed Portion= 58.00%), (d= D/A= Dividend Portion or Payout= 30.00%), and (i= I/S= Interest Portion= 1.00%), are known, then it's (v= Variable Portion planned) is:

$$v = 1-f-(\$i+H/\{d[1-y][1-t]\})/\{\$'[1+s]\}$$
$$= 1-0.5800-(51,294*0.0100+1,659/\{0.3000[1-0.3000][1-0.3000]\})/\{48,851[1+0.0500]\}$$
$$= 19.00\%$$

Law-3266:

If both ($\$'$= Sales of Past Year= 48,851 millions USD), (**H**= Home Taken Dividend= 1,659 millions USD), (**s**= [S/S']-1= Sales Growth= 5.00%), (**t**= T/B= Tax Rate= 30.00%), (**y**= Y/D= Yielding Tax Rate of Dividend= 30.00%), ($\$$= Sales or Revenues= 51,294 millions USD), (**v**= V/S= Variable Portion= 19.00%), (**d**= D/A= Dividend Portion or Payout= 30.00%), and (**i**= I/S= Interest Portion= 1.00%), are known, then it's (**f**= Fixed Portion planned) is:

f= 1-**v**-($\$$**i**+**H**/{**d**[1-**y**][1-**t**]})/{$\$'$[1+**s**]}
   = 1-0.1900-(51,294*0.0100+1,659
      /{0.3000[1-0.3000]
      [1-0.3000]})/{48,851
      [1+0.0500]}
   = 58.00%

Law-3267:

If both ($\$'$= Sales of Past Year= 48,851 millions USD), (**H**= Home Taken Dividend= 1,659 millions USD), (**s**= [S/S']-1= Sales Growth= 5.00%), (**t**= T/B= Tax Rate= 30.00%), (**y**= Y/D= Yielding Tax Rate of Dividend= 30.00%), (**f**= F/S= Fixed Portion= 58.00%), (**v**= V/S= Variable Portion= 19.00%), (**d**= D/A= Dividend Portion or Payout= 30.00%), and (**i**= I/S= Interest Portion= 1.00%), are known, then it's ($\$$= Sales or Revenues planned) is:

$\$ = (\$'[1+s][1-v-f]-H/\{d[1-y][1-t]\})/i$
  $= (48,851[1+0.0500][1-0.1900-0.5800]$
    $-1,659/\{0.3000[1-0.3000]$
    $[1-0.3000]\})/0.0100$
  $= \underline{51,294}$ millions USD

Law-3268:

If both ($\$'$= Sales of Past Year= 48,851 millions USD), (**H**= Home Taken Dividend= 1,659 millions USD), (**s**= [S/S']-1= Sales Growth= 5.00%), (**t**= T/B= Tax Rate= 30.00%), (**y**= Y/D= Yielding Tax Rate of Dividend= 30.00%), (**f**= F/S= Fixed Portion= 58.00%), (**v**= V/S= Variable Portion= 19.00%), (**d**= D/A= Dividend Portion or Payout= 30.00%), and ($\$$= Sales or Revenues= 51,294 millions USD), are known, then it's (**i**= Interest Portion planned) is:

$i = (\$'[1+s][1-v-f]-H/\{d[1-y][1-t]\})/\$$
  $= (48,851[1+0.0500][1-0.1900-0.5800]$
    $-1,659/\{0.3000[1-0.3000]$
    $[1-0.3000]\})/51,294$
  $= \underline{1.00\%}$

Law-3269:

If both (S'= Sales of Past Year= 48,851 millions USD), (i= I/S= Interest Portion= 1.00%), (s= [S/S']-1= Sales Growth= 5.00%), (t= T/B= Tax Rate= 30.00%), (y= Y/D= Yielding Tax Rate of Dividend= 30.00%), (f= F/S= Fixed Portion= 58.00%), (v= V/S= Variable Portion= 19.00%), (d= D/A= Dividend Portion or Payout= 30.00%), and (S= Sales or Revenues= 51,294 millions USD), are known, then it's (H= Home Taken Dividend planned) is:

$H = d[1-y][1-t]\{S'[1+s]-S'v[1+s]-S'f[1+s]$
$\quad -S'i[1+s]\} = d[1-y][1-t]\{S'[1+s][1-v-f-i]\}$
$\quad = 0.3000[1-0.3000][1-0.3000]\{48.851$
$\qquad [1+0.0500][1-0.1900-0.5800$
$\qquad -0.0100]\}$
$\quad = 1,659$ millions USD

Law-3270:

If both ($'= Sales of Past Year= 48,851 millions USD), (i= I/S= Interest Portion= 1.00%), (s= [S/S']-1= Sales Growth= 5.00%), (t= T/B= Tax Rate= 30.00%), (y= Y/D= Yielding Tax Rate of Dividend= 30.00%), (f= F/S= Fixed Portion= 58.00%), (v= V/S= Variable Portion= 19.00%), (H= Home Taken Dividend= 1,659 millions USD), and ($= Sales or Revenues= 51,294 millions USD), are known, then it's (d= Dividend Portion or Payout planned) is:

$$d = H/([1-y][1-t]\{\$'[1+s][1-v-f-i]\})$$
$$= 1,659/([1-0.3000][1-0.3000]\{48,851[1+0.0500][1-0.1900-0.5800-0.0100]\})$$
$$= 30.00\%$$

Law-3271:

If both ($'= Sales of Past Year= 48,851 millions USD), (i= I/S= Interest Portion= 1.00%), (s= [S/S']-1= Sales Growth= 5.00%), (t= T/B= Tax Rate= 30.00%), (d= D/A= Dividend Portion or Payout= 30.00%), (f= F/S= Fixed Portion= 58.00%), (v= V/S= Variable Portion= 19.00%), (H= Home Taken Dividend= 1,659 millions USD), and ($= Sales or Revenues= 51,294 millions USD), are known, then it's (y= Yielding Tax Rate of Dividend planned) is:

$$y = 1 - H/(d[1-t]\{\$'[1+s][1-v-f-i]\})$$
$$= 1 - 1,659/(0.3000[1-0.3000]\{[1+0.0500][1-0.1900-0.5800-0.0100]\})$$
$$= 30.00\%$$

Law-3272:
If both ($'$ = Sales of Past Year= 48,851 millions USD), (i= I/S= Interest Portion= 1.00%), (s= [S/S']-1= Sales Growth= 5.00%), (y= Y/D= Yielding Tax Rate of Dividend= 30.00%), (d= D/A= Dividend Portion or Payout= 30.00%), (f= F/S= Fixed Portion= 58.00%), (v= V/S= Variable Portion= 19.00%), (H= Home Taken Dividend= 1,659 millions USD), and (S= Sales or Revenues= 51,294 millions USD), are known, then it's (t= Tax Rate planned) is:

$$t = 1-H/(d[1-y]\{S'[1+s][1-v-f-i]\})$$
$$= 1-1,659/(0.3000[1-0.3000]$$
$$\{[1+0.0500][1-0.1900-0.5800$$
$$-0.0100]\})$$
$$= 30.00\%$$

Law-3273:

If both (**t**= T/B= Tax Rate= 30.00%), (**i**= I/S= Interest Portion= 1.00%), (**s**= [S/S']-1= Sales Growth= 5.00%), (**y**= Y/D= Yielding Tax Rate of Dividend= 30.00%), (**d**= D/A= Dividend Portion or Payout= 30.00%), (**f**= F/S= Fixed Portion= 58.00%), (**v**= V/S= Variable Portion= 19.00%), (**H**= Home Taken Dividend= 1,659 millions USD), and (**$**= Sales or Revenues= 51,294 millions USD), are known, then it's (**$'**= Sales Past) must be:

$'= (**H**/{**d**[1-**y**][1-**t**]})/{[1+**s**][1-**v**-**f**-**i**]}
 = (1,659/{0.3000[1-0.3000]
  [1-0.3000]})/{[1+0.0500]
  [1-0.1900-0.5800-0.0100]})
 = 48,851 millions USD

Law-3274:

If both (**t**= T/B= Tax Rate= 30.00%), (**i**= I/S= Interest Portion= 1.00%), (**$'**= Sales of Past Year= 48,851 millions USD), (**y**= Y/D= Yielding Tax Rate of Dividend= 30.00%), (**d**= D/A= Dividend Portion or Payout= 30.00%), (**f**= F/S= Fixed Portion= 58.00%), (**v**= V/S= Variable Portion= 19.00%), (**H**= Home Taken Dividend= 1,659 millions USD),and (**$**= Sales or Revenues= 51,294 millions USD), are known, then it's (**s**= Sales Growth planned) is:

$s = (H/\{d[1-y][1-t]\})/\{\$'[1-v-f-i]\}-1$
$= (1,659/\{0.3000[1-0.3000]$
$[1-0.3000]\})/\{48,851[1-0.1900$
$-0.5800-0.0100]\})-1$
$= 5.00\%$

Law-3275:

If both (**t**= T/B= Tax Rate= 30.00%), (**i**= I/S= Interest Portion= 1.00%), (**$'**= Sales of Past Year= 48,851 millions USD), (**y**= Y/D= Yielding Tax Rate of Dividend= 30.00%), (**d**= D/A= Dividend Portion or Payout= 30.00%), (**f**= F/S= Fixed Portion= 58.00%), (**s**= [S/S']-1= Sales Growth= 5.00%), (**H**= Home Taken Dividend= 1,659 millions USD),and (**$**= Sales or Revenues= 51,294 millions USD), are known, then it's (**v**= Variable Portion planned) is:

$$v = 1\text{-}f\text{-}i\text{-}(H/\{d[1\text{-}y][1\text{-}t]\})/\{\$'[1+s]\}$$
$$= 1\text{-}0.5800\text{-}0.0100\text{-}(1,659/\{0.3000 \\ [1\text{-}0.3000][1\text{-}0.3000]\}) \\ /\{48,851[1+0.0500]\})$$
$$= 19.00\%$$

Law-3276:

If both (**t**= T/B= Tax Rate= 30.00%), (**i**= I/S= Interest Portion= 1.00%), (**$'**= Sales of Past Year= 48,851 millions USD), (**y**= Y/D= Yielding Tax Rate of Dividend= 30.00%), (**d**= D/A= Dividend Portion or Payout= 30.00%), (**v**= V/S= Variable Portion= 19.00%), (**s**= [S/S']-1= Sales Growth= 5.00%), (**H**= Home Taken Dividend= 1,659 millions USD),and (**$**= Sales or Revenues= 51,294 millions USD), are known, then it's (**f**= Fix Portion planned) is:

$$f = 1\text{-}v\text{-}i\text{-}(H/\{d[1\text{-}y][1\text{-}t]\})/\{\$'[1+s]\}$$
$$= 1\text{-}0.1900\text{-}0.0100\text{-}(1,659/\{0.3000 \\ [1\text{-}0.3000][1\text{-}0.3000]\}) \\ /\{48,851[1+0.0500]\})$$
$$= 58.00\%$$

Law-3277:

If both (**t**= T/B= Tax Rate= 30.00%), (**f**= F/S= Fixed Portion= 58.00%), (**$'**= Sales of Past Year= 48,851 millions USD), (**y**= Y/D= Yielding Tax Rate of Dividend= 30.00%), (**d**= D/A= Dividend Portion or Payout= 30.00%), (**v**= V/S= Variable Portion= 19.00%), (**s**= [S/S']-1= Sales Growth= 5.00%), (**H**= Home Taken Dividend= 1,659 millions USD),and (**$**= Sales or Revenues= 51,294 millions USD), are known, then it's (**i**= Interest Portion planned) is:

$$i = 1 - v - f - (H/\{d[1-y][1-t]\})/\{\$'[1+s]\}$$
$$= 1 - 0.1900 - 0.5800 - (1,659/\{0.3000[1-0.3000][1-0.3000]\})/\{48,851[1+0.0500]\})$$
$$= 1.00\%$$

CHAPTER-13:
R= RETAINED EARNINGS Optimization

Based on 12/31/15 Pfizer Annual Report
(http://www.nasdaq.com/symbol/pfe/financials?query=income-statement)

Law-3278:
  If both (**A**= After Tax Income= 7,899 millions USD), and (**D**= Dividend Paid= 2,370 millions USD), are known, then it's (**R**= Retained Earnings planned), is:
  R= **A-D**
  $= 7,899-2,370$
  $= \underline{5,529}$ millions USD

Law-3279:
  If both (**R**= Retained Earnings= 5,529 millions USD), and (**D**= Dividend Paid= 2,370 millions USD), are known, then it's (**A**= After Tax Income planned), is:
  A= **R+D**
  $= 5,529+2,370$
  $= \underline{7,899}$ millions USD

Law-3280:
   If both (**R**= Retained Earnings= 5,529 millions USD), and (**A**= After Tax Income= 7,899 millions USD), are known, then it's (**D**= Dividend Paid planned), is:
   D= **A**-**R**
   = 7,899-5,529
   = 2,370 millions USD

Law-3281:
   If both (**d**= D/A= Dividend Portion or Payout= 30.00%), and (**A**= After Tax Income= 7,899 millions USD), are known, then it's (**R**= Retained Earnings planned), is:
   R= **A**-**Ad**= **A**[1-**d**]
   = 7,899[1-0.3000]
   = 5,529 millions USD

Law-3282:
   If both (**d**= D/A= Dividend Portion or Payout= 30.00%), and (**R**= Retained Earnings= 5,529 millions USD), are known, then it's (**A**= After Tax Income planned), is:
   A= **R**/[1-**d**]
   = 5,529/[1-0.3000]
   = 7,899 millions USD

Law-3283:

If both (**A**= After Tax Income= 7,899 millions USD), and (**R**= Retained Earnings= 5,529 millions USD), are known, then it's (**d**= Dividend Portion or Payout planned), is:

$$\begin{aligned}\mathbf{d} &= 1\text{-}\mathbf{R}/\mathbf{A} \\ &= 1\text{-}5{,}529/7{,}899 \\ &= 30.00\%\end{aligned}$$

Law-3284:

If both (**d**= D/A= Dividend Portion or Payout= 30.00%), (**T**= Tax Paid= 3,385 millions USD), and (**B**= Before Tax Income= 11,285 millions USD), are known, then it's (**R**= Retained Earnings planned), is:

$$\begin{aligned}\mathbf{R} &= [1\text{-}\mathbf{d}][\mathbf{B}\text{-}\mathbf{T}] \\ &= [1\text{-}0.3000][11{,}285\text{-}3{,}385] \\ &= 5{,}529 \text{ millions USD}\end{aligned}$$

Law-3285:

If both (**d**= D/A= Dividend Portion or Payout= 30.00%), (**T**= Tax Paid= 3,385 millions USD), and (**R**= Retained Earnings= 5,529 millions USD), are known, then it's (**B**= Before Tax Income planned), is:

$$\begin{aligned}\mathbf{B} &= \mathbf{T}+\mathbf{R}/[1\text{-}\mathbf{d}] \\ &= 3{,}385+5{,}529/[1\text{-}0.3000] \\ &= 11{,}285 \text{ millions USD}\end{aligned}$$

Law-3286:
   If both (**d**= D/A= Dividend Portion or Payout= 30.00%), (**B**= Before Tax Income= 11,285 millions USD), and (**R**= Retained Earnings= 5,529 millions USD), are known, then it's (**T**= Tax Paid planned), is:

   $T = B - R/[1-d]$
   $= 11,285 - 5,529/[1-0.3000]$
   = 3,385 millions USD

Law-3287:
   If both (**T**= Tax Paid= 3,385 millions USD), (**B**= Before Tax Income= 11,285 millions USD), and (**R**= Retained Earnings= 5,529 millions USD), are known, then it's (**d**= Dividend Portion or Payout planned), is:

   $d = 1 - R/[B-T]$
   $= 1 - 5,529/[11,285 - 3,385]$
   = 30.00%

Law-3288:
   If both (**B**= Before Tax Income= 11,285 millions USD), (**d**= D/A= Dividend Portion or Payout= 30.00%), and (**t**= T/B= Tax Rate= 30.00%), are known, then it's (**R**= Retained Earnings planned), is:
   $R = [B - Bt][1-d] = B[1-t][1-d]$
   $= 11,285[1-0.3000][1-0.3000]$
   = 5,529 millions USD

Law-3289:

If both (**R**= Retained Earnings= 5,529 millions USD), (**d**= D/A= Dividend Portion or Payout= 30.00%), and (**t**= T/B= Tax Rate= 30.00%), are known, then it's (**B**= Before Tax Income planned), is:

$$B = R/[1-t][1-d]$$
$$= 5,529/[1-0.3000][1-0.3000]$$
$$= \underline{11,285} \text{ millions USD}$$

Law-3290:

If both (**R**= Retained Earnings= 5,529 millions USD), (**d**= D/A= Dividend Portion or Payout= 30.00%), and (**B**= Before Tax Income= 11,285 millions USD), are known, then it's (**t**= Tax Rate planned), is:

$$t = 1 - R/\{B[1-d]\}$$
$$= 1 - 5,529/\{11,285[1-0.3000]\}$$
$$= \underline{30.00\%}$$

Law-3291:

If both (**R**= Retained Earnings= 5,529 millions USD), (**t**= T/B= Tax Rate= 30.00%), and (**B**= Before Tax Income= 11,285 millions USD), are known, then it's (**d**= Dividend Portion or Payout planned), is:

$$d = 1 - R/\{B[1-t]\}$$
$$= 1 - 5,529/\{11,285[1-0.3000]\}$$
$$= \underline{30.00\%}$$

Law-3292:
   If both (**R**= Retained Earnings= 5,529 millions USD), (**t**= T/B= Tax Rate= 30.00%), (**I**= Interest Expense= 513 millions USD), and (**O**= Operational Surplus= 11,798 millions USD), are known, then it's (**R**= Retained Earnings planned), is:
   R= [1-**t**][1-**d**][**O**-**I**]
   $\quad\quad$ = [1-0.3000][1-0.3000][11,798-513]
   $\quad\quad$ = 5,529 millions USD

Law-3293:
   If both (**R**= Retained Earnings= 5,529 millions USD), (**d**= D/A= Dividend Portion or Payout= 30.00%), (**I**= Interest Expense= 513 millions USD), and (**O**= Operational Surplus= 11,798 millions USD), are known, then it's (**t**= Tax Rate planned), is:
   t= 1-**R**/{[1-**d**][**O**-**I**]}
   $\quad\quad$ = 1-5,529/{[1-0.3000][11,798-513]}
   $\quad\quad$ = 30.00%

Law-3294:
   If both (**R**= Retained Earnings= 5,529 millions USD), (**t**= T/B= Tax Rate= 30.00%), (**I**= Interest Expense= 513 millions USD), and (**O**= Operational Surplus= 11,798 millions USD), are known, then it's (**d**= Dividend Portion or Payout planned), is:
   d= 1-**R**/{[1-**t**][**O**-**I**]}
   $\quad\quad$ = 1-5,529/{[1-0.3000][11,798-513]}
   $\quad\quad$ = 30.00%

Law-3295:

If both (**R**= Retained Earnings= 5,529 millions USD), (**t**= T/B= Tax Rate= 30.00%), (**I**= Interest Expense= 513 millions USD), and (**d**= D/A= Dividend Portion or Payout= 30.00%), are known, then it's (**O**= Operational Surplus planned), is:

O= **I**+**R**/{[1-**t**][1-**d**]}
= 513+5,529/{[1-0.3000][1-0.3000]}
= 11,798 millions USD

Law-3296:

If both (**R**= Retained Earnings= 5,529 millions USD), (**t**= T/B= Tax Rate= 30.00%), (**O**= Operational Surplus= 11,798 millions USD), and (**d**= D/A= Dividend Portion or Payout= 30.00%), are known, then it's (**I**= Interest Expense planned), is:

I= **O**-**R**/{[1-**t**][1-**d**]}
= 11,798-5,529/{[1-0.3000][1-0.3000]}
= 513 millions USD

Law-3297:
If both ($= Sales or Revenues= 51,294 millions USD), (i= I/S= Interest Portion= 1.00%), (t= T/B= Tax Rate= 30.00%), (O= Operational Surplus= 11,798 millions USD), and (d= D/A= Dividend Portion or Payout= 30.00%), are known, then it's (R= Retained Earnings planned), is:
$$R = [1-t][1-d][O-\$i]$$
$$= [1-0.3000][1-0.3000][11,798 - 51,294*0.0100]$$
$$= 5,529 \text{ millions USD}$$

Law-3298:
If both (i= I/S= Interest Portion= 1.00%), ($= Sales or Revenues= 51,294 millions USD), (R= Retained Earnings= 5,529 millions USD), (O= Operational Surplus= 11,798 millions USD), and (d= D/A= Dividend Portion or Payout= 30.00%), are known, then it's (t= Tax Rate planned), is:
$$t = 1 - R/\{[1-d][O-\$i]\}$$
$$= 513-5,529/\{[1-0.3000][11,7978 - 51,294*0.0100]\}$$
$$= 30.00\%$$

Law-3299:

If both (i= I/S= Interest Portion= 1.00%), (S= Sales or Revenues= 51,294 millions USD), (R= Retained Earnings= 5,529 millions USD), (O= Operational Surplus= 11,798 millions USD), and (t= T/B= Tax Rate= 30.00%), are known, then it's (d= Dividend Portion or Payout planned), is:

$$d = 1 - R/\{[1-t][O-Si]\}$$
$$= 1 - 5{,}529/\{[1-0.3000][11{,}7978 - 51{,}294*0.0100]\}$$
$$= 30.00\%$$

Law-3300:

If both (i= I/S= Interest Portion= 1.00%), (S= Sales or Revenues= 51,294 millions USD), (R= Retained Earnings= 5,529 millions USD), (d= D/A= Dividend Portion or Payout= 30.00%), and (t= T/B= Tax Rate= 30.00%), are known, then it's (O= Operational Surplus planned), is:

$$O = Si + R/\{[1-t][1-d]\}$$
$$= 51{,}294*0.0100 + 5{,}529/\{[1-0.3000][1-0.3000]\}$$
$$= 11{,}798 \text{ millions USD}$$

Law-3301:

If both (i= I/S= Interest Portion= 1.00%), (O= Operational Surplus= 11,798 millions USD), (R= Retained Earnings= 5,529 millions USD), (d= D/A= Dividend Portion or Payout= 30.00%), and (t= T/B= Tax Rate= 30.00%), are known, then it's (S= Sales or Revenues planned), is:

$S = (O-R/\{[1-t][1-d]\})/i$
$= (11,798-5,529/\{[1-0.3000]$
$[1-0.3000]\})/ 0.0100$
$= 51,294$ millions USD

Law-3302:

If both (S= Sales or Revenues= 51,294 millions USD), (O= Operational Surplus= 11,798 millions USD), (R= Retained Earnings= 5,529 millions USD), (d= D/A= Dividend Portion or Payout= 30.00%), and (t= T/B= Tax Rate= 30.00%), are known, then it's (i= Interest Portion planned), is:

$i = (O-R/\{[1-t][1-d]\})/S$
$= (11,798-5,529/\{[1-0.3000]$
$[1-0.3000]\})/51,294$
$= 1.00\%$

Law-3303:

If both ($ $'$ = Sales of Past Year= 48,851 millions USD), (s= [S/S']-1= Sales Growth= 5.00%), (O= Operational Surplus= 11,798 millions USD), (i= I/S= Interest Portion= 1.00%), (d= D/A= Dividend Portion or Payout= 30.00%), and (t= T/B= Tax Rate= 30.00%), are known, then it's (R= Retained Earnings planned), is:

R= [1-t][1-d]{O-$ $'$ i[1+s]}
   = [1-0.3000][1-0.3000]{11,798
         -48,851*0.0100[1+0.0500]
   = 5,529 millions USD

Law-3304:

If both ($ $'$ = Sales of Past Year= 48,851 millions USD), (s= [S/S']-1= Sales Growth= 5.00%), (R= Retained Earnings= 5,529 millions USD), (i= I/S= Interest Portion= 1.00%), (d= D/A= Dividend Portion or Payout= 30.00%), and (t= T/B= Tax Rate= 30.00%), are known, then it's (O= Operational Surplus planned), is:

O= $ $'$ i[1+s]+R/{[1-t][1-d]}
   = 48,851*0.0100[1+0.0500]+5,529
          /{[1-0.3000][1-0.3000]}
   = 11,798 millions USD

Law-3305:

If both (**O**= Operational Surplus= 11,798 millions USD), (**s**= [S/S']-1= Sales Growth= 5.00%), (**R**= Retained Earnings= 5,529 millions USD), (**i**= I/S= Interest Portion= 1.00%), (**d**= D/A= Dividend Portion or Payout= 30.00%), and (**t**= T/B= Tax Rate= 30.00%), are known, then it's (**S'**= Sales Past), must be:

$S' = (O-R/\{[1-t][1-d]\})/\{i[1+s]\}$
$= 11{,}798 - 5{,}529/\{[1-0.3000]$
$[1-0.3000]\})/\{0.0100$
$[1+0.0500]\}$
$= 48{,}851$ millions USD

Law-3306:

If both (**O**= Operational Surplus= 11,798 millions USD), (**s**= [S/S']-1= Sales Growth= 5.00%), (**R**= Retained Earnings= 5,529 millions USD), (**S'**= Sales of Past Year= 48,851 millions USD), (**d**= D/A= Dividend Portion or Payout= 30.00%), and (**t**= T/B= Tax Rate= 30.00%), are known, then it's (**i**= Interest Portion planned), is:

$i = (O-R/\{[1-t][1-d]\})/\{S'[1+s]\}$
$= 11{,}798 - 5{,}529/\{[1-0.3000]$
$[1-0.3000]\})/\{48{,}851$
$[1+0.0500]\}$
$= 1.00\%$

Law-3307:

If both (**O**= Operational Surplus= 11,798 millions USD), (**i**= I/S= Interest Portion= 1.00%), (**R**= Retained Earnings= 5,529 millions USD), (**$'**= Sales of Past Year= 48,851 millions USD), (**d**= D/A= Dividend Portion or Payout= 30.00%), and (**t**= T/B= Tax Rate= 30.00%), are known, then it's (**s**= Sales Growth planned), is:

$$s = (O-R/\{[1-t][1-d]\})/[\$'i]-1$$
$$= 11{,}798-5{,}529/\{[1-0.3000][1-0.3000]\})/\{48{,}851 * 0.0100\}-1$$
$$= \underline{5.00\%}$$

Law-3308:

If both (**O**= Operational Surplus= 11,798 millions USD), (**i**= I/S= Interest Portion= 1.00%), (**R**= Retained Earnings= 5,529 millions USD), (**$'**= Sales of Past Year= 48,851 millions USD), (**d**= D/A= Dividend Portion or Payout= 30.00%), and (**s**= [S/S']-1= Sales Growth= 5.00%), are known, then it's (**t**= Tax Rate planned), is:

$$t = 1-R/([1-d]\{O-\$'i[1+s]\})$$
$$= 1-5{,}529/([1-0.3000]\{11{,}798-48{,}851[1+0.0500]\})$$
$$= \underline{30.00\%}$$

Law-3309:

If both (O= Operational Surplus= 11,798 millions USD), (t= T/B= Tax Rate= 30.00%), (i= I/S= Interest Portion= 1.00%), (R= Retained Earnings= 5,529 millions USD), (S'= Sales of Past Year= 48,851 millions USD), and (s= [S/S']-1= Sales Growth= 5.00%), are known, then it's (d= Dividend Portion or Payout planned), is:

$$d = 1-R/([1-t]\{O-S'i[1+s]\})$$
$$= 1-5,529/([1-0.3000]\{11,798 -48,851[1+0.0500]\})$$
$$= 30.00\%$$

Law-3310:

If both (M= Margin of Contribution= 41,548 millions USD), (I= Interest Expense= 513 millions USD), (d= D/A= Dividend Portion or Payout= 30.00%), and (F= Fixed Cost= 29,750 millions USD), (t= T/B= Tax Rate= 30.00%), are known, then it's (R= Retained Earnings planned), is:

$$R = [1-t][1-d][M-F-I]$$
$$= [1-0.3000][1-0.3000][41,548-29,750 -513]$$
$$= 5,529 \text{ millions USD}$$

Law-3311:

If both (**R**= Retained Earnings= 5,529 millions USD), (**d**= D/A= Dividend Portion or Payout= 30.00%), (**I**= Interest Expense= 513 millions USD), (**F**= Fixed Cost= 29,750 millions USD), and (**t**= T/B= Tax Rate= 30.00%), are known, then it's (**M**= Margin of Contribution planned), is:

$$M = F+I+R/\{[1-t][1-d]\}$$
$$= 29{,}750+513+5{,}529/\{[1-0.3000][1-0.3000]\}$$
$$= \underline{41{,}548} \text{ millions USD}$$

Law-3312:

If both (**R**= Retained Earnings= 5,529 millions USD), (**I**= Interest Expense= 513 millions USD), (**d**= D/A= Dividend Portion or Payout= 30.00%), (**M**= Margin of Contribution= 41,548 millions USD), and (**t**= T/B= Tax Rate= 30.00%), are known, then it's (**F**= Fixed Cost planned), is:

$$F = M-I-R/\{[1-t][1-d]\}$$
$$= 41{,}548-513-5{,}529/\{[1-0.3000][1-0.3000]\}$$
$$= \underline{29{,}750} \text{ millions USD}$$

Law-3313:

If both (**R**= Retained Earnings= 5,529 millions USD), (**F**= Fixed Cost= 29,750 millions USD), (**d**= D/A= Dividend Portion or Payout= 30.00%), (**M**= Margin of Contribution= 41,548 millions USD), and (**t**= T/B= Tax Rate= 30.00%), are known, then it's (**I**= Interest Expense planned), is:

$$I = M-F-R/\{[1-t][1-d]\}$$
$$= 41,548-29,750-5,529/\{[1-0.3000][1-0.3000]\}$$
$$= \underline{513} \text{ millions USD}$$

Law-3314:

If both (**R**= Retained Earnings= 5,529 millions USD), (**F**= Fixed Cost= 29,750 millions USD), (**d**= D/A= Dividend Portion or Payout= 30.00%), (**M**= Margin of Contribution= 41,548 millions USD), and (**I**= Interest Expense= 513 millions USD), are known, then it's (**t**= Tax Rate planned), is:

$$t = 1-R/\{[1-d][M-F-I]\}$$
$$= 1-5,529/\{[1-0.3000][41,548-29,750-513]\}$$
$$= \underline{30.00}\%$$

Law-3315:

If both (**R**= Retained Earnings= 5,529 millions USD), (**F**= Fixed Cost= 29,750 millions USD), (**t**= T/B= Tax Rate= 30.00%), (**M**= Margin of Contribution= 41,548 millions USD), and (**I**= Interest Expense= 513 millions USD), are known, then it's (**d**= Dividend Portion or Payout planned), is:

$$d = 1 - R/\{[1-t][M-F-I]\}$$
$$= 1 - 5,529/\{[1-0.3000][41,548-29,750-513]\}$$
$$= 30.00\%$$

Law-3316:

If both (**d**= D/A= Dividend Portion or Payout= 30.00%), (**F**= Fixed Cost= 29,750 millions USD), (**t**= T/B= Tax Rate= 30.00%), (**S**= Sales or Revenues= 51,294 millions USD), (**M**= Margin of Contribution= 41,548 millions USD), and (**i**= I/S= Interest Portion= 1.00%), are known, then it's (**R**= Retained Earnings planned), is:

$$R = [1-t][1-d][M-F-Si]$$
$$= [1-0.3000][1-0.3000][41,548-29,750-51,294*0.0100]$$
$$= 5,529 \text{ millions USD}$$

Law-3317:
If both (**d**= D/A= Dividend Portion or Payout= 30.00%), (**F**= Fixed Cost= 29,750 millions USD), (**t**= T/B= Tax Rate= 30.00%), (**S**= Sales or Revenues= 51,294 millions USD), (**R**= Retained Earnings= 5,529 millions USD), and (**i**= I/S= Interest Portion= 1.00%), are known, then it's (**M**= Margin of Contribution planned), is:

M= **F**+**Si**+**R**/{[1-**t**][1-**d**]}
= 29,750+51,294*0.0100+5,529
/{[1-0.3000][1-0.3000]}
= 41,548 millions USD

Law-3318:
If both (**d**= D/A= Dividend Portion or Payout= 30.00%), (**M**= Margin of Contribution= 41,548 millions USD), (**t**= T/B= Tax Rate= 30.00%), (**S**= Sales or Revenues= 51,294 millions USD), (**R**= Retained Earnings= 5,529 millions USD), and (**i**= I/S= Interest Portion= 1.00%), are known, then it's (**M**= Margin of Contribution planned), is:

F= **M**-**Si**-**R**/{[1-**t**][1-**d**]}
= 41,548-51,294*0.0100-5,529
/{[1-0.3000][1-0.3000]}
= 29,750 millions USD

Law-3319:
If both (**d**= D/A= Dividend Portion or Payout= 30.00%), (**M**= Margin of Contribution= 41,548 millions USD), (**t**= T/B= Tax Rate= 30.00%), (**F**= Fixed Cost= 29,750 millions USD), (**R**= Retained Earnings= 5,529 millions USD), and (**i**= I/S= Interest Portion= 1.00%), are known, then it's (**$**= Sales or Revenues planned), is:

$\$ = (M-F-R/\{[1-t][1-d]\})/i$
$= 41{,}548 - 29{,}750 - 5{,}529/\{[1-0.3000][1-0.3000]\}/0.0100$
$= \underline{51{,}294}$ millions USD

Law-3320:
If both (**d**= D/A= Dividend Portion or Payout= 30.00%), (**M**= Margin of Contribution= 41,548 millions USD), (**t**= T/B= Tax Rate= 30.00%), (**F**= Fixed Cost= 29,750 millions USD), (**R**= Retained Earnings= 5,529 millions USD), and (**$**= Sales or Revenues= 51,294 millions USD), are known, then it's (**i**= Interest Portion planned), is:

$i = (M-F-R/\{[1-t][1-d]\})/\$$
$= 41{,}548 - 29{,}750 - 5{,}529/\{[1-0.3000][1-0.3000]\}/51{,}294$
$= \underline{1.00\%}$

Law-3321:

If both (**d**= D/A= Dividend Portion or Payout= 30.00%), (**M**= Margin of Contribution= 41,548 millions USD), (**i**= I/S= Interest Portion= 1.00%), (**F**= Fixed Cost= 29,750 millions USD), (**R**= Retained Earnings= 5,529 millions USD), and (**S**= Sales or Revenues= 51,294 millions USD), are known, then it's (**t**= Tax Rate planned), is:

$$\mathbf{t} = 1 - \mathbf{R}/\{[1-\mathbf{d}][\mathbf{M}-\mathbf{F}-\mathbf{Si}]\}$$
$$= 1 - 5{,}529/\{[1-0.3000][41{,}548-29{,}750 -51{,}294*0.0100]\}$$
$$= 30.00\%$$

Law-3322:

If both (**t**= T/B= Tax Rate= 30.00%), (**M**= Margin of Contribution= 41,548 millions USD), (**i**= I/S= Interest Portion= 1.00%), (**F**= Fixed Cost= 29,750 millions USD), (**R**= Retained Earnings= 5,529 millions USD), and (**S**= Sales or Revenues= 51,294 millions USD), are known, then it's (**d**= Dividend Portion or Payout planned), is:

$$\mathbf{d} = 1 - \mathbf{R}/\{[1-\mathbf{t}][\mathbf{M}-\mathbf{F}-\mathbf{Si}]\}$$
$$= 1 - 5{,}529/\{[1-0.3000][41{,}548-29{,}750 -51{,}294*0.0100]\}$$
$$= 30.00\%$$

Law-3323:

If both (**t**= T/B= Tax Rate= 30.00%), (**M**= Margin of Contribution= 41,548 millions USD), (**i**= I/S= Interest Portion= 1.00%), (**F**= Fixed Cost= 29,750 millions USD), (**d**= D/A= Dividend Portion or Payout= 30.00%), (**$'**= Sales of Past Year= 48,851 millions USD), and (**s**= [S/S']-1= Sales Growth= 5.00%), are known, then it's (**R**= Retained Earnings planned), is:

$$\begin{aligned}
\mathbf{R} &= [1\text{-}\mathbf{t}][1\text{-}\mathbf{d}]\{\mathbf{M}\text{-}\mathbf{F}\text{-}\mathbf{\$'i}[1+\mathbf{s}] \\
&= [1\text{-}0.3000][1\text{-}0.3000]\{41{,}548\text{-}29{,}750 \\
&\quad \text{-}48{,}851*0.0100[1+0.0500]\} \\
&= \underline{5{,}529} \text{ millions USD}
\end{aligned}$$

Law-3324:

If both (**t**= T/B= Tax Rate= 30.00%), (**s**= [S/S']-1= Sales Growth= 5.00%), (**i**= I/S= Interest Portion= 1.00%), (**d**= D/A= Dividend Portion or Payout= 30.00%), (**$'**= Sales of Past Year= 48,851 millions USD), (**R**= Retained Earnings= 5,529 millions USD), and (**F**= Fixed Cost= 29,750 millions USD), are known, then it's (**M**= Margin of Contribution planned), is:

$$\begin{aligned}
\mathbf{M} &= \mathbf{F}+\mathbf{\$'i}[1+\mathbf{s}]+\mathbf{R}/\{[1\text{-}\mathbf{t}][1\text{-}\mathbf{d}]\} \\
&= 29{,}750+48{,}851*0.0100[1+0.0500] \\
&\quad +5{,}529/\{[1\text{-}0.3000][1\text{-}0.3000]\} \\
&= \underline{41{,}548} \text{ millions USD}
\end{aligned}$$

Law-3325:
If both (t= T/B= Tax Rate= 30.00%), (s= [S/S']-1= Sales Growth= 5.00%), (i= I/S= Interest Portion= 1.00%), (d= D/A= Dividend Portion or Payout= 30.00%), (S'= Sales of Past Year= 48,851 millions USD), (R= Retained Earnings= 5,529 millions USD), and (M= Margin of Contribution= 41,548 millions USD), are known, then it's (F= Fixed Cost planned), is:

$F = M - S'i[1+s] - R/\{[1-t][1-d]\}$
   = 41,548 - 48,851*0.0100[1+0.0500]
      -5,529/{[1-0.3000]1-0.3000]}
   = <u>29,750</u> millions USD

Law-3326:
If both (F= Fixed Cost= 29,750 millions USD), (t= T/B= Tax Rate= 30.00%), (s= [S/S']-1= Sales Growth= 5.00%), (i= I/S= Interest Portion= 1.00%), (d= D/A= Dividend Portion or Payout= 30.00%), (R= Retained Earnings= 5,529 millions USD), and (M= Margin of Contribution= 41,548 millions USD), are known, then it's (S'= Sales Past), must be:

$S' = (M - F - R/\{[1-t][1-d]\})/\{i[1+s]\}$
   = (41,548 - 29,750 - 5,529/{[1-0.3000]
      [1-0.3000]})
      /{0.0100[1+0.0500]}
   = <u>48,851</u> millions USD

Law-3327:
If both (**F**= Fixed Cost= 29,750 millions USD), (**t**= T/B= Tax Rate= 30.00%), (**s**= [S/S']-1= Sales Growth= 5.00%), (**S'**= Sales of Past Year= 48,851 millions USD), (**d**= D/A= Dividend Portion or Payout= 30.00%), (**R**= Retained Earnings= 5,529 millions USD), and (**M**= Margin of Contribution= 41,548 millions USD), are known, then it's (**i**= Interest Portion planned), is:

$$i = M-F-R/\{[1-t][1-d]\}/\{S'[1+s]\}$$
$$= 41,548-29,750-5,529/\{[1-0.3000][1-0.3000]\}$$
$$/\{48,851[1+0.0500]\}$$
$$= \underline{1.00\%}$$

Law-3328:
If both (**F**= Fixed Cost= 29,750 millions USD), (**t**= T/B= Tax Rate= 30.00%), (**i**= I/S= Interest Portion= 1.00%), (**S'**= Sales of Past Year= 48,851 millions USD), (**d**= D/A= Dividend Portion or Payout= 30.00%), (**R**= Retained Earnings= 5,529 millions USD), and (**M**= Margin of Contribution= 41,548 millions USD), are known, then it's (**s**= Sales Growth planned), is:

$$s = (M-F-R/\{[1-t][1-d]\})/[S'i]-1$$
$$= (41,548-29,750-5,529/\{[1-0.3000][1-0.3000]\})$$
$$/[48,851*0.0100]-1$$
$$= \underline{5.00\%}$$

Law-3329:
If both (**F**= Fixed Cost= 29,750 millions USD), (**s**= [S/S']-1= Sales Growth= 5.00%), (**i**= I/S= Interest Portion= 1.00%), (**$'**= Sales of Past Year= 48,851 millions USD), (**d**= D/A= Dividend Portion or Payout= 30.00%), (**R**= Retained Earnings= 5,529 millions USD), and (**M**= Margin of Contribution= 41,548 millions USD), are known, then it's (**t**= Tax Rate planned), is:

$$t = 1-R/([1-d]\{M-F-\$'i[1+s]\})$$
$$= 1-5,529/(1-0.3000]\{41,548-29,750 -48,851*0.0100[1+0.0500]\})$$
$$= 30.00\%$$

Law-3330:
If both (**F**= Fixed Cost= 29,750 millions USD), (**s**= [S/S']-1= Sales Growth= 5.00%), (**i**= I/S= Interest Portion= 1.00%), (**$'**= Sales of Past Year= 48,851 millions USD), (**t**= T/B= Tax Rate= 30.00%), (**R**= Retained Earnings= 5,529 millions USD), and (**M**= Margin of Contribution= 41,548 millions USD), are known, then it's (**d**= Dividend Portion or Payout planned), is:

$$d = 1-R/([1-t]\{M-F-\$'i[1+s]\})$$
$$= 1-5,529/(1-0.3000]\{41,548-29,750 -48,851*0.0100[1+0.0500]\})$$
$$= 30.00\%$$

Law-3331:

If both (**I**= Interest Expense= 513 millions USD), (**f**= F/S= Fixed Portion= 58.00%), (**t**= T/B= Tax Rate= 30.00%), (**$**= Sales or Revenues= 51,294 millions USD), (**d**= D/A= Dividend Portion or Payout= 30.00%), and (**M**= Margin of Contribution= 41,548 millions USD), are known, then it's (**R**= Retained Earnings planned), is:

$$R = [1-t][1-d][M-\$f-I]$$
$$= [1-0.3000][-0.3000][41,548-51,294*0.5800-513]$$
$$= 5,529 \text{ millions USD}$$

Law-3332:

If both (**I**= Interest Expense= 513 millions USD), (**f**= F/S= Fixed Portion= 58.00%), (**t**= T/B= Tax Rate= 30.00%), (**$**= Sales or Revenues= 51,294 millions USD), (**d**= D/A= Dividend Portion or Payout= 30.00%), and (**R**= Retained Earnings= 5,529 millions USD), are known, then it's (**M**= Margin of Contribution planned), is:

$$M = \$f + I + R / \{[1-t][1-d]\}$$
$$= 51,294*0.5800+513+5,529 / \{[1-0.3000][-0.3000]\}$$
$$= 41,548 \text{ millions USD}$$

Law-3333:
If both (**I**= Interest Expense= 513 millions USD), (**f**= F/S= Fixed Portion= 58.00%), (**t**= T/B= Tax Rate= 30.00%), (**M**= Margin of Contribution= 41,548 millions USD), (**d**= D/A= Dividend Portion or Payout= 30.00%), and (**R**= Retained Earnings= 5,529 millions USD), are known, then it's (**$**= Sales or Revenues planned), is:

$$\$ = (\mathbf{M}\text{-}\mathbf{I}\text{-}\mathbf{R}/\{[1\text{-}\mathbf{t}][1\text{-}\mathbf{d}]\})/\mathbf{f}$$
$$= (41{,}548\text{-}513\text{-}5{,}529/\{[1\text{-}0.3000][1\text{-}0.3000]\})/0.5800$$
$$= 51{,}294 \text{ millions USD}$$

Law-3334:
If both (**I**= Interest Expense= 513 millions USD), (**$**= Sales or Revenues= 51,294 millions USD), (**t**= T/B= Tax Rate= 30.00%), (**M**= Margin of Contribution= 41,548 millions USD), (**d**= D/A= Dividend Portion or Payout= 30.00%), and (**R**= Retained Earnings= 5,529 millions USD), are known, then it's (**f**= Fixed Portion planned), is:

$$\mathbf{f} = (\mathbf{M}\text{-}\mathbf{I}\text{-}\mathbf{R}/\{[1\text{-}\mathbf{t}][1\text{-}\mathbf{d}]\})/\$$$
$$= (41{,}548\text{-}513\text{-}5{,}529/\{[1\text{-}0.3000][1\text{-}0.3000]\})/51{,}294$$
$$= 58.00\%$$

Law-3335:
If both (f= F/S= Fixed Portion= 58.00%), ($\$$= Sales or Revenues= 51,294 millions USD), (t= T/B= Tax Rate= 30.00%), (M= Margin of Contribution= 41,548 millions USD), (d= D/A= Dividend Portion or Payout= 30.00%), and (R= Retained Earnings= 5,529 millions USD), are known, then it's (I= Interest Expense planned), is:

$$I = M\text{-}\$f\text{-}R/\{[1\text{-}t][1\text{-}d]\}$$
$$= 41{,}548\text{-}51{,}294*0.5800\text{-}5{,}529 /\{[1\text{-}0.3000][\text{-}0.3000]\}$$
$$= 513 \text{ millions USD}$$

Law-3336:
If both (f= F/S=Fixed Portion= 58.00%), ($\$$= Sales or Revenues= 51,294 millions USD), (I= Interest Expense= 513 millions USD), (M= Margin of Contribution= 41,548 millions USD), (d= D/A= Dividend Portion or Payout= 30.00%), and (R= Retained Earnings= 5,529 millions USD), are known, then it's (t= Tax Rate planned), is:

$$t = 1\text{-}R/\{[1\text{-}d][M\text{-}\$f\text{-}I]\}$$
$$= 1\text{-}5{,}529/\{[1\text{-}0.3000][41{,}548 \text{-}51{,}294*0.5800\text{-}513]\}$$
$$= 30.00\%$$

Law-3337:

If both (**f**= F/S= Fixed Portion= 58.00%), (**S**= Sales or Revenues= 51,294 millions USD), (**I**= Interest Expense= 513 millions USD), (**M**= Margin of Contribution= 41,548 millions USD), (**t**= T/B= Tax Rate= 30.00%), and (**R**= Retained Earnings= 5,529 millions USD), are known, then it's (**d**= Dividend Portion or Payout planned), is:

$$d = 1 - R/\{[1-t][M-Sf-I]\}$$
$$= 1 - 5,529/\{[1-0.3000][41,548 -51,294*0.5800-513]\}$$
$$= 30.00\%$$

Law-3338:

If both (**f**= F/S= Fixed Portion= 58.00%), (**S**= Sales or Revenues= 51,294 millions USD), (**i**= I/S= Interest Portion= 1.00%), (**M**= Margin of Contribution= 41,548 millions USD), (**t**= T/B= Tax Rate= 30.00%), and (**d**= D/A= Dividend Portion or Payout= 30.00%), are known, then it's (**R**= Retained Earnings planned), is:

$$R = [1-t][1-d][M-Sf-Si] = [1-t][1-d]\{M-S[f+i]\}$$
$$= [1-0.3000][1-0.3000]\{41,548 -51,294[0.5800+0.0100]\}$$
$$= 5,529 \text{ millions USD}$$

Law-3339:
If both (**f**= F/S= Fixed Portion= 58.00%), (**$**= Sales or Revenues= 51,294 millions USD), (**i**= I/S= Interest Portion= 1.00%), (**R**= Retained Earnings= 5,529 millions USD), (**t**= T/B= Tax Rate= 30.00%), and (**d**= D/A= Dividend Portion or Payout= 30.00%), are known, then it's (**M**= Margin of Contribution planned), is:

$$M= \$[f+i]+R/\{[1-t][1-d]\}$$
$$= 51{,}294[0.5800+0.0100]+5{,}529 /\{[1-0.3000][1-0.3000]\}$$
$$= 41{,}548 \text{ millions USD}$$

Law-3340:
If both (**f**= F/S= Fixed Portion= 58.00%), (**M**= Margin of Contribution= 41,548 millions USD), (**i**= I/S= Interest Portion= 1.00%), (**R**= Retained Earnings= 5,529 millions USD), (**t**= T/B= Tax Rate= 30.00%), and (**d**= D/A= Dividend Portion or Payout= 30.00%), are known, then it's (**$**= Sales or Revenues planned), is:

$$\$= (M-R/\{[1-t][1-d]\})/[f+i]$$
$$= (41{,}548-5{,}529/\{[1-0.3000][1-0.3000]\})/ [0.5800+0.0100]$$
$$= 51{,}294 \text{ millions USD}$$

Law-3341:
If both ($= Sales or Revenues= 51,294 millions USD), (M= Margin of Contribution= 41,548 millions USD), (i= I/S= Interest Portion= 1.00%), (R= Retained Earnings= 5,529 millions USD), (t= T/B= Tax Rate= 30.00%), and (d= D/A= Dividend Portion or Payout= 30.00%), are known, then it's (f= Fixed Portion planned), is:

$$f = (M-R/\{[1-t][1-d]\})/\$-i$$
$$= (41{,}548-5{,}529/\{[1-0.3000][1-0.3000]\})/51{,}294-0.0100$$
$$= 58.00\%$$

Law-3342:
If both ($= Sales or Revenues= 51,294 millions USD), (M= Margin of Contribution= 41,548 millions USD), (f= F/S= Fixed Portion= 58.00%), (R= Retained Earnings= 5,529 millions USD), (t= T/B= Tax Rate= 30.00%), and (d= D/A= Dividend Portion or Payout= 30.00%), are known, then it's (i= Interest Portion planned), is:

$$i = (M-R/\{[1-t][1-d]\})/\$-f$$
$$= (41{,}548-5{,}529/\{[1-0.3000][1-0.3000]\})/51{,}294-0.5800$$
$$= 1.00\%$$

Law-3343:
If both ($= Sales or Revenues= 51,294 millions USD), (M= Margin of Contribution= 41,548 millions USD), (f= F/S= Fixed Portion= 58.00%), (R= Retained Earnings= 5,529 millions USD), (i= I/S= Interest Portion= 1.00%), and (d= D/A= Dividend Portion or Payout= 30.00%), are known, then it's (t= Tax Rate planned), is:

$$t = 1-R/(1-d)\{M-\$[f+i]\})$$
$$= 1-5,529/([1-0.3000]\{41,548-51,294\ [0.5800+0.0100]\})$$
$$= 30.00\%$$

Law-3344:
If both ($= Sales or Revenues= 51,294 millions USD), (M= Margin of Contribution= 41,548 millions USD), (f= F/S= Fixed Portion= 58.00%), (R= Retained Earnings= 5,529 millions USD), (i= I/S= Interest Portion= 1.00%), and (t= T/B= Tax Rate= 30.00%), are known, then it's (d= Dividend Portion or Payout planned), is:

$$d = 1-R/([1-t]\{M-\$[f+i]\})$$
$$= 1-5,529/([1-0.3000]\{41,548-51,294\ [0.5800+0.0100]\})$$
$$= 30.00\%$$

Law-3345:

If both ($= Sales or Revenues= 51,294 millions USD), ($'= Sales of Past Year= 48,851 millions USD), (s= [S/S']-1= Sales Growth= 5.00%), (M= Margin of Contribution= 41,548 millions USD), (f= F/S= Fixed Portion= 58.00%), (d= D/A= Dividend Portion or Payout= 30.00%), (i= I/S= Interest Portion= 1.00%), and (t= T/B= Tax Rate= 30.00%), are known, then it's (R= Retained Earnings planned), is:

$$R = [1-t][1-d]\{M-\$f-\$'i[1+s]\}$$
$$= [1-0.3000][1-0.3000]\{41,548-51,294$$
$$*0.5800-48,851*0.0100$$
$$[1+0.0500]\}$$
$$= 5,529 \text{ millions USD}$$

Law-3346:

If both ($= Sales or Revenues= 51,294 millions USD), ($'= Sales of Past Year= 48,851 millions USD), (s= [S/S']-1= Sales Growth= 5.00%), (R= Retained Earnings= 5,529 millions USD), (f= F/S– Fixed Portion= 58.00%), (d= D/A= Dividend Portion or Payout= 30.00%), (i= I/S= Interest Portion= 1.00%), and (t= T/B= Tax Rate= 30.00%), are known, then it's (M= Margin of Contribution planned), is:

$$M = \$f+\$'i[1+s]+R/\{[1-t][1-d]\}$$
$$= 51,294*0.5800+48,851*0.0100$$
$$[1+0.0500]+5,529/\{[1-0.3000]$$
$$[1-0.3000]\}$$
$$= 41,548 \text{ millions USD}$$

Law-3347:
If both (**M**= Margin of Contribution= 41,548 millions USD), (**$'**= Sales of Past Year= 48,851 millions USD), (**s**= [S/S']-1= Sales Growth= 5.00%), (**R**= Retained Earnings= 5,529 millions USD), (**f**= F/S= Fixed Portion= 58.00%), (**d**= D/A= Dividend Portion or Payout= 30.00%), (**i**= I/S= Interest Portion= 1.00%), and (**t**= T/B= Tax Rate= 30.00%), are known, then it's (**$**= Sales or Revenues planned), is:

$$\$ = (M-\$'i[1+s]-R/\{[1-t][1-d]\})/f$$
$$= (41{,}548-48{,}851*0.0100[1+0.0500]$$
$$-5{,}529/\{[1-0.3000][1-0.3000]\})$$
$$/0.5800$$
$$= 51{,}294 \text{ millions USD}$$

Law-3348:
If both (**M**= Margin of Contribution= 41,548 millions USD), (**$'**= Sales of Past Year= 48,851 millions USD), (**s**= [S/S']-1= Sales Growth= 5.00%), (**R**= Retained Earnings= 5,529 millions USD), (**$**= Sales or Revenues= 51,294 millions USD), (**d**= D/A= Dividend Portion or Payout= 30.00%), (**i**= I/S= Interest Portion= 1.00%), and (**t**= T/B= Tax Rate= 30.00%), are known, then it's (**f**= Fixed Portion planned), is:

$$f = (M-\$'i[1+s]-R/\{[1-t][1-d]\})/\$$$
$$= (41{,}548-48{,}851*0.0100[1+0.0500]$$
$$-5{,}529/\{[1-0.3000][1-0.3000]\})$$
$$/51{,}294$$
$$= 58.00\%$$

Law-3349:

If both (**M**= Margin of Contribution= 41,548 millions USD), (**f**= F/S= Fixed Portion= 58.00%), (**s**= [S/S']-1= Sales Growth= 5.00%), (**R**= Retained Earnings= 5,529 millions USD), (**S**= Sales or Revenues= 51,294 millions USD), (**d**= D/A= Dividend Portion or Payout= 30.00%), (**i**= I/S= Interest Portion= 1.00%), and (**t**= T/B= Tax Rate= 30.00%), are known, then it's (**S'**= Sales Past), must be:

$$S' = (M-Sf-R/\{[1-t][1-d]\})/\{i[1+s]\}$$
$$= (41,548-51,294*0.5800-5,529$$
$$/\{[1-0.3000][1-0.3000]\})$$
$$/\{0.0100[1+0.0500]\}$$
$$= 48,851 \text{ millions USD}$$

Law-3350:

If both (**M**= Margin of Contribution= 41,548 millions USD), (**f**= F/S= Fixed Portion= 58.00%), (**s**= [S/S']-1= Sales Growth= 5.00%), (**R**= Retained Earnings= 5,529 millions USD), (**S**= Sales or Revenues= 51,294 millions USD), (**d**= D/A= Dividend Portion or Payout= 30.00%), (**S'**= Sales of Past Year= 48,851 millions USD), and (**t**= T/B= Tax Rate= 30.00%), are known, then it's (**i**= Interest Portion planned), is:

$$i = (M-Sf-R/\{[1-t][1-d]\})/\{S'[1+s]\}$$
$$= (41,548-51,294*0.5800-5,529$$
$$/\{[1-0.3000][1-0.3000]\})$$
$$/\{48,851[1+0.0500]\}$$
$$= 1.00\%$$

Law-3351:

If both (**M**= Margin of Contribution= 41,548 millions USD), (**f**= F/S= Fixed Portion= 58.00%), (**R**= Retained Earnings= 5,529 millions USD), (**$**= Sales or Revenues= 51,294 millions USD), (**d**= D/A= Dividend Portion or Payout= 30.00%), (**$'**= Sales of Past Year= 48,851 millions USD), and (**t**= T/B= Tax Rate= 30.00%), are known, then it's (**s**= Sales Growth planned), is:

$$s= (M-\$f-R/\{[1-t][1-d]\})/[\$'i]-1$$
$$= (41,548-51,294*0.5800-5,529 /\{[1-0.3000][1-0.3000]\}) /[48,851*0.0100]-1$$
$$= 5.00\%$$

Law-3352:

If both (**M**= Margin of Contribution= 41,548 millions USD), (**f**= F/S= Fixed Portion= 58.00%), (**i**= I/S= Interest Portion= 1.00%), (**R**= Retained Earnings= 5,529 millions USD), (**$**= Sales or Revenues= 51,294 millions USD), (**d**= D/A= Dividend Portion or Payout= 30.00%), (**$'**= Sales of Past Year= 48,851 millions USD), and (**s**= [S/S']-1= Sales Growth= 5.00%), are known, then it's (**t**= Tax Rate planned), is:

$$t= 1-R/([1-d]\{M-\$f-\$'i[1+s]\})$$
$$= 1-5,529/([1-0.3000]\{41,548-51,294 *0.5800-48,851*0.0100 [1+0.0500]\})$$
$$= 30.00\%$$

Law-3353:

If both (**M**= Margin of Contribution= 41,548 millions USD), (**f**= F/S=Fixed Portion= 58.00%), (**i**= I/S= Interest Portion= 1.00%), (**R**= Retained Earnings= 5,529 millions USD), (**S**= Sales or Revenues= 51,294 millions USD), (**t**= T/B= Tax Rate= 30.00%), (**S'**= Sales of Past Year= 48,851 millions USD), and (**s**= [S/S']-1= Sales Growth= 5.00%), are known, then it's (**d**= Dividend Portion or Payout planned), is:

$$d = 1 - R/([1-t]\{M-Sf-S'i[1+s]\})$$
$$= 1 - 5,529/([1-0.3000]\{41,548 - 51,294$$
$$*0.5800 - 48,851*0.0100$$
$$[1+0.0500]\})$$
$$= 30.00\%$$

Law-3354:

If both (**M**= Margin of Contribution= 41,548 millions USD), (**f**= F/S= Fixed Portion= 58.00%), (**I**= Interest Expense= 513 millions USD), (**d**= D/A= Dividend Portion or Payout= 30.00%), (**S**= Sales or Revenues= 51,294 millions USD), (**t**= T/B= Tax Rate= 30.00%), (**S'**= Sales of Past Year= 48,851 millions USD), and (**s**= [S/S']-1= Sales Growth= 5.00%), are known, then it's (**R**= Retained Earnings planned), is:

$$R = [1-t][1-d]\{M-I-S'f[1+s]\}$$
$$= [1-0.3000][1-0.3000]\{41,548-513$$
$$-48,851*0.5800[1+0.0500]\})$$
$$= 5,529 \text{ millions USD}$$

Law-3355:
   If both (**R**= Retained Earnings= 5,529 millions USD), (**f**= F/S= Fixed Portion= 58.00%), (**I**= Interest Expense= 513 millions USD), (**d**= D/A= Dividend Portion or Payout= 30.00%), (**S**= Sales or Revenues= 51,294 millions USD), (**t**= T/B= Tax Rate= 30.00%), (**S'**= Sales of Past Year= 48,851 millions USD), and (**s**= [S/S']-1= Sales Growth= 5.00%), are known, then it's (**M**= Retained Earnings planned), is:
   M= **S'f**[1+**s**]+**I**+**R**/{[1-**t**][1-**d**]}
   = 48,851*0.5800[1+0.0500]+513
   +5,529/{[1-0.3000][1-0.3000]}
   = 41,548 millions USD

Law-3356:
   If both (**R**= Retained Earnings= 5,529 millions USD), (**f**= F/S= Fixed Portion= 58.00%), (**I**= Interest Expense= 513 millions USD), (**d**= D/A= Dividend Portion or Payout= 30.00%), (**S**= Sales or Revenues= 51,294 millions USD), (**t**= T/B= Tax Rate= 30.00%), (**M**= Margin of Contribution= 41,548 millions USD), and (**s**= [S/S']-1= Sales Growth= 5.00%), are known, then it's (**S'**= Sales Past), must be:
   S'= **M-I-R**/{[1-**t**][1-**d**]}/{**f**[1+**s**]}
   = 41,548-513-5,529/{[1-0.3000]
   [1-0.3000]}
   /{0.5800[1+0.0500]}
   = 48,851 millions USD

Law-3357:
If both (**R**= Retained Earnings= 5,529 millions USD), (**$'**= Sales of Past Year= 48,851 millions USD), (**I**= Interest Expense= 513 millions USD), (**d**= D/A= Dividend Portion or Payout= 30.00%), (**$**= Sales or Revenues= 51,294 millions USD), (**t**= T/B= Tax Rate= 30.00%), (**M**= Margin of Contribution= 41,548 millions USD), and (**s**= [S/S']-1= Sales Growth= 5.00%), are known, then it's (**f**= Fixed Portion planned), must be:

f= **M-I-R**/{[1-**t**][1-**d**]}/{**$'**[1+**s**]}
= 41,548-513-5,529/{[1-0.3000]
[1-0.3000]}
/{48,851[1+0.0500]}
= 58.00%

Law-3358:
If both (**R**= Retained Earnings= 5,529 millions USD), (**$'**= Sales of Past Year= 48,851 millions USD), (**I**= Interest Expense= 513 millions USD), (**d**= D/A= Dividend Portion or Payout= 30.00%), (**$**= Sales or Revenues= 51,294 millions USD), (**t**= T/B= Tax Rate= 30.00%), (**M**= Margin of Contribution= 41,548 millions USD), and (**f**= F/S= Fixed Portion= 58.00%), are known, then it's (**s**= Sales Growth planned), is:

s= **M-I-R**/{[1-**t**][1-**d**]}/[**$'f**]-1
= 41,548-513-5,529/{[1-0.3000]
[1-0.3000]}/[48,851*0.5800]-1
= 5.00%

Law-3359:

If both (**R**= Retained Earnings= 5,529 millions USD), (**$'**= Sales of Past Year= 48,851 millions USD), (**s**= [S/S']-1= Sales Growth= 5.00%), (**d**= D/A= Dividend Portion or Payout= 30.00%), (**$**= Sales or Revenues= 51,294 millions USD), (**t**= T/B= Tax Rate= 30.00%), (**M**= Margin of Contribution= 41,548 millions USD), and (**f**= F/S= Fixed Portion= 58.00%), are known, then it's (**I**= Interest Expense planned), is:

$$\begin{aligned}
\mathbf{I} &= \mathbf{M}\text{-}\mathbf{\$'f}[1+\mathbf{s}]\text{-}\mathbf{R}/\{[1\text{-}\mathbf{t}][1\text{-}\mathbf{d}]\} \\
&= 41{,}548\text{-}48{,}851*0.5800[1+0.0500] \\
&\quad -5{,}529/\{[1\text{-}0.3000][1\text{-}0.3000]\} \\
&= \underline{513} \text{ millions USD}
\end{aligned}$$

Law-3360:

If both (**R**= Retained Earnings= 5,529 millions USD), (**$'**= Sales of Past Year= 48,851 millions USD), (**s**= [S/S']-1= Sales Growth= 5.00%), (**d**= D/A= Dividend Portion or Payout= 30.00%), (**$**= Sales or Revenues= 51,294 millions USD), (**I**= Interest Expense= 513 millions USD), (**M**= Margin of Contribution= 41,548 millions USD), and (**f**= F/S= Fixed Portion= 58.00%), are known, then it's (**t**= Tax Rate planned), is:

$$\begin{aligned}
\mathbf{t} &= 1\text{-}\mathbf{R}/([1\text{-}\mathbf{d}][\mathbf{M}\text{-}\mathbf{\$'f}[1+\mathbf{s}]\text{-}\mathbf{I}]) \\
&= 1\text{-}5{,}529/([1\text{-}0.3000]\{41{,}548\text{-}48{,}851 \\
&\quad *0.5800[1+0.0500]\text{-}513\}) \\
&= \underline{30.00\%}
\end{aligned}$$

Law-3361:
If both (**R**= Retained Earnings= 5,529 millions USD), (**S'**= Sales of Past Year= 48,851 millions USD), (**s**= [S/S']-1= Sales Growth= 5.00%), (**t**= T/B= Tax Rate= 30.00%), (**S**= Sales or Revenues= 51,294 millions USD), (**I**= Interest Expense= 513 millions USD), (**M**= Margin of Contribution= 41,548 millions USD), and (**f**= F/S= Fixed Portion= 58.00%), are known, then it's (**d**= Dividend Portion or Payout planned), is:

$$d= 1-R/([1-t][M-S'f[1+s]-I])$$
$$= 1-5,529/([1-0.3000]\{41,548-48,851$$
$$*0.5800[1+0.0500]-513\})$$
$$= 30.00\%$$

Law-3362:
If both (**d**= D/A= Dividend Portion or Payout= 30.00%), (**S'**= Sales of Past Year= 48,851 millions USD), (**s**= [S/S']-1= Sales Growth= 5.00%), (**t**= T/B= Tax Rate= 30.00%), (**S**= Sales or Revenues= 51,294 millions USD), (**i**= I/S= Interest Portion= 1.00%), (**M**= Margin of Contribution= 41,548 millions USD), and (**f**= F/S= Fixed Portion= 58.00%), are known, then it's (**R**= Retained Earnings planned), is:

$$R= [1-t][1-d][M-S'f[1+s]-Si\}$$
$$= [1-0.3000][1-0.3000]\{41,548-48,851$$
$$*0.5800[1+0.0500]$$
$$-51,294*0.0100\}$$
$$= 5,529 \text{ millions USD}$$

Law-3363:
If both (**d**= D/A= Dividend Portion or Payout= 30.00%), (**$'**= Sales of Past Year= 48,851 millions USD), (**s**= [S/S']-1= Sales Growth= 5.00%), (**t**= T/B= Tax Rate= 30.00%), (**$**= Sales or Revenues= 51,294 millions USD), (**i**= I/S= Interest Portion= 1.00%), (**R**= Retained Earnings= 5,529 millions USD), and (**f**= F/S= Fixed Portion= 58.00%), are known, then it's (**M**= Margin of Contribution planned), is:

$$M = \$'f[1+s] + \$i + R/\{[1-t][1-d]\}$$
$$= 48,851*0.5800[1+0.0500] - 51,294*0.0100 + 5,529/\{[1-0.3000][1-0.3000]\}$$
$$= \underline{41,548} \text{ millions USD}$$

Law-3364:
If both (**d**= D/A= Dividend Portion or Payout= 30.00%), (**M**= Margin of Contribution= 41,548 millions USD), (**s**= [S/S']-1= Sales Growth= 5.00%), (**t**= T/B= Tax Rate= 30.00%), (**$**= Sales or Revenues= 51,294 millions USD), (**i**= I/S= Interest Portion= 1.00%), (**R**= Retained Earnings= 5,529 millions USD), and (**f**= F/S= Fixed Portion= 58.00%), are known, then it's (**$'**= Sales Past), must be:

$$\$' = (M - \$i - R/\{[1-t][1-d]\})/\{f[1+s]\}$$
$$= (41,548 - 51,294*0.0100 - 5,529/\{[1-0.3000][1-0.3000]\})/\{0.5800[1+0.0500]\}$$
$$= \underline{48,851} \text{ millions USD}$$

Law-3365:
If both (**d**= D/A= Dividend Portion or Payout= 30.00%), (**M**= Margin of Contribution= 41,548 millions USD), (**s**= [S/S']-1= Sales Growth= 5.00%), (**t**= T/B= Tax Rate= 30.00%), (**S**= Sales or Revenues= 51,294 millions USD), (**i**= I/S= Interest Portion= 1.00%), (**R**= Retained Earnings= 5,529 millions USD), and (**S'**= Sales of Past Year= 48,851 millions USD), are known, then it's (**f**= Fixed Portion planned), is:

$f = (M-Si-R/\{[1-t][1-d]\})/\{S'[1+s]\}$
$= (41{,}548-51{,}294*0.0100-5{,}529$
$\quad /\{[1-0.3000][1-0.3000]\})$
$\quad /\{48{,}851[1+0.0500]\}$
$= 58.00\%$

Law-3366:
If both (**d**= D/A= Dividend Portion or Payout= 30.00%), (**M**= Margin of Contribution= 41,548 millions USD), (**f**= F/S= Fixed Portion= 58.00%), (**t**= T/B= Tax Rate= 30.00%), (**S**=Sales or Revenues= 51,294 millions USD), (**i**= I/S= Interest Portion= 1.00%), (**R**= Retained Earnings= 5,529 millions USD), and (**S'**= Sales of Past Year= 48,851 millions USD), are known, then it's (**s**= Sales Growth planned), is:

$s = (M-Si-R/\{[1-t][1-d]\})/[S'f]-1$
$= (41{,}548-51{,}294*0.0100-5{,}529$
$\quad /\{[1-0.3000][1-0.3000]\})$
$\quad /[48{,}851*0.5800]-1$
$= 5.00\%$

Law-3367:

If both (**d**= D/A= Dividend Portion or Payout= 30.00%), (**M**= Margin of Contribution= 41,548 millions USD), (**f**= F/S= Fixed Portion= 58.00%), (**t**= T/B= Tax Rate= 30.00%), (**s**= [S/S']-1= Sales Growth= 5.00%), (**i**= I/S= Interest Portion= 1.00%), (**R**= Retained Earnings= 5,529 millions USD), and (**S'**= Sales of Past Year= 48,851 millions USD), are known, then it's (**S**= Sales or Revenues planned), is:

$$S= (M-S'f[1+s]-R/\{[1-t][1-d]\})/i$$
$$= (41{,}548-48{,}851*0.5800[1+0.0500]$$
$$-5{,}529/\{[1-0.3000][1-0.3000]\}$$
$$/0.0100$$
$$= \underline{51{,}294} \text{ millions USD}$$

Law-3368:

If both (**d**= D/A= Dividend Portion or Payout= 30.00%), (**M**= Margin of Contribution= 41,548 millions USD), (**f**= F/S= Fixed Portion= 58.00%), (**t**= T/B= Tax Rate= 30.00%), (**s**= [S/S']-1= Sales Growth= 5.00%), (**S**= Sales or Revenues= 51,294 millions USD), (**R**= Retained Earnings= 5,529 millions USD), and (**S'**= Sales of Past Year= 48,851 millions USD), are known, then it's (**i**= Interest Portion planned), is:

$$i= (M-S'f[1+s]-R/\{[1-t][1-d]\})/S$$
$$= (41{,}548-48{,}851*0.5800[1+0.0500]$$
$$-5{,}529/\{[1-0.3000][1-0.3000]\}$$
$$/51{,}294$$
$$= \underline{1.00\%}$$

Law-3369:

If both (**d**= D/A= Dividend Portion or Payout= 30.00%), (**M**= Margin of Contribution= 41,548 millions USD), (**f**= F/S= Fixed Portion= 58.00%), (**i**= I/S= Interest Portion= 1.00%), **s**= [S/S']-1= Sales Growth= 5.00%), (**$**= Sales or Revenues= 51,294 millions USD), (**R**= Retained Earnings= 5,529 millions USD), and (**$'**= Sales of Past Year= 48,851 millions USD), are known, then it's (**t**= Tax Rate planned), is:

$$\mathbf{d}= 1-\mathbf{R}/([1-\mathbf{d}]\{\mathbf{M}-\mathbf{\$'f}[1+\mathbf{s}]-\mathbf{\$i}\})$$
$$= 1-5,529/([1-0.3000]\{41,548-48,851$$
$$*0.5800[1+0.0500]$$
$$-51,294*0.0100\})$$
$$= 30.00\%$$

Law-3370:

If both (**t**= T/B= Tax Rate= 30.00%), (**M**= Margin of Contribution= 41,548 millions USD), (**f**= F/S= Fixed Portion= 58.00%), (**i**= I/S= Interest Portion= 1.00%), **s**= [S/S']-1= Sales Growth= 5.00%), (**$**= Sales or Revenues= 51,294 millions USD), (**R**= Retained Earnings= 5,529 millions USD), and (**$'**= Sales of Past Year= 48,851 millions USD), are known, then it's (**d**= Dividend Portion or Payout planned), is:

$$\mathbf{d}= 1-\mathbf{R}/([1-\mathbf{t}][1-\mathbf{d}]\{\mathbf{M}-\mathbf{\$'f}[1+\mathbf{s}]-\mathbf{\$i}\})$$
$$= 1-5,529/([1-0.3000]\{41,548-48,851$$
$$*0.5800[1+0.0500]$$
$$-51,294*0.0100\})$$
$$= 30.00\%$$

Law-3371:

If both (**t**= T/B= Tax Rate= 30.00%), (**M**= Margin of Contribution= 41,548 millions USD), (**f**= F/S= Fixed Portion= 58.00%), (**i**= I/S= Interest Portion= 1.00%), **s**= [S/S']-1= Sales Growth= 5.00%), (**$**= Sales or Revenues= 51,294 millions USD), (**d**= D/A= Dividend Portion or Payout= 30.00%), and (**$'**= Sales of Past Year= 48,851 millions USD), are known, then it's (**R**= Retained Earnings planned), is:

$$R = [1-t][1-d]\{M-S'f[1+s]-S'i[1+s]\}$$
$$= [1-t][1-d]\{M-S'f[1+s][f+i]\}$$
$$= [1-0.3000][1-0.3000]\{41,548-48,851$$
$$[1+0.0500][0.5800+0.0100]\}$$
$$= \underline{5,529} \text{ millions USD}$$

Law-3372:

If both (**t**= T/B= Tax Rate= 30.00%), (**R**= Retained Earnings= 5,529 millions USD), (**f**= F/S= Fixed Portion= 58.00%), (**i**= I/S= Interest Portion= 1.00%), **s**= [S/S']-1= Sales Growth= 5.00%), (**$**= Sales or Revenues= 51,294 millions USD), (**d**= D/A= Dividend Portion or Payout= 30.00%), and (**$'**= Sales of Past Year= 48,851 millions USD), are known, then it's (**M**= Margin of Contribution planned), is:

$$M = S'[1+s][f+i] + R/\{[1-t][1-d]\}$$
$$= 48,851[1+0.0500][0.5800+0.0100]$$
$$+ 5,529/\{[1-0.3000][1-0.3000]\}$$
$$= \underline{41,548} \text{ millions USD}$$

Law-3373:
If both (**t**= T/B= Tax Rate= 30.00%), (**R**= Retained Earnings= 5,529 millions USD), (**f**= F/S= Fixed Portion= 58.00%), (**i**= I/S= Interest Portion= 1.00%), **s**= [S/S']-1= Sales Growth= 5.00%), (**S**= Sales or Revenues= 51,294 millions USD), (**d**= D/A= Dividend Portion or Payout= 30.00%), and (**M**= Margin of Contribution= 41,548 millions USD), are known, then it's (**S'**= Sales Past), must be:

$$S' = (M-R/\{[1-t][1-d]\})/\{[1+s][f+i]\}$$
$$= (41{,}548 - 5{,}529/\{[1-0.3000][1-0.3000]\})/\{[1+0.0500][0.5800+0.0100]\}$$
$$= 48{,}851 \text{ millions USD}$$

Law-3374:
If both (**t**= T/B= Tax Rate= 30.00%), (**R**= Retained Earnings= 5,529 millions USD), (**f**= F/S= Fixed Portion= 58.00%), (**i**= I/S= Interest Portion= 1.00%), (**S'**= Sales of Past Year= 48,851 millions USD), (**S**= Sales or Revenues= 51,294 millions USD), (**d**= D/A= Dividend Portion or Payout= 30.00%), and (**M**= Margin of Contribution= 41,548 millions USD), are known, then it's (**s**= Sales Growth planned), is:

$$s = (M-R/\{[1-t][1-d]\})/\{S'[f+i]\} - 1$$
$$= (41{,}548 - 5{,}529/\{[1-0.3000][1-0.3000]\})/\{48{,}851[0.5800+0.0100]\} - 1$$
$$= 5.00\%$$

Law-3375:

If both (**t**= T/B= Tax Rate= 30.00%), (**R**= Retained Earnings= 5,529 millions USD), (**s**= [S/S']-1= Sales Growth= 5.00%), (**i**= I/S= Interest Portion= 1.00%), (**$'**= Sales of Past Year= 48,851 millions USD), (**$**= Sales or Revenues= 51,294 millions USD), (**d**= D/A= Dividend Portion or Payout= 30.00%), and (**M**= Margin of Contribution= 41,548 millions USD), are known, then it's (**f**= Fixed Portion planned), is:

$$f= (M-R/\{[1-t][1-d]\})/\{\$'[1+s]\}-i$$
$$= (41,548-5,529/\{[1-0.3000][1-0.3000]\})/\{48,851[1+0.0500]-0.0100\}$$
$$= 58.00\%$$

Law-3376:

If both (**t**= T/B= Tax Rate= 30.00%), (**R**= Retained Earnings= 5,529 millions USD), (**s**= [S/S']-1= Sales Growth= 5.00%), (**f**= F/S= Fixed Portion= 58.00%), (**$'**= Sales of Past Year= 48,851 millions USD), (**$**= Sales or Revenues= 51,294 millions USD), (**d**= D/A= Dividend Portion or Payout= 30.00%), and (**M**= Margin of Contribution= 41,548 millions USD), are known, then it's (**i**= Interest Portion planned), is:

$$i= (M-R/\{[1-t][1-d]\})/\{\$'[1+s]\}-f$$
$$= (41,548-5,529/\{[1-0.3000][1-0.3000]\})/\{48,851[1+0.0500]-0.5800]$$
$$= 1.00\%$$

Law-3377:

If both (**i**= I/S= Interest Portion= 1.00%), (**R**= Retained Earnings= 5,529 millions USD), (**s**= [S/S']-1= Sales Growth= 5.00%), (**f**= F/S= Fixed Portion= 58.00%), (**S'**= Sales of Past Year= 48,851 millions USD), (**S**= Sales or Revenues= 51,294 millions USD), (**d**= D/A= Dividend Portion or Payout= 30.00%), and (**M**= Margin of Contribution= 41,548 millions USD), are known, then it's (**t**= Tax Rate planned), is:

t= 1-**R**/([1-**d**]{**M**-**S'**[1+**s**][**f**+**i**]})
   = 1-5,529/([1-0.3000]{41,548
       -48,851[1+0.0500][0.5800
       +0.0100]})
   = 30.00%

Law-3378:

If both (**i**= I/S= Interest Portion= 1.00%), (**R**= Retained Earnings= 5,529 millions USD), (**s**= [S/S']-1= Sales Growth= 5.00%), (**f**= F/S= Fixed Portion= 58.00%), (**S'**= Sales of Past Year= 48,851 millions USD), (**S**= Sales or Revenues= 51,294 millions USD), (**t**= T/B= Tax Rate= 30.00%), and (**M**= Margin of Contribution= 41,548 millions USD), are known, then it's (**d**= Dividend Portion or Payout planned), is:

d= 1-**R**/([1-**t**]{**M**-**S'**[1+**s**][**f**+**i**]})
   = 1-5,529/([1-0.3000]{41,548-48,851
       [1+0.0500][0.5800+0.0100]})
   = 30.00%

Law-3379:
If both (**I**= Interest Expense= 513 millions USD), (**d**= D/A= Dividend Portion or Payout= 30.00%), (**F**= Fixed Cost= 29,750 millions USD), (**$**= Sales or Revenues= 51,294 millions USD), (**t**= T/B= Tax Rate= 30.00%), and (**V**= Variable Cost= 9,746 millions USD), are known, then it's (**R**= Retained Earnings planned), is:

$$R = [1-t][1-d][\$-V-F-I]$$
$$= [1-0.3000][1-0.3000][51,294-9,746-29,750-513]\})$$
$$= \underline{5,529} \text{ millions USD}$$

Law-3380:
If both (**I**= Interest Expense= 513 millions USD), (**d**= D/A= Dividend Portion or Payout= 30.00%), (**F**= Fixed Cost= 29,750 millions USD), (**R**= Retained Earnings= 5,529 millions USD), (**t**= T/B= Tax Rate= 30.00%), and (**V**= Variable Cost= 9,746 millions USD), are known, then it's (**$**= Sales or Revenues planned), is:

$$\$ = V+F+I+R/\{[1-t]1-d]\}$$
$$= 9,746+29,750+513+5,529 /\{[1-0.3000][1-0.3000]\}$$
$$= \underline{51,294} \text{ millions USD}$$

Law-3381:
If both (**I**= Interest Expense= 513 millions USD), (**d**= D/A= Dividend Portion or Payout= 30.00%), (**F**= Fixed Cost= 29,750 millions USD), (**R**= Retained Earnings= 5,529 millions USD), (**t**= T/B= Tax Rate= 30.00%), and (**$**= Sales or Revenues= 51,294 millions USD), are known, then it's (**V**= Variable Cost planned), is:

$$V = \$-F-I-R/\{[1-t]1-d]\}$$
$$= 51{,}294-29{,}750-513-5{,}529/\{[1-0.3000][1-0.3000]\}$$
$$= 9{,}746 \text{ millions USD}$$

Law-3382:
If both (**I**= Interest Expense= 513 millions USD), (**d**= D/A= Dividend Portion or Payout= 30.00%), (**V**= Variable Cost= 9,746 millions USD), (**R**= Retained Earnings= 5,529 millions USD), (**t**= T/B= Tax Rate= 30.00%), and (**$**= Sales or Revenues= 51,294 millions USD), are known, then it's (**F**= Fixed Cost planned), is:

$$F = \$-V-I-R/\{[1-t]1-d]\}$$
$$= 51{,}294-9{,}746-513-5{,}529/\{[1-0.3000][1-0.3000]\}$$
$$= 29{,}750 \text{ millions USD}$$

Law-3383:

If both (**F**= Fixed Cost= 29,750 millions USD), (**d**= D/A= Dividend Portion or Payout= 30.00%), (**V**= Variable Cost= 9,746 millions USD), (**R**= Retained Earnings= 5,529 millions USD), (**t**= T/B= Tax Rate= 30.00%), and (**$**= Sales or Revenues= 51,294 millions USD), are known, then it's (**I**= Interest Expense planned), is:

$$I = \$-V-F-R/\{[1-t]1-d]\}$$
$$= 51{,}294 - 9{,}746 - 29{,}750 - 5{,}529$$
$$\quad /\{[1-0.3000][1-0.3000]\}$$
$$= \underline{513} \text{ millions USD}$$

Law-3384:

If both (**F**= Fixed Cost= 29,750 millions USD), (**d**= D/A= Dividend Portion or Payout= 30.00%), (**V**= Variable Cost= 9,746 millions USD), (**R**= Retained Earnings= 5,529 millions USD), (**I**= Interest Expense= 513 millions USD), and (**$**= Sales or Revenues= 51,294 millions USD), are known, then it's (**t**= Tax Rate planned), is:

$$t = 1 - R/\{[1-d][\$-V-F-I]\}$$
$$= 1 - 5{,}529/\{[1-0.3000][51{,}294 - 9{,}746$$
$$\quad -29{,}750 - 513]\}$$
$$= \underline{30.00\%}$$

Law-3385:
If both (**F**= Fixed Cost= 29,750 millions USD), (**t**= T/B= Tax Rate= 30.00%), (**V**= Variable Cost= 9,746 millions USD), (**R**= Retained Earnings= 5,529 millions USD), (**I**= Interest Expense= 513 millions USD), and (**S**= Sales or Revenues= 51,294 millions USD), are known, then it's (**d**= Dividend Portion or Payout planned), is:

$$\mathbf{d}= 1-\mathbf{R}/\{[1-\mathbf{t}][\mathbf{S}-\mathbf{V}-\mathbf{F}-\mathbf{I}]\}$$
$$= 1-5,529/\{[1-0.3000][51,294-9,746-29,750-513]\}$$
$$= 30.00\%$$

Law-3386:
If both (**F**= Fixed Cost= 29,750 millions USD), (**t**= T/B= Tax Rate= 30.00%), (**V**= Variable Cost= 9,746 millions USD), (**i**= I/S= Interest Portion= 1.00%), (**d**= D/A= Dividend Portion or Payout= 30.00%), and (**S**= Sales or Revenues= 51,294 millions USD), are known, then it's (**R**= Retained Earnings planned), is:

$$\mathbf{R}= [1-\mathbf{t}][1-\mathbf{d}][\mathbf{S}-\mathbf{V}-\mathbf{F}-\mathbf{Si}]$$
$$= [1-\mathbf{t}][1-\mathbf{d}]\{[\mathbf{S}[1-\mathbf{i}]-\mathbf{V}-\mathbf{F}\}$$
$$= [1-0.3000][1-0.3000]\{51,294[1-0.0100]-9,746-29,750\}$$
$$= 5,529 \text{ millions USD}$$

Law-3387:
If both (**F**= Fixed Cost= 29,750 millions USD), (**t**= T/B= Tax Rate= 30.00%), (**V**= Variable Cost= 9,746 millions USD), (**R**= Retained Earnings= 5,529 millions USD), (**i**= I/S= Interest Portion= 1.00%), and (**d**= D/A= Dividend Portion or Payout= 30.00%), are known, then it's (**$**= Sales or Revenues planned), is:

$$\$= (V+F+R/\{[1-t][1-d]\})/[1-i]$$
$$= (9,746+29,750+5,529/\{[1-0.3000][1-0.3000]\})/[1-0.0100]$$
$$= 51,294 \text{ millions USD}$$

Law-3388:
If both (**F**= Fixed Cost= 29,750 millions USD), (**t**= T/B= Tax Rate= 30.00%), (**V**= Variable Cost= 9,746 millions USD), (**R**= Retained Earnings= 5,529 millions USD), (**$**= Sales or Revenues= 51,294 millions USD), and (**d**= D/A= Dividend Portion or Payout= 30.00%), are known, then it's (**i**= Interest Portion planned), is:

$$i= 1-(V+F+R/\{[1-t][1-d]\})/\$$$
$$= 1-(9,746+29,750+5,529/\{[1-0.3000][1-0.3000]\})/51,294$$
$$= 1.00\%$$

Law-3389:

If both (**F**= Fixed Cost= 29,750 millions USD), (**t**= T/B= Tax Rate= 30.00%), (**i**= I/S= Interest Portion= 1.00%), (**R**= Retained Earnings= 5,529 millions USD), (**$**= Sales or Revenues= 51,294 millions USD), and (**d**= D/A= Dividend Portion or Payout= 30.00%), are known, then it's (**V**= Variable Cost planned), is:

$$V = \$[1-i] - F - R / \{[1-t][1-d]\}$$
$$= 51,294[1-0.0100] - 29,750 - 5,529 / \{[1-0.3000][1-0.3000]\}$$
$$= 9,746 \text{ millions USD}$$

Law-3390:

If both (**t**= T/B= Tax Rate= 30.00%), (**i**= I/S= Interest Portion= 1.00%), (**V**= Variable Cost= 9,746 millions USD), (**R**= Retained Earnings= 5,529 millions USD), (**$**= Sales or Revenues= 51,294 millions USD), and (**d**= D/A= Dividend Portion or Payout= 30.00%), are known, then it's (**F**= Fixed Cost planned), is:

$$F = \$[1-i] - V - R / \{[1-t][1-d]\}$$
$$= 51,294[1-0.0100] - 9,746 - 5,529 / \{[1-0.3000][1-0.3000]\}$$
$$= 29,750 \text{ millions USD}$$

Law-3391:
If both (**i**= I/S= Interest Portion= 1.00%), (**V**= Variable Cost= 9,746 millions USD), (**R**= Retained Earnings= 5,529 millions USD), (**$**= Sales or Revenues= 51,294 millions USD), (**d**= D/A= Dividend Portion or Payout= 30.00%), and (**F**= Fixed Cost= 29,750 millions USD), are known, then it's (**t**= Tax Rate planned), is:

$$t = 1\text{-}R/\{[1\text{-}d]\{\$[1\text{-}i]\text{-}V\text{-}F\}$$
$$= 1\text{-}5{,}529/\{[1\text{-}0.3000]\{51{,}294[1\text{-}0.0100]\text{-}9{,}746\text{-}29{,}750\}$$
$$= \underline{30.00\%}$$

Law-3392:
If both (**i**= I/S= Interest Portion= 1.00%), (**V**= Variable Cost= 9,746 millions USD), (**R**= Retained Earnings= 5,529 millions USD), (**$**= Sales or Revenues= 51,294 millions USD), (**t**= T/B= Tax Rate= 30.00%), and (**F**= Fixed Cost= 29,750 millions USD), are known, then it's (**d**= Dividend Portion or Payout planned), is:

$$d = 1\text{-}R/([1\text{-}t]\{\$[1\text{-}i]\text{-}V\text{-}F\})$$
$$= 1\text{-}5{,}529/([1\text{-}0.3000]\{51{,}294[1\text{-}0.0100]\text{-}9{,}746\text{-}29{,}750\})$$
$$= \underline{30.00\%}$$

Law-3393:

If both ($'=$ Sales of Past Year= 48,851 millions USD), ($s=$ [S/S']-1= Sales Growth= 5.00%), ($i=$ I/S= Interest Portion= 1.00%), ($V=$ Variable Cost= 9,746 millions USD), ($d=$ D/A= Dividend Portion or Payout= 30.00%), ($S=$ Sales or Revenues= 51,294 millions USD), ($t=$ T/B= Tax Rate= 30.00%), and ($F=$ Fixed Cost= 29,750 millions USD), are known, then it's ($R=$ Retained Earnings planned), is:

$R = [1-t][1-d]\{S-V-F-Si[1+s]\}$
$= [1-0.3000][1-0.3000]\{51,294-9,746$
$-29,750-48,851*0.0100$
$[1+0.0500]\}$
$= 5,529$ millions USD

Law-3394:

If both ($'=$ Sales of Past Year= 48,851 millions USD), ($s=$ [S/S']-1= Sales Growth= 5.00%), ($i=$ I/S= Interest Portion= 1.00%),($R=$ Retained Earnings= 5,529 millions USD), ($V=$ Variable Cost= 9,746 millions USD), ($d=$ D/A= Dividend Portion or Payout= 30.00%), ($t=$ T/B= Tax Rate= 30.00%), and ($F=$ Fixed Cost= 29,750 millions USD), are known, then it's ($S=$ Sales or Revenues planned), is:

$S = V+F+Si[1+s]+R/\{[1-t][1-d]\}$
$= 9,746+29,750+48,851*0.0100$
$[1+0.0500]+5,529$
$/\{[1-0.3000][1-0.3000]\}$
$= 51,294$ millions USD

Law-3395:
If both ($'$= Sales of Past Year= 48,851 millions USD), (s= [S/S']-1= Sales Growth= 5.00%), (i= I/S= Interest Portion= 1.00%), (S= Sales or Revenues= 51,294 millions USD), (R= Retained Earnings= 5,529 millions USD), (d= D/A= Dividend Portion or Payout= 30.00%), (t= T/B= Tax Rate= 30.00%), and (F= Fixed Cost= 29,750 millions USD), are known, then it's (V= Variable Cost planned), is:

$$V = S-F-Si[1+s]-R/\{[1-t][1-d]\}$$
$$= 51,294-29,750-48,851*0.0100$$
$$[1+0.0500]-5,529/\{[1-0.3000]$$
$$[1-0.3000]\}$$
$$= \underline{9,746} \text{ millions USD}$$

Law-3396:
If both ($'$= Sales of Past Year= 48,851 millions USD), (s= [S/S']-1= Sales Growth= 5.00%), (i= I/S= Interest Portion= 1.00%), (S= Sales or Revenues= 51,294 millions USD), (R= Retained Earnings= 5,529 millions USD), (d= D/A= Dividend Portion or Payout= 30.00%), (t= T/B= Tax Rate= 30.00%), and (V= Variable Cost= 9,746 millions USD), are known, then it's (F= Fixed Cost planned), is:

$$F = S-V-Si[1+s]-R/\{[1-t][1-d]\}$$
$$= 51,294-9,746-48,851*0.0100$$
$$[1+0.0500]-5,529$$
$$/\{[1-0.3000][1-0.3000]\}$$
$$= \underline{29,750} \text{ millions USD}$$

Law-3397:
If both (**s**= [S/S']-1= Sales Growth= 5.00%), (**i**= I/S= Interest Portion= 1.00%), (**$**= Sales or Revenues= 51,294 millions USD), (**R**= Retained Earnings= 5,529 millions USD), (**F**= Fixed Cost= 29,750 millions USD), (**d**= D/A= Dividend Portion or Payout= 30.00%), (**t**= T/B= Tax Rate= 30.00%), and (**V**= Variable Cost= 9,746 millions USD), are known, then it's (**$'**= Sales Past), must be:
$$\$' = (\$ \text{-} V \text{-} F \text{-} R / \{[1\text{-}t][1\text{-}d]\}) / \{i[1+s]\}$$
$$= (51{,}294\text{-}9{,}746\text{-}29{,}750\text{-}5{,}529$$
$$/\{[1\text{-}0.3000][1\text{-}0.3000]\})$$
$$/\{0.0100[1+0.0500]\}$$
$$= 48{,}851 \text{ millions USD}$$

Law-3398:
If both (**s**= [S/S']-1= Sales Growth= 5.00%), (**$**= Sales or Revenues= 51,294 millions USD), (**R**= Retained Earnings= 5,529 millions USD), (**F**= Fixed Cost= 29,750 millions USD), (**d**= D/A= Dividend Portion or Payout= 30.00%), (**$'**= Sales of Past Year= 48,851 millions USD), (**t**= T/B= Tax Rate= 30.00%), and (**V**= Variable Cost= 9,746 millions USD), are known, then it's (**i**= Interest Portion planned), is:
$$i = (\$ \text{-} V \text{-} F \text{-} R / \{[1\text{-}t][1\text{-}d]\}) / \{\$'[1+s]\}$$
$$= (51{,}294\text{-}9{,}746\text{-}29{,}750\text{-}5{,}529$$
$$/\{[1\text{-}0.3000][1\text{-}0.3000]\})$$
$$/\{48{,}851[1+0.0500]\}$$
$$= 1.00\%$$

Law-3399:

If both (i= I/S= Interest Portion= 1.00%), ($\$$= Sales or Revenues= 51,294 millions USD), (R= Retained Earnings= 5,529 millions USD), (F= Fixed Cost= 29,750 millions USD), (d= D/A= Dividend Portion or Payout= 30.00%), ($\$$'= Sales of Past Year= 48,851 millions USD), (t= T/B= Tax Rate= 30.00%), and (V= Variable Cost= 9,746 millions USD), are known, then it's (s= Sales Growth planned), is:

$$s = (\$-V-F-R/\{[1-t][1-d]\})/[\$'i]-1$$
$$= (51,294-9,746-29,750-5,529$$
$$/\{[1-0.3000][1-0.3000]\})$$
$$/[48,851*0.0100]-1$$
$$= 5.00\%$$

Law-3400:

If both (i= I/S= Interest Portion= 1.00%), ($\$$= Sales or Revenues= 51,294 millions USD), (R= Retained Earnings= 5,529 millions USD), (F= Fixed Cost= 29,750 millions USD), (d= D/A=Dividend Portion or Payout= 30.00%), (s= [S/S']-1= Sales Growth= 5.00%), ($\$$'= Sales of Past Year= 48,851 millions USD), and (V= Variable Cost= 9,746 millions USD), are known, then it's (t= Tax Rate planned), is:

$$t = 1-R/([1-d]\{\$-V-F-\$'i[1+s]\})$$
$$= 1-5,529/([1-0.3000]\{51,294-9,746$$
$$-29,750-48,851*0.0100$$
$$[1+0.0500]\})$$
$$= 30.00\%$$

Law-3401:

If both (i= I/S= Interest Portion= 1.00%), (S= Sales or Revenues= 51,294 millions USD), (R= Retained Earnings= 5,529 millions USD), (F= Fixed Cost= 29,750 millions USD), (t= T/B= Tax Rate= 30.00%), (s= [S/S']-1= Sales Growth= 5.00%), (S'= Sales of Past Year= 48,851 millions USD), and (V= Variable Cost= 9,746 millions USD), are known, then it's (d= Dividend Portion or Payout planned), is:

$$d = 1-R/([1-t]\{S-V-F-S'i[1+s]\})$$
$$= 1-5{,}529/([1-0.3000]\{51{,}294-9{,}746$$
$$-29{,}750-48{,}851*0.0100$$
$$[1+0.0500]\})$$
$$= 30.00\%$$

Law-3402:

If both (I= Interest Expense= 513 millions USD), (f= F/S= Fixed Portion= 58.00%), (S= Sales or Revenues= 51,294 millions USD), (d= D/A= Dividend Portion or Payout= 30.00%), (t= T/B= Tax Rate= 30.00%), and (V= Variable Cost= 9,746 millions USD), are known, then it's (R= Retained Earnings planned), is:

$$R = [1-t][1-d][S-V-Sf-I]$$
$$= [1-t][1-d]\{S[1-f]-V-I\}$$
$$= [1-0.3000][1-0.3000]\{51{,}294$$
$$[1-0.5800]-9{,}746-513\}$$
$$= 5{,}529 \text{ millions USD}$$

Law-3403:
If both (**I**= Interest Expense= 513 millions USD), (**f**= F/S= Fixed Portion= 58.00%), (**R**= Retained Earnings= 5,529 millions USD), (**d**= D/A= Dividend Portion or Payout= 30.00%), (**t**= T/B= Tax Rate= 30.00%), and (**V**= Variable Cost= 9,746 millions USD), are known, then it's (**$**= Sales or Revenues planned), is:

$= (**V**+**I**+**R**/{[1-**t**][1-**d**]})/[1-**f**]
 = (9,746+513+5,529[1-0.3000]
     [1-0.3000]+513}
 = 51,294 millions USD

Law-3404:
If both (**I**= Interest Expense= 513 millions USD), (**$**= Sales or Revenues= 51,294 millions USD), (**R**= Retained Earnings= 5,529 millions USD), (**d**= D/A= Dividend Portion or Payout= 30.00%), (**t**= T/B= Tax Rate= 30.00%), and (**V**= Variable Cost= 9,746 millions USD), are known, then it's (**f**= Fixed Portion planned), is:

f= 1-(**V**+**I**+**R**/{[1-**t**][1-**d**]})/**$**
 = 1-(9,746+513+5,529[1-0.3000]
     [1-0.3000]+513}/51,294
 = 58.00%

Law-3405:
If both (**I**= Interest Expense= 513 millions USD), (**f**= F/S= Fixed Portion= 58.00%), (**R**= Retained Earnings= 5,529 millions USD), (**$**= Sales or Revenues= 51,294 millions USD), (**d**= D/A= Dividend Portion or Payout= 30.00%), (**t**= T/B= Tax Rate= 30.00%), and are known, then it's (**V**= Variable Cost planned), is:

$$V = \$[1-f]-I-R/\{[1-t][1-d]\}$$
$$= 51{,}294[1-0.5800]-513-5{,}529/\{[1-0.3000][1-0.3000]\}$$
$$= 9{,}746 \text{ millions USD}$$

Law-3406:
If both (**f**= F/S= Fixed Portion= 58.00%), (**R**= Retained Earnings= 5,529 millions USD), (**$**= Sales or Revenues= 51,294 millions USD), (**d**= D/A= Dividend Portion or Payout= 30.00%), (**t**= T/B= Tax Rate= 30.00%), and (**V**= Variable Cost= 9,746 millions USD), are known, then it's (**I**= Interest Expense planned), is:

$$I = \$[1-f]-V-R/\{[1-t][1-d]\}$$
$$= 51{,}294[1-0.5800]-9{,}746-5{,}529/\{[1-0.3000][1-0.3000]\}$$
$$= 513 \text{ millions USD}$$

Law-3407:

If both (**f**= F/S= Fixed Portion= 58.00%), (**R**= Retained Earnings= 5,529 millions USD), (**$**=Sales or Revenues= 51,294 millions USD), (**I**= Interest Expense= 513 millions USD), (**d**= D/A= Dividend Portion or Payout= 30.00%), and (**V**= Variable Cost= 9,746 millions USD), are known, then it's (**t**= Tax Rate planned), is:

$$\mathbf{t}= 1-\mathbf{R}/([1-\mathbf{d}]\{\mathbf{\$}[1-\mathbf{f}]-\mathbf{V}-\mathbf{I}\})$$
$$=1-5,529/([1-0.3000][51,294[1-0.5800]-9,746-513\})$$
$$= \underline{30.00\%}$$

Law-3408:

If both (**f**= F/S=Fixed Portion= 58.00%), (**t**= T/B= Tax Rate= 30.00%), (**R**= Retained Earnings= 5,529 millions USD), (**$**=Sales or Revenues= 51,294 millions USD), (**I**= Interest Expense= 513 millions USD), and (**V**= Variable Cost= 9,746 millions USD), are known, then it's (**d**= Dividend Portion or Payout planned), is:

$$\mathbf{d}= 1-\mathbf{R}/([1-\mathbf{t}]\{\mathbf{\$}[1-\mathbf{f}]-\mathbf{V}-\mathbf{I}\})$$
$$=1-5,529/([1-0.3000]\{51,294[1-0.5800]-9,746-513\})$$
$$= \underline{30.00\%}$$

Law-3409:

If both (**f**= F/S= Fixed Portion= 58.00%), (**t**= T/B= Tax Rate= 30.00%), (**$**=Sales or Revenues= 51,294 millions USD), (**i**= I/S= Interest Portion= 1.00%), (**d**= D/A= Dividend Portion or Payout= 30.00%), and (**V**= Variable Cost= 9,746 millions USD), are known, then it's (**R**= Retained Earnings planned), is:

$$R = [1-t][1-d]\{\$-V-\$f-\$i\}$$
$$= [1-t][1-d]\{\$[1-f-i]-V\}$$
$$= [1-0.3000][1-0.3000]\{51,294$$
$$[1-0.5800-0.0100]- 9,746\}$$
$$= 5,529 \text{ millions USD}$$

Law-3410:

If both (**f**= F/S= Fixed Portion= 58.00%), (**t**= T/B= Tax Rate= 30.00%), (**i**= I/S= Interest Portion= 1.00%), (**d**= D/A= Dividend Portion or Payout= 30.00%), (**R**= Retained Earnings= 5,529 millions USD), and (**V**= Variable Cost= 9,746 millions USD), are known, then it's (**$**= Sales or Revenues planned), is:

$$\$ = (V+R/\{[1-t][1-d]\})[1-f-i]$$
$$= (9,746+5,529/\{[1-0.3000]$$
$$[1-0.3000]\})$$
$$/[1-0.5800-0.0100]$$
$$= 51,294 \text{ millions USD}$$

Law-3411:

If both (**t**= T/B= Tax Rate= 30.00%), (**i**= I/S= Interest Portion= 1.00%), (**d**= D/A= Dividend Portion or Payout= 30.00%), (**$**=Sales or Revenues= 51,294 millions USD), (**R**= Retained Earnings= 5,529 millions USD), and (**V**= Variable Cost= 9,746 millions USD), are known, then it's (**f**= Fixed Portion planned), is:

$$f = 1-i-(V+R/\{[1-t][1-d]\})/\$$$
$$= 1-0.0100-(9,746+5,529/\{[1-0.3000][1-0.3000]\})/51,294$$
$$= \underline{58.00\%}$$

Law-3412:

If both (**t**= T/B= Tax Rate= 30.00%), (**V**= Variable Cost= 9,746 millions USD), (**d**= D/A= Dividend Portion or Payout= 30.00%), (**$**=Sales or Revenues= 51,294 millions USD), (**R**= Retained Earnings= 5,529 millions USD), and (**f**= F/S= Fixed Portion= 58.00%), are known, then it's (**i**= Interest Portion planned), is:

$$i = 1-f-(V+R/\{[1-t][1-d]\})/\$$$
$$= 1-0.5800-(9,746+5,529/\{[1-0.3000][1-0.3000]\})/51,294$$
$$= \underline{1.00\%}$$

Law-3413:

If both (**t**= T/B= Tax Rate= 30.00%), (**i**= I/S= Interest Portion= 1.00%), (**d**= D/A= Dividend Portion or Payout= 30.00%), (**$**= Sales or Revenues= 51,294 millions USD), (**R**= Retained Earnings= 5,529 millions USD), and (**f**= F/S= Fixed Portion= 58.00%), are known, then it's (**V**= Interest Portion planned), is:

$$V = \$[1-f-i]-R/\{[1-t][1-d]\}$$
$$= 51,294[1-0.5800-0.0100]-5,529$$
$$/\{[1-0.3000][1-0.3000]\}$$
$$= 9,746 \text{ millions USD}$$

Law-3414:

If both (**i**= I/S= Interest Portion= 1.00%), (**V**= Variable Cost= 9,746 millions USD), (**d**= D/A= Dividend Portion or Payout= 30.00%), (**$**=Sales or Revenues= 51,294 millions USD), (**R**= Retained Earnings= 5,529 millions USD), and (**f**= F/S= Fixed Portion= 58.00%), are known, then it's (**t**= Tax Rate planned), is:

$$t = 1-R/([1-d]\{\$[1-f-i]-V\})$$
$$= 1- 5,529/([1-0.3000]\{51,294$$
$$[1-0.5800-0.0100-9,746\})$$
$$= 30.00\%$$

Law-3415:

If both (**i**= I/S= Interest Portion= 1.00%), (**V**= Variable Cost= 9,746 millions USD), (**t**= T/B= Tax Rate= 30.00%), (**$**=Sales or Revenues= 51,294 millions USD), (**R**= Retained Earnings= 5,529 millions USD), and (**f**= F/S= Fixed Portion= 58.00%), are known, then it's (**d**= Dividend Portion or Payout planned), is:

$$d = 1\text{-}R/([1\text{-}t]\{\$[1\text{-}f\text{-}i]\text{-}V\})$$
$$= 1 - 5{,}529/([1\text{-}0.3000]\{51{,}294$$
$$[1\text{-}0.5800\text{-}0.0100]\text{-}9{,}746\})$$
$$= \underline{30.00\%}$$

Law-3416:

If both (**i**= I/S= Interest Portion= 1.00%), (**V**= Variable Cost= 9,746 millions USD), (**$'**= Sales of Past Year= 48,851 millions USD), (**s**= [S/S']-1= Sales Growth= 5.00%), (**t**= T/B= Tax Rate= 30.00%), (**$**=Sales or Revenues= 51,294 millions USD), (**d**= D/A=Dividend Portion or Payout= 30.00%), and (**f**= F/S=Fixed Portion= 58.00%), are known, then it's (**R**= Retained Earnings planned), is:

$$R = [1\text{-}t][1\text{-}d]\{\$\text{-}V\text{-}\$f\text{-}\$'i[1+s]\}$$
$$= [1\text{-}t][1\text{-}d]\{\$[1\text{-}f]\text{-}V\text{-}\$'i[1+s]\}$$
$$= [1\text{-}0.3000][1\text{-}0.3000]\{51{,}294$$
$$[1\text{-}0.5800]\text{-}9{,}746\text{-}48{,}851$$
$$*0.0100[1+0.0500]\}$$
$$= \underline{30.00\%}$$

Law-3417:
If both (i= I/S= Interest Portion= 1.00%), (V= Variable Cost= 9,746 millions USD), (S'= Sales of Past Year= 48,851 millions USD), (s= [S/S']-1= Sales Growth= 5.00%), (t= T/B= Tax Rate= 30.00%), (R= Retained Earnings= 5,529 millions USD), (d= D/A= Dividend Portion or Payout= 30.00%), and (f= F/S= Fixed Portion= 58.00%), are known, then it's (S= Sales or Revenues planned), is:

$S = (V + S'i[1+s] + R/\{[1-t][1-d]\})/[1-f]$
$ = (9,746 + 48,851*0.0100[1+0.0500]$
$ + 5,529/\{[1-0.3000][1-0.3000]\})$
$ /[1-0.5800]$
$ = 51,294$ millions USD

Law-3418:
If both (i= I/S= Interest Portion= 1.00%), (S= Sales or Revenues= 51,294 millions USD), (S'= Sales of Past Year= 48,851 millions USD), (s= [S/S']-1= Sales Growth= 5.00%), (t= T/B= Tax Rate= 30.00%), (R= Retained Earnings= 5,529 millions USD), (d= D/A= Dividend Portion or Payout= 30.00%), and (f= F/S= Fixed Portion= 58.00%), are known, then it's (V= Variable Cost planned), is:

$V = S[1-f] - S'i[1+s] - R/\{[1-t][1-d]\}$
$ = 51,294[1-0.5800] - 48,851*0.0100$
$ [1+0.0500] - 5,529/\{[1-0.3000]$
$ [1-0.3000]\}$
$ = 9,746$ millions USD

Law-3419:

If both (i= I/S= Interest Portion= 1.00%), ($\$$= Sales or Revenues= 51,294 millions USD), (V= Variable Cost= 9,746 millions USD), (s= [S/S']-1= Sales Growth= 5.00%), (t= T/B= Tax Rate= 30.00%), (R= Retained Earnings= 5,529 millions USD), (d= D/A= Dividend Portion or Payout= 30.00%), and (f= F/S= Fixed Portion= 58.00%), are known, then it's ($\$'$= Sales Past), must be:

$$\$' = (\$[1-f]-V-R/\{[1-t][1-d]\})/\{i[1+s]\}$$
$$= (51,294[1-0.5800]-9,746-5,529$$
$$/\{[1-0.3000][1-0.3000]\})$$
$$/\{0.0100[1+0.0500]\}$$
$$= 48,851 \text{ millions USD}$$

Law-3420:

If both (i= I/S= Interest Portion= 1.00%), ($\$$= Sales or Revenues= 51,294 millions USD), (V= Variable Cost= 9,746 millions USD), (s= [S/S']-1= Sales Growth= 5.00%), (t= T/B= Tax Rate= 30.00%), (R= Retained Earnings= 5,529 millions USD), (d= D/A= Dividend Portion or Payout= 30.00%), and ($\$'$= Sales of Past Year= 48,851 millions USD), are known, then it's (f= Fixed Portion planned), is:

$$f = 1-(V+\$'i[1+s]+R/\{[1-t][1-d]\})/\$$$
$$= 1-(9,746+48,851*0.0100[1+0.0500]$$
$$+5,529/\{[1-0.3000]$$
$$[1-0.3000]\})/51,294$$
$$= 58.00\%$$

Law-3421:

If both (**i**= I/S= Interest Portion= 1.00%), (**$**= Sales or Revenues= 51,294 millions USD), (**f**= F/S= Fixed Portion= 58.00%), (**V**= Variable Cost= 9,746 millions USD), (**t**= T/B= Tax Rate= 30.00%), (**R**= Retained Earnings= 5,529 millions USD), (**d**= D/A= Dividend Portion or Payout= 30.00%), and (**$'**= Sales of Past Year= 48,851 millions USD), are known, then it's (**s**= Sales Growth planned), is:

$$s = (\$[1-f]-V-R/\{[1-t][1-d]\})/[\$'i]-1$$
$$= (51,294[1-0.5800]-9,746-5,529$$
$$/\{[1-0.3000][1-0.3000]\})$$
$$/[48,851*0.0100]-1$$
$$= 5.00\%$$

Law-3422:

If both (**s**= [S/S']-1= Sales Growth= 5.00%), (**$**= Sales or Revenues= 51,294 millions USD), (**f**= F/S=Fixed Portion= 58.00%), (**V**= Variable Cost= 9,746 millions USD), (**t**= T/B= Tax Rate= 30.00%), (**R**= Retained Earnings= 5,529 millions USD), (**d**= D/A=Dividend Portion or Payout= 30.00%), and (**$'**= Sales of Past Year= 48,851 millions USD), are known, then it's (**i**= Interest Portion planned), is:

$$i = (\$[1-f]-V-R/\{[1-t][1-d]\})/\{\$'[1+s]\}$$
$$= (51,294[1-0.5800]-9,746-5,529$$
$$/\{[1-0.3000][1-0.3000]\})$$
$$/\{48,851[1+0.0500]\}$$
$$= 1.00\%$$

Law-3423:
If both ($=$ [S/S']-1= Sales Growth= 5.00%), ($= Sales or Revenues= 51,294 millions USD), (f= F/S= Fixed Portion= 58.00%), (V= Variable Cost= 9,746 millions USD), (i= I/S= Interest Portion= 1.00%), (R= Retained Earnings= 5,529 millions USD), (d= D/A= Dividend Portion or Payout= 30.00%), and ($'= Sales of Past Year= 48,851 millions USD), are known, then it's (t= Tax Rate planned), is:

$t = 1-R/([1-d]\{\$[1-f]-V-\$'i[1+s]\})$
$= 1-5,529/([1-0.3000]\{51,294$
$[1-0.5800]-9,746- 48,851$
$*0.0100[1+0.0500]\}$
$= 30.00\%$

Law-3424:
If both ($=$ [S/S']-1= Sales Growth= 5.00%), ($= Sales or Revenues= 51,294 millions USD), (f= F/S= Fixed Portion= 58.00%), (V= Variable Cost= 9,746 millions USD), (i= I/S= Interest Portion= 1.00%), (R= Retained Earnings= 5,529 millions USD), (t= T/B= Tax Rate= 30.00%), and ($'= Sales of Past Year= 48,851 millions USD), are known, then it's (d= Dividend Portion or Payout planned), is:

$d = 1-R/([1-t]\{\$[1-f]-V-\$'i[1+s]\})$
$= 1-5,529/([1-0.3000]\{51,294$
$[1-0.5800]-9,746- 48,851$
$*0.0100[1+0.0500]\}$
$= 30.00\%$

Finance Construction-4, *Tim Asikin, Steve Asikin, Indra Senihardja*

Law-3425:

If both (s= [S/S']-1= Sales Growth= 5.00%), (S= Sales or Revenues= 51,294 millions USD), (f= F/S=Fixed Portion= 58.00%), (V= Variable Cost= 9,746 millions USD), (I= Interest Expense= 513 millions USD), (d= D/A= Dividend Portion or Payout= 30.00%), (t= T/B= Tax Rate= 30.00%), and (S'= Sales of Past Year= 48,851 millions USD), are known, then it's (R= Retained Earnings planned), is:

R= [1-t][1-d]{S-V-I-$S'f$[1+s]})
    = [1-0.3000][1-0.3000]{51,294-9,746
        - 513-48,851*0.5800
        [1+0.0500]}
    = 5,529 millions USD

Law-3426:

If both (s= [S/S']-1= Sales Growth= 5.00%), (f= F/S= Fixed Portion= 58.00%), (V= Variable Cost= 9,746 millions USD), (R= Retained Earnings= 5,529 millions USD), (I= Interest Expense= 513 millions USD), (d= D/A= Dividend Portion or Payout= 30.00%), (t= T/B= Tax Rate= 30.00%), and (S'= Sales of Past Year= 48,851 millions USD), are known, then it's (S= Sales or Revenues planned), is:

S= V+I+$S'f$[1+s]+R/{[1-t][1-d]}
    = 9,746+513+48,851*0.5800
        [1+0.0500]+5,529
        /{[1-0.3000][1-0.3000]}
    = 51,294 millions USD

Law-3427:
If both (s= [S/S']-1= Sales Growth= 5.00%), (f= F/S= Fixed Portion= 58.00%), (S= Sales or Revenues= 51,294 millions USD), (R= Retained Earnings= 5,529 millions USD), (I= Interest Expense= 513 millions USD), (d= D/A= Dividend Portion or Payout= 30.00%), (t= T/B= Tax Rate= 30.00%), and (S'= Sales of Past Year= 48,851 millions USD), are known, then it's (V= Variable Cost planned), is:

$$V = S-I-S'f[1+s]-R/\{[1-t][1-d]\}$$
$$= 51{,}294-513-48{,}851*0.5800$$
$$[1+0.0500]-5{,}529$$
$$/\{[1-0.3000][1-0.3000]\}$$
$$= 9{,}746 \text{ millions USD}$$

Law-3428:
If both (s= [S/S']-1= Sales Growth= 5.00%), (f= F/S=Fixed Portion= 58.00%), (S= Sales or Revenues= 51,294 millions USD), (R= Retained Earnings= 5,529 millions USD), (I= Interest Expense= 513 millions USD), (d= D/A= Dividend Portion or Payout= 30.00%), (t= T/B= Tax Rate= 30.00%), and (V= Variable Cost= 9,746 millions USD), are known, then it's (S'= Sales Past), must be:

$$S' = (S-V-I-R/\{[1-t][1-d]\})/\{f[1+s]\}$$
$$= (51{,}294-9{,}746-513-5{,}529/\{[1-0.3000]$$
$$[1-0.3000]\})$$
$$/\{0.5800[1+0.0500]\}$$
$$= 48{,}851 \text{ millions USD}$$

Law-3429:

If both (s= [S/S']-1= Sales Growth= 5.00%), (S'= Sales of Past Year= 48,851 millions USD), (S= Sales or Revenues= 51,294 millions USD), (R= Retained Earnings= 5,529 millions USD), (I= Interest Expense= 513 millions USD), (d= D/A= Dividend Portion or Payout= 30.00%), (t= T/B= Tax Rate= 30.00%), and (V= Variable Cost= 9,746 millions USD), are known, then it's (f= Fixed Portion planned), is:

f= (S-V-I-R/{[1-t][1-d]})/{S'[1+s]}
= (51,294-9,746-513-5,529/{[1-0.3000][1-0.3000]})
/{48,851[1+0.0500]}
= 58.00%

Law-3430:

If both (f= F/S= Fixed Portion= 58.00%), (S'= Sales of Past Year= 48,851 millions USD), (S= Sales or Revenues= 51,294 millions USD), (R= Retained Earnings= 5,529 millions USD), (I= Interest Expense= 513 millions USD), (d= D/A= Dividend Portion or Payout= 30.00%), (t= T/B= Tax Rate= 30.00%), and (V= Variable Cost= 9,746 millions USD), are known, then it's (s= Sales Growth planned), is:

s= (S-V-I-R/{[1-t][1-d]})/[$S'f$]-1
= (51,294-9,746-513-5,529/{[1-0.3000][1-0.3000]})
/[48,851*0.5800]-1
= 5.00%

Law-3431:

If both (**f**= F/S= Fixed Portion= 58.00%), (**$'**= Sales of Past Year= 48,851 millions USD), (**$**= Sales or Revenues= 51,294 millions USD), (**R**= Retained Earnings= 5,529 millions USD), (**s**= [S/S']-1= Sales Growth= 5.00%), (**d**= D/A= Dividend Portion or Payout= 30.00%), (**t**= T/B= Tax Rate= 30.00%), and (**V**= Variable Cost= 9,746 millions USD), are known, then it's (**I**= Sales Growth planned), is:

I= **$-V-$'f**[1+**s**]-**R**/{[1-**t**][1-**d**]}
 = 51,294-9,746-48,851*0.5800
  [1+0.0500]-5,529/{[1-0.3000]
  [1-0.3000]}
 = 513 millions USD

Law-3432:

If both (**f**= F/S= Fixed Portion= 58.00%), (**$'**= Sales of Past Year= 48,851 millions USD), (**$**= Sales or Revenues= 51,294 millions USD), (**R**= Retained Earnings= 5,529 millions USD), (**s**= [S/S']-1= Sales Growth= 5.00%), (**d**= D/A= Dividend Portion or Payout= 30.00%), (**I**= Interest Expense= 513 millions USD), and (**V**= Variable Cost= 9,746 millions USD), are known, then it's (**t**= Tax Rate planned), is:

t= 1-**R**/([1-**d**]{**$-V-$'f**[1+**s**]-**I**}
 = 1-5,529/([1-0.3000]{51,294-9,746
  -48,851*0.5800[1+0.0500]
  -513}
 = 30.00%

Law-3433:

If both (**f**= F/S= Fixed Portion= 58.00%), (**S'**= Sales of Past Year= 48,851 millions USD), (**S**= Sales or Revenues= 51,294 millions USD), (**R**= Retained Earnings= 5,529 millions USD), (**s**= [S/S']-1= Sales Growth= 5.00%), (**t**= T/B= Tax Rate= 30.00%), (**I**= Interest Expense= 513 millions USD), and (**V**= Variable Cost= 9,746 millions USD), are known, then it's (**d**= Dividend Portion or Payout planned), is:

$$\mathbf{d} = 1 - \mathbf{R}/([1-\mathbf{t}]\{\mathbf{S} - \mathbf{V} - \mathbf{S'f}[1+\mathbf{s}] - \mathbf{I}\})$$
$$= 1 - 5{,}529/([1-0.3000]\{51{,}294 - 9{,}746 - 48{,}851 * 0.5800[1+0.0500] - 513\})$$
$$= 30.00\%$$

Law-3434:

If both (**f**= F/S= Fixed Portion= 58.00%), (**S'**= Sales of Past Year= 48,851 millions USD), (**S**= Sales or Revenues= 51,294 millions USD), (**d**= D/A= Dividend Portion or Payout= 30.00%), (**s**= [S/S']-1= Sales Growth= 5.00%), (**t**= T/B= Tax Rate= 30.00%), (**i**= I/S= Interest Portion= 1.00%), and (**V**= Variable Cost= 9,746 millions USD), are known, then it's (**R**= Retained Earnings planned), is:

$$\mathbf{R} = [1-\mathbf{t}][1-\mathbf{d}]\{\mathbf{S} - \mathbf{V} - \mathbf{S'f}[1+\mathbf{s}] - \mathbf{Si}\}$$
$$= [1-\mathbf{t}][1-\mathbf{d}]\{\mathbf{S}[1-\mathbf{i}] - \mathbf{V} - \mathbf{S'f}[1+\mathbf{s}]\}$$
$$= [1-0.3000][1-0.3000]\{51{,}294[1-0.0100] - 9{,}746 - 48{,}851 * 0.5800[1+0.0500]\}$$
$$= 5{,}529 \text{ millions USD}$$

Law-3435:

If both (**f**= F/S= Fixed Portion= 58.00%), (**S'**= Sales of Past Year= 48,851 millions USD), (**R**= Retained Earnings= 5,529 millions USD), (**d**= D/A= Dividend Portion or Payout= 30.00%), (**s**= [S/S']-1= Sales Growth= 5.00%), (**t**= T/B= Tax Rate= 30.00%), (**i**= I/S= Interest Portion= 1.00%), and (**V**= Variable Cost= 9,746 millions USD), are known, then it's (**S**= Sales or Revenues planned), is:

$$S = (V+S'f[1+s]+R/\{[1-t][1-d]\})/[1-i]$$
$$= (9,746+48,851*0.5800[1+0.0500]$$
$$+5,529/\{[1-0.3000]$$
$$[1-0.3000]\})/[1-0.0100]$$
$$= 51,294 \text{ millions USD}$$

Law-3436:

If both (**f**= F/S=Fixed Portion= 58.00%), (**S'**= Sales of Past Year= 48,851 millions USD), (**R**= Retained Earnings= 5,529 millions USD), (**d**= D/A= Dividend Portion or Payout= 30.00%), (**s**= [S/S']-1= Sales Growth= 5.00%), (**t**= T/B= Tax Rate= 30.00%), (**S**= Sales or Revenues= 51,294 millions USD), and (**V**= Variable Cost= 9,746 millions USD), are known, then it's (**i**= Interest Portion planned), is:

$$i = 1-(V+S'f[1+s]+R/\{[1-t][1-d]\})/S$$
$$= 1-(9,746+48,851*0.5800[1+0.0500]$$
$$+5,529/\{[1-0.3000]$$
$$[1-0.3000]\})/51,294$$
$$= 1.00\%$$

Law-3437:
If both (**f**= F/S=Fixed Portion= 58.00%), (**$'**= Sales of Past Year= 48,851 millions USD), (**R**= Retained Earnings= 5,529 millions USD), (**d**= D/A= Dividend Portion or Payout= 30.00%), (**s**= [S/S']-1= Sales Growth= 5.00%), (**t**= T/B= Tax Rate= 30.00%), (**$**= Sales or Revenues= 51,294 millions USD), and (**i**= I/S= Interest Portion= 1.00%), are known, then it's (**V**= Variable Cost planned), is:

$$V = \$[1-i] - \$'f[1+s] - R/\{[1-t][1-d]\}$$
$$= 51{,}294[1-0.0100] - 48{,}851*0.5800$$
$$[1+0.0500] - 5{,}529/\{[1-0.3000]$$
$$[1-0.3000]\}$$
$$= \underline{9{,}746} \text{ millions USD}$$

Law-3438:
If both (**f**= F/S=Fixed Portion= 58.00%), (**V**= Variable Cost= 9,746 millions USD), (**R**= Retained Earnings= 5,529 millions USD), (**d**= D/A= Dividend Portion or Payout= 30.00%), (**s**= [S/S']-1= Sales Growth= 5.00%), (**t**= T/B= Tax Rate= 30.00%), (**$**= Sales or Revenues= 51,294 millions USD), and (**i**= I/S= Interest Portion= 1.00%), are known, then it's (**$'**= Sales Past), must be:

$$\$' = (\$[1-i] - V - R/\{[1-t][1-d]\})/\{f[1+s]\}$$
$$= (51{,}294[1-0.0100] - 9{,}746 - 5{,}529$$
$$/\{[1-0.3000][1-0.3000]\})$$
$$/\{0.5800[1+0.0500]\}$$
$$= \underline{48{,}851} \text{ millions USD}$$

Law-3439:

If both ($\$'$= Sales of Past Year= 48,851 millions USD), (**V**= Variable Cost= 9,746 millions USD), (**R**= Retained Earnings= 5,529 millions USD), (**d**= D/A= Dividend Portion or Payout= 30.00%), (**s**= [S/S']-1= Sales Growth= 5.00%), (**t**= T/B= Tax Rate= 30.00%), ($\$$= Sales or Revenues= 51,294 millions USD), and (**i**= I/S= Interest Portion= 1.00%), are known, then it's (**f**= Fixed Portion planned), is:

$$f = (\$[1-i]-V-R/\{[1-t][1-d]\})/\{\$'[1+s]\}$$
$$= (51,294[1-0.0100]-9,746-5,529$$
$$/\{[1-0.3000][1-0.3000]\})$$
$$/\{48,851[1+0.0500]\}$$
$$= 58.00\%$$

Law-3440:

If both ($\$'$= Sales of Past Year= 48,851 millions USD), (**V**= Variable Cost= 9,746 millions USD), (**R**= Retained Earnings= 5,529 millions USD), (**d**= D/A= Dividend Portion or Payout= 30.00%), (**f**= F/S=Fixed Portion= 58.00%), (**t**= T/B= Tax Rate= 30.00%), ($\$$= Sales or Revenues= 51,294 millions USD), and (**i**= I/S= Interest Portion= 1.00%), are known, then it's (**s**= Sales Growth planned), is:

$$s = \$[1-i]-V-R/\{[1-t][1-d]\})/[\$'f]-1$$
$$= (51,294[1-0.0100]-9,746-5,529$$
$$/\{[1-0.3000][1-0.3000]\})$$
$$/[48,851*0.5800]-1$$
$$= 5.00\%$$

Law-3441:

If both ($'$= Sales of Past Year= 48,851 millions USD), (V= Variable Cost= 9,746 millions USD), (R= Retained Earnings= 5,529 millions USD), (d= D/A= Dividend Portion or Payout= 30.00%), (f= F/S= Fixed Portion= 58.00%), (s= [S/S']-1= Sales Growth= 5.00%), ($\$$= Sales or Revenues= 51,294 millions USD), and (i= I/S= Interest Portion= 1.00%), are known, then it's (t= Tax Rate planned), is:

t= 1-R/([1-d]{$\$$[1-i]-V-$\$'f$[1+$s$]
= 1-5,529/([1-0.3000]{51,294[1-0.0100]
-9,746-48,851*0.5800
[1+0.0500]})
= 30.00%

Law-3442:

If both ($'$= Sales of Past Year= 48,851 millions USD), (V= Variable Cost= 9,746 millions USD), (R= Retained Earnings= 5,529 millions USD), (d= D/A= Dividend Portion or Payout= 30.00%), (t= T/R= Tax Rate= 30.00%), (f= F/S= Fixed Portion= 58.00%), (s= [S/S']-1= Sales Growth= 5.00%), ($\$$= Sales or Revenues= 51,294 millions USD), and (i= I/S= Interest Portion= 1.00%), are known, then it's (d= Dividend Portion or Payout planned), is:

d= 1-R/([1-t]{$\$$[1-i]-V-$\$'f$[1+$s$]
= 1-5,529/([1-0.3000]{51,294[1-0.0100]
-9,746-48,851*0.5800
[1+0.0500]})
= 30.00%

Law-3443:

If both ($\$'$= Sales of Past Year= 48,851 millions USD), (V= Variable Cost= 9,746 millions USD), (t= T/B= Tax Rate= 30.00%), (d= D/A= Dividend Portion or Payout= 30.00%), (f= F/S= Fixed Portion= 58.00%), (s= [S/S']-1= Sales Growth= 5.00%), ($\$$= Sales or Revenues= 51,294 millions USD), and (i= I/S= Interest Portion= 1.00%), are known, then it's (R= Retained Earnings planned), is:

$$R = [1-t][1-d]\{\$-V-\$'f[1+s]-\$'i[1+s]\}$$
$$= [1-t][1-d]\{\$-V-\$'[1+s][f+i]\}$$
$$= [1-0.3000][1-0.3000]\{51,294-9,746 -48,851\,[1+0.0500] [0.5800+0.0100]\}$$
$$= 5,529 \text{ millions USD}$$

Law-3444:

If both ($\$'$= Sales of Past Year= 48,851 millions USD), (V= Variable Cost= 9,746 millions USD), (t= T/B= Tax Rate= 30.00%), (d= D/A= Dividend Portion or Payout= 30.00%), (f= F/S= Fixed Portion= 58.00%), (s= [S/S']-1= Sales Growth= 5.00%), (R= Retained Earnings= 5,529 millions USD), and (i= I/S= Interest Portion= 1.00%), are known, then it's ($\$$= Sales or Revenues planned), is:

$$\$ = V+\$'[1+s][f+i]+R/\{[1-t][1-d]\}$$
$$= 9,746+48,851\,[1+0.0500][0.5800 +0.0100]+5,529/\{[1-0.3000] [1-0.3000]\}$$
$$= 51,294 \text{ millions USD}$$

Law-3445:
If both ($'= Sales of Past Year= 48,851 millions USD), ($= Sales or Revenues= 51,294 millions USD), (t= T/B= Tax Rate= 30.00%), (d= D/A= Dividend Portion or Payout= 30.00%), (f= F/S= Fixed Portion= 58.00%), (s= [S/S']-1= Sales Growth= 5.00%), (R= Retained Earnings= 5,529 millions USD), and (i= I/S= Interest Portion= 1.00%), are known, then it's (V= Variable Cost planned), is:

$$V = \$ - \$'[1+s][f+i] - R/\{[1-t][1-d]\}$$
$$= 51{,}294 - 48{,}851\,[1+0.0500][0.5800 + 0.0100] - 5{,}529/\{[1-0.3000][1-0.3000]\}$$
$$= \underline{9{,}746} \text{ millions USD}$$

Law-3446:
If both (V= Variable Cost= 9,746 millions USD), ($= Sales or Revenues= 51,294 millions USD), (t= T/B= Tax Rate= 30.00%), (d= D/A= Dividend Portion or Payout= 30.00%), (f= F/S= Fixed Portion= 58.00%), (s= [S/S']-1= Sales Growth= 5.00%), (R= Retained Earnings= 5,529 millions USD), and (i= I/S= Interest Portion= 1.00%), are known, then it's ($'= Sales Past), must be:

$$\$' = (\$ - V - R/\{[1-t][1-d]\}) / \{[1+s][f+i]\}$$
$$= (51{,}294 - 9{,}746 - 5{,}529/\{[1-0.3000][1-0.3000]\}) / \{[1+0.0500][0.5800+0.0100]\}$$
$$= \underline{48{,}851} \text{ millions USD}$$

Law-3447:

If both (**V**= Variable Cost= 9,746 millions USD), (**$**= Sales or Revenues= 51,294 millions USD), (**t**= T/B= Tax Rate= 30.00%), (**d**= D/A= Dividend Portion or Payout= 30.00%), (**f**= F/S= Fixed Portion= 58.00%), (**$'**= Sales of Past Year= 48,851 millions USD), (**R**= Retained Earnings= 5,529 millions USD), and (**i**= I/S= Interest Portion= 1.00%), are known, then it's (**s**= Sales Gorowth planned), is:

$$s = (\$-V-R/\{[1-t][1-d]\})/\{\$'[f+i]\}-1$$
$$= (51{,}294-9{,}746-5{,}529/\{[1-0.3000][1-0.3000]\})/\{48{,}851[0.5800+0.0100]\}-1$$
$$= 5.00\%$$

Law-3448:

If both (**V**= Variable Cost= 9,746 millions USD), (**$**= Sales or Revenues= 51,294 millions USD), (**t**= T/B= Tax Rate= 30.00%), (**d**= D/A= Dividend Portion or Payout= 30.00%), (**s**= [S/S']-1= Sales Growth= 5.00%), (**$'**= Sales of Past Year= 48,851 millions USD), (**R**= Retained Earnings= 5,529 millions USD), and (**i**= I/S= Interest Portion= 1.00%), are known, then it's (**f**= Fixed Portion planned), is:

$$f = (\$-V-R/\{[1-t][1-d]\})/\{\$'[1+s]\}-i$$
$$= (51{,}294-9{,}746-5{,}529/\{[1-0.3000][1-0.3000]\})/\{48{,}851[1+0.0500]\}-0.0100$$
$$= 58.00\%$$

Law-3449:
If both (**V**= Variable Cost= 9,746 millions USD), (**$**= Sales or Revenues= 51,294 millions USD), (**t**= T/B= Tax Rate= 30.00%), (**d**= D/A= Dividend Portion or Payout= 30.00%), (**s**= [S/S']-1= Sales Growth= 5.00%), (**$'**= Sales of Past Year= 48,851 millions USD), (**R**= Retained Earnings= 5,529 millions USD), and (**f**= F/S= Fixed Portion= 58.00%), are known, then it's (**i**= Interest Portion planned), is:

$$i = (\$-V-R/\{[1-t][1-d]\})/\{\$'[1+s]\}-v$$
$$= (51,294-9,746-5,529/\{[1-0.3000][1-0.3000]\})/\{48,851[1+0.0500]\}-0.1900$$
$$= 1.00\%$$

Law-3450:
If both (**V**= Variable Cost= 9,746 millions USD), (**$**= Sales or Revenues= 51,294 millions USD), (**i**= I/S= Interest Portion= 1.00%), (**d**= D/A= Dividend Portion or Payout= 30.00%), (**s**= [S/S']-1= Sales Growth= 5.00%), (**$'**= Sales of Past Year= 48,851 millions USD), (**R**= Retained Earnings= 5,529 millions USD), and (**f**= F/S= Fixed Portion= 58.00%), are known, then it's (**t**= Tax Rate planned), is:

$$t = 1 - R/([1-d]\{\$-V-\$'[1+s][f+i]\})$$
$$= 1-5,529/([1-0.3000]\{51,294-9,746 -48,851[1+0.0500][0.5800+0.0100]\})$$
$$= 30.00\%$$

Law-3451:

If both (**V**= Variable Cost= 9,746 millions USD), (**S**= Sales or Revenues= 51,294 millions USD), (**i**= I/S= Interest Portion= 1.00%), (**t**= T/B= Tax Rate= 30.00%), (**s**= [S/S']-1= Sales Growth= 5.00%), (**S'**= Sales of Past Year= 48,851 millions USD), (**R**= Retained Earnings= 5,529 millions USD), and (**f**= F/S=Fixed Portion= 58.00%), are known, then it's (**d**= Dividend Portion or Payout planned), is:

$$d = 1-R/([1-t]\{S-V-S'[1+s][f+i]\})$$
$$= 1-5,529/([1-0.3000]\{51,294-9,746$$
$$-48,851[1+0.0500]$$
$$[0.5800+0.0100]\}$$
$$= \underline{30.00\%}$$

Law-3452:

If both (**I**= Interest Expense= 513 millions USD), (**v**= V/S= Variable Portion= 19.00%), (**S**= Sales or Revenues= 51,294 millions USD), (**t**= T/B= Tax Rate= 30.00%), (**d**= D/A= Dividend Portion or Payout= 30.00%), and (**F**= Fixed Cost= 29,750 millions USD), are known, then it's (**R**= Retained Earnings planned), is:

$$R = [1-t][1-d][S-Sv-F-I] = [1-t][1-d]\{S[1-v]-F-I\}$$
$$= [1-0.3000][1-0.3000]\{51,294$$
$$[1-0.1900]-29,750-513\}$$
$$= \underline{5,529} \text{ millions USD}$$

Law-3453:

If both (**I**= Interest Expense= 513 millions USD), (**v**= V/S= Variable Portion= 19.00%), (**R**= Retained Earnings= 5,529 millions USD), (**t**= T/B= Tax Rate= 30.00%), (**d**= D/A= Dividend Portion or Payout= 30.00%), and (**F**= Fixed Cost= 29,750 millions USD), are known, then it's (**S**= Sales or Revenues planned), is:

$$S = (F+I+R/\{[1-t][1-d]\})/[1-v]$$
$$= (29{,}750+513+5{,}529/\{[1-0.3000][1-0.3000]\})/[1-0.1900]$$
$$= 51{,}294 \text{ millions USD}$$

Law-3454:

If both (**I**= Interest Expense= 513 millions USD), (**S**= Sales or Revenues= 51,294 millions USD), (**R**= Retained Earnings= 5,529 millions USD), (**t**= T/B= Tax Rate= 30.00%), (**d**= D/A= Dividend Portion or Payout= 30.00%), and (**F**= Fixed Cost= 29,750 millions USD), are known, then it's (**v**= Variable Portion planned), is:

$$v = 1-(F+I+R/\{[1-t][1-d]\})/S$$
$$= 1-(29{,}750+513+5{,}529/\{[1-0.3000][1-0.3000]\})/51{,}294$$
$$= 19.00\%$$

Law-3455:

If both (**I**= Interest Expense= 513 millions USD), (**$**= Sales or Revenues= 51,294 millions USD), (**R**= Retained Earnings= 5,529 millions USD), (**t**= T/B= Tax Rate= 30.00%), (**d**= D/A= Dividend Portion or Payout= 30.00%), and (**v**= V/S= Variable Portion= 19.00%), are known, then it's (**F**= Fixed Cost planned), is:

$$F = \$[1-v] - I - R / \{[1-t][1-d]\}$$
$$= 51{,}294[1-0.1900] - 513 - 5{,}529 / \{[1-0.3000][1-0.3000]\}$$
$$= \underline{29{,}750} \text{ millions USD}$$

Law-3456:

If both (**F**= Fixed Cost= 29,750 millions USD), (**$**= Sales or Revenues= 51,294 millions USD), (**R**= Retained Earnings= 5,529 millions USD), (**t**= T/B= Tax Rate= 30.00%), (**d**= D/A= Dividend Portion or Payout= 30.00%), and (**v**= V/S= Variable Portion= 19.00%), are known, then it's (**I**= Interest Expense planned), is:

$$I = \$[1-v] - F - R / \{[1-t][1-d]\}$$
$$= 51{,}294[1-0.1900] - 29{,}750 - 5{,}529 / \{[1-0.3000][1-0.3000]\}$$
$$= \underline{513} \text{ millions USD}$$

Law-3457:

If both (**F**= Fixed Cost= 29,750 millions USD), (**$**= Sales or Revenues= 51,294 millions USD), (**R**= Retained Earnings= 5,529 millions USD), (**I**= Interest Expense= 513 millions USD), (**d**= D/A= Dividend Portion or Payout= 30.00%), and (**v**= V/S= Variable Portion= 19.00%), are known, then it's (**t**= Tax Rate planned), is:

$$t = 1 - R/([1-d]\{\$[1-v]-F-I\})$$
$$= 1 - 5{,}529/([1-0.3000]\{51{,}294$$
$$[1-0.1900]-29{,}750-513\})$$
$$= 30.00\%$$

Law-3458:

If both (**F**= Fixed Cost= 29,750 millions USD), (**$**= Sales or Revenues= 51,294 millions USD), (**R**= Retained Earnings= 5,529 millions USD), (**I**= Interest Expense= 513 millions USD), (**t**= T/B= Tax Rate= 30.00%), and (**v**= V/S= Variable Portion= 19.00%), are known, then it's (**d**= Dividend Portion or Payout planned), is:

$$d = 1 - R/([1-t]\{\$[1-v]-F-I\})$$
$$= 1 - 5{,}529/([1-0.3000]\{51{,}294$$
$$[1-0.1900]-29{,}750-513\})$$
$$= 30.00\%$$

Law-3459:

If both (**F**= Fixed Cost= 29,750 millions USD), (**$**= Sales or Revenues= 51,294 millions USD), (**d**= D/A= Dividend Portion or Payout= 30.00%), (**i**= I/S= Interest Portion= 1.00%), (**t**= T/B= Tax Rate= 30.00%), and (**v**= V/S= Variable Portion= 19.00%), are known, then it's (**R**= Retained Earnings planned), is:

$$\begin{aligned}
\mathbf{R} &= [1\text{-}\mathbf{t}][1\text{-}\mathbf{d}][\mathbf{\$}\text{-}\mathbf{\$v}\text{-}\mathbf{F}\text{-}\mathbf{\$i}] \\
&= [1\text{-}\mathbf{t}][1\text{-}\mathbf{d}][\mathbf{\$}[1\text{-}\mathbf{v}\text{-}\mathbf{i}]\text{-}\mathbf{F}\} \\
&= [1\text{-}0.3000][1\text{-}0.3000]\{51{,}294 \\
&\qquad [1\text{-}0.1900\text{-}0.0100]\text{-}29{,}750\} \\
&= 5{,}529 \text{ millions USD}
\end{aligned}$$

Law-3460:

If both (**F**= Fixed Cost= 29,750 millions USD), (**R**= Retained Earnings= 5,529 millions USD), (**d**= D/A= Dividend Portion or Payout= 30.00%), (**i**= I/S= Interest Portion= 1.00%), (**t**= T/B= Tax Rate= 30.00%), and (**v**= V/S= Variable Portion= 19.00%), are known, then it's (**$**= Sales or Revenues planned), is:

$$\begin{aligned}
\mathbf{\$} &= (\mathbf{F}+\mathbf{R}/\{[1\text{-}\mathbf{t}][1\text{-}\mathbf{d}]\}) / [1\text{-}\mathbf{v}\text{-}\mathbf{i}] \\
&= (29{,}750+5{,}529/\{[1\text{-}0.3000] \\
&\qquad [1\text{-}0.3000]\}) \\
&\qquad /\{[1\text{-}0.1900\text{-}0.0100]\} \\
&= 51{,}294 \text{ millions USD}
\end{aligned}$$

Law-3461:

If both (**F**= Fixed Cost= 29,750 millions USD), (**R**= Retained Earnings= 5,529 millions USD), (**d**= D/A= Dividend Portion or Payout= 30.00%), (**i**= I/S= Interest Portion= 1.00%), (**t**= T/B= Tax Rate= 30.00%), and (**$**= Sales or Revenues= 51,294 millions USD), are known, then it's (**v**= Variable Portion planned), is:

$$v = 1 - (F+R/\{[1-t][1-d]\})/\$-i$$
$$= 1-(29,750+5,529/\{[1-0.3000]$$
$$[1-0.3000]\})/51,294-0.0100$$
$$= 19.00\%$$

Law-3462:

If both (**F**= Fixed Cost= 29,750 millions USD), (**R**= Retained Earnings= 5,529 millions USD), (**d**= D/A= Dividend Portion or Payout= 30.00%), (**v**= V/S= Variable Portion= 19.00%), (**t**= T/B= Tax Rate= 30.00%), and (**$**= Sales or Revenues= 51,294 millions USD), are known, then it's (**i**= Interest Portion planned), is:

$$i = 1 - (F+R/\{[1-t][1-d]\})/\$-v$$
$$= 1-(29,750+5,529/\{[1-0.3000]$$
$$[1-0.3000]\})/51,294-0.1900$$
$$= 1.00\%$$

Law-3463:

If both (i= I/S= Interest Portion= 1.00%), (R= Retained Earnings= 5,529 millions USD), (d= D/A=Dividend Portion or Payout= 30.00%), (v= V/S= Variable Portion= 19.00%), (t= T/B= Tax Rate= 30.00%), and ($\$$= Sales or Revenues= 51,294 millions USD), are known, then it's (F= Fixed Cost planned), is:

$$F= \$[1-v-i]-R/\{[1-t][1-d]\}$$
$$= 51,294[1-0.1900-0.0100]-5,529$$
$$/\{[1-0.3000][1-0.3000]\}$$
$$= \underline{29,750} \text{ millions USD}$$

Law-3464:

If both (i= I/S= Interest Portion= 1.00%), (R= Retained Earnings= 5,529 millions USD), (d= D/A= Dividend Portion or Payout= 30.00%), (v= V/S= Variable Portion= 19.00%), (F= Fixed Cost= 29,750 millions USD), and ($\$$= Sales or Revenues= 51,294 millions USD), are known, then it's (t= Tax Rate planned), is:

$$t= 1-R/([1-d]\{\$[1-v-i]-F\})$$
$$= 1-5,529/([1-0.3000]\{51,294[1-0.1900$$
$$-0.0100]-29,750\})$$
$$= \underline{30.00\%}$$

Law-3465:
If both (**i**= I/S= Interest Portion= 1.00%), (**R**= Retained Earnings= 5,529 millions USD), (**t**= T/B= Tax Rate= 30.00%), (**v**= V/S= Variable Portion= 19.00%), (**F**= Fixed Cost= 29,750 millions USD), and (**$**= Sales or Revenues= 51,294 millions USD), are known, then it's (**d**= Dividend Portion or Payout planned), is:

$$d= 1\text{-}R/([1\text{-}t]\{\$[1\text{-}v\text{-}i]\text{-}F\})$$
$$= 1\text{-}5,529/([1\text{-}0.3000]\{51,294 [1\text{-}0.1900\text{-}0.0100]\text{-}F\})$$
$$= 30.00\%$$

Law-3466:
If both (**i**= I/S= Interest Portion= 1.00%), (**$'**= Sales of Past Year= 48,851 millions USD), (**s**= [S/S']-1= Sales Growth= 5.00%), (**d**= D/A= Dividend Portion or Payout= 30.00%), (**t**= T/B= Tax Rate= 30.00%), (**v**= V/S= Variable Portion= 19.00%), (**F**= Fixed Cost= 29,750 millions USD), and (**$**= Sales or Revenues= 51,294 millions USD), are known, then it's (**R**= Retained Earnings planned), is:

$$R= [1\text{-}t][1\text{-}d]\{\$\text{-}\$v\text{-}F\text{-}\$'i[1+s]\}$$
$$= [1\text{-}t][1\text{-}d]\{\$[1\text{-}v]\text{-}F\text{-}\$'i[1+s]\}$$
$$= [1\text{-}0,3000][1\text{-}0.3000]\{51,294 [1\text{-}0.1900]\text{-}29,750\text{-}48,851 *0,0100[1+0.0500]\}$$
$$= 5,529 \text{ millions USD}$$

Law-3467:

If both (**i**= I/S= Interest Portion= 1.00%), (**$'**= Sales of Past Year= 48,851 millions USD), (**s**= [S/S']-1= Sales Growth= 5.00%), (**d**= D/A= Dividend Portion or Payout= 30.00%), (**t**= T/B= Tax Rate= 30.00%), (**v**= V/S= Variable Portion= 19.00%), (**F**= Fixed Cost= 29,750 millions USD), and (**R**= Retained Earnings= 5,529 millions USD), are known, then it's (**$**= Retained Earnings planned), is:

$$\$ = (F+\$'i[1+s]+R/[[1-t][1-d]]/[1-v]$$
$$= (29{,}750+48{,}851*0{,}0100[1+0{,}0500]$$
$$+ 5{,}529/\{[1-0{,}3000][1-0{,}3000]\}$$
$$/ [1-0{,}1900]$$
$$= 51{,}294 \text{ millions USD}$$

Law-3468:

If both (**i**= I/S= Interest Portion= 1.00%), (**$'**= Sales of Past Year= 48,851 millions USD), (**s**= [S/S']-1= Sales Growth= 5.00%), (**d**= D/A= Dividend Portion or Payout= 30.00%), (**t**= T/B= Tax Rate= 30.00%), (**$**= Sales or Revenues= 51,294 millions USD), (**F**= Fixed Cost= 29,750 millions USD), and (**R**= Retained Earnings= 5,529 millions USD), are known, then it's (**v**= Retained Earnings planned), is:

$$v = 1-(F+\$'i[1+s]+R/\{[1-t][1-d]\}/\$$$
$$= 1-(29{,}750+48{,}851*0{,}0100[1+0{,}0500]$$
$$+5{,}529/\{[1-0{,}3000][1-0{,}3000]\}$$
$$/51{,}294$$
$$= 19.00\%$$

Law-3469:

If both (i= I/S= Interest Portion= 1.00%), ($\$'$= Sales of Past Year= 48,851 millions USD), (s= [S/S']-1= Sales Growth= 5.00%), (d= D/A= Dividend Portion or Payout= 30.00%), (t= T/B= Tax Rate= 30.00%), ($\$$= Sales or Revenues= 51,294 millions USD), (v= V/S= Variable Portion= 19.00%), and (R= Retained Earnings= 5,529 millions USD), are known, then it's (F= Fixed Cost planned), is:

$$F = \$[1-v]-\$'i[1+s]-R/\{[[1-t][1-d]]\}$$
$$= 51,294[1-0.1900]-48,851*0.0100$$
$$[1+0.0500] -5,529$$
$$/\{[1-0.3000][1-0.3000]\}$$
$$= 29,750 \text{ millions USD}$$

Law-3470:

If both (i= I/S= Interest Portion= 1.00%), (F= Fixed Cost= 29,750 millions USD), (s= [S/S']-1= Sales Growth= 5.00%), (d= D/A= Dividend Portion or Payout= 30.00%), (t= T/B= Tax Rate= 30.00%), ($\$$= Sales or Revenues= 51,294 millions USD), (v= V/S= Variable Portion= 19.00%), and (R= Retained Earnings= 5,529 millions USD), are known, then it's ($\$'$= Sales Past), must be:

$$\$' = (\$[1-v]-\$'i[1+s]-R/\{[[1-t][1-d]]\})/\{i[1+s]\}$$
$$= (51,294[1-0.1900]-29,750 -5,529$$
$$/\{[1-0.3000][1-0.3000]\})$$
$$/\{0.0100[1+0.0500]\}$$
$$= 48,851 \text{ millions USD}$$

Law-3471:

If both ($'$= Sales of Past Year= 48,851 millions USD), (**F**= Fixed Cost= 29,750 millions USD), (**s**= [S/S']-1= Sales Growth= 5.00%), (**d**= D/A= Dividend Portion or Payout= 30.00%), (**t**= T/B= Tax Rate= 30.00%), ($= Sales or Revenues= 51,294 millions USD), (**v**= V/S= Variable Portion= 19.00%), and (**R**= Retained Earnings= 5,529 millions USD), are known, then it's (**i**= Interest Portion planned), is:

$$i = (\$[1-v] - \$'i[1+s] - R/\{[[1-t][1-d]]\})/\{\$'[1+s]\}$$
$$= (51,294[1-0.1900] - 29,750 - 5,529$$
$$/\{[1-0.3000][1-0.3000]\})$$
$$/\{48,851[1+0.0500]\}$$
$$= \underline{1.00\%}$$

Law-3472:

If both ($'$= Sales of Past Year= 48,851 millions USD), (**F**= Fixed Cost= 29,750 millions USD), (**i**= I/S= Interest Portion= 1.00%), (**d**= D/A= Dividend Portion or Payout= 30.00%), (**t**= T/B= Tax Rate= 30.00%), ($= Sales or Revenues= 51,294 millions USD), (**v**= V/S= Variable Portion= 19.00%), and (**R**= Retained Earnings= 5,529 millions USD), are known, then it's (**s**= Sales Growth planned), is:

$$s = (\$[1-v] - \$'i[1+s] - R/\{[[1-t][1-d]]\})/[\$'i] - 1$$
$$= (51,294[1-0.1900] - 29,750 - 5,529$$
$$/\{[1-0.3000][1-0.3000]\})$$
$$/[48,851*0.0100] - 1$$
$$= \underline{5.00\%}$$

Law-3473:

If both ($**'**= Sales of Past Year= 48,851 millions USD), (**F**= Fixed Cost= 29,750 millions USD), (**i**= I/S= Interest Portion= 1.00%), (**d**= D/A= Dividend Portion or Payout= 30.00%), (**s**= [S/S']-1= Sales Growth= 5.00%), ($= Sales or Revenues= 51,294 millions USD), (**v**= V/S= Variable Portion= 19.00%), and (**R**= Retained Earnings= 5,529 millions USD), are known, then it's (**t**= Tax Rate planned), is:

$\mathbf{t}$= 1-**R**/([1-**d**]{$[1-**v**]-**F**-$'**i**[1+**s**]})
= 1-5,529/([1-0.3000]{51,294[1-0.1900]
-29,750 -48,851*0.0100
[1+0.0500]})
= 30.00%

Law-3474:

If both ($**'**= Sales of Past Year= 48,851 millions USD), (**F**= Fixed Cost= 29,750 millions USD), (**i**= I/S= Interest Portion= 1.00%), (**t**= T/B= Tax Rate= 30.00%), (**s**= [S/S']-1= Sales Growth= 5.00%), ($= Sales or Revenues= 51,294 millions USD), (**v**= V/S= Variable Portion= 19.00%), and (**R**= Retained Earnings= 5,529 millions USD), are known, then it's (**d**= Dividend Portion or Payout planned), is:

$\mathbf{d}$= 1-**R**/([1-**t**]{$[1-**v**]-**F**-$'**i**[1+**s**]})
= 1-5,529/([1-0.3000]{51,294[1-0.1900]
-29,750 -48,851*0.0100
[1+0.0500]})
= 30.00%

Law-3475:

If both (**I**= Interest Expense= 513 millions USD), (**f**= F/S= Fixed Portion= 58.00%), (**t**= T/B= Tax Rate= 30.00%), (**S**= Sales or Revenues= 51,294 millions USD), (**v**= V/S= Variable Portion= 19.00%), and (**d**= D/A= Dividend Portion or Payout= 30.00%), are known, then it's (**R**= Retained Earnings planned), is:

$$R = [1-t][1-d]\{S-Sv-Sf-I\} = [1-t][1-d]\{S[1-v-f]-I\}$$
$$= [1-0.3000][1-0.3000]\{51,294[1-0.1900-0.5800]-513\}$$
$$= \underline{5,529} \text{ millions USD}$$

Law-3476:

If both (**I**= Interest Expense= 513 millions USD), (**f**= F/S= Fixed Portion= 58.00%), (**t**= T/B= Tax Rate= 30.00%), (**R**= Retained Earnings= 5,529 millions USD), (**v**= V/S= Variable Portion= 19.00%), and (**d**= D/A= Dividend Portion or Payout= 30.00%), are known, then it's (**S**= Sales or Revenues planned), is:

$$S = (I+R/\{[1-t][1-d]\})/[1-v-f]$$
$$= (513+5,529/\{[1-0.3000][1-0.3000]\})/[1-0.1900-0.5800]$$
$$= \underline{51,294} \text{ millions USD}$$

Law-3477:

If both (**I**= Interest Expense= 513 millions USD), (**f**= F/S= Fixed Portion= 58.00%), (**t**= T/B= Tax Rate= 30.00%), (**R**= Retained Earnings= 5,529 millions USD), (**S**= Sales or Revenues= 51,294 millions USD), and (**d**= D/A= Dividend Portion or Payout= 30.00%), are known, then it's (**v**= Variable Portion planned), is:

$$v = 1-f-(I+R/\{[1-t][1-d]\})/S$$
$$= 1-0.5800-(513+5,529/\{[1-0.3000][1-0.3000]\}/51,294$$
$$= 19.00\%$$

Law-3478:

If both (**I**= Interest Expense= 513 millions USD), (**v**= V/S= Variable Portion= 19.00%), (**t**= T/B= Tax Rate= 30.00%), (**R**= Retained Earnings= 5,529 millions USD), (**S**= Sales or Revenues= 51,294 millions USD), and (**d**= D/A= Dividend Portion or Payout= 30.00%), are known, then it's (**f**= Fixed Portion planned), is:

$$f = 1-v-(I+R/\{[1-t][1-d]\})/S$$
$$= 1-0.1900-(513+5,529/\{[1-0.3000][1-0.3000]\}/51,294$$
$$= 58.00\%$$

Law-3479:

If both (v= V/S= Variable Portion= 19.00%), (t= T/B= Tax Rate= 30.00%), (R= Retained Earnings= 5,529 millions USD), (f= F/S= Fixed Portion= 58.00%), (S= Sales or Revenues= 51,294 millions USD), and (d= D/A= Dividend Portion or Payout= 30.00%), are known, then it's (I= Interest Expense planned), is:

$I = S[1-v-f] - R/\{[1-t][1-d]\}$
   $= 51,294[1-0.1900-0.5800] - 5,529$
   $/\{[1-0.3000][1-0.3000]\}$
   $= \underline{513}$ millions USD

Law-3480:

If both (v= V/S= Variable Portion= 19.00%), (I= Interest Expense= 513 millions USD), (R= Retained Earnings= 5,529 millions USD), (f= F/S= Fixed Portion= 58.00%), (S= Sales or Revenues= 51,294 millions USD), and (d= D/A= Dividend Portion or Payout= 30.00%), are known, then it's (t= Tax Rate planned), is:

$t = 1 - R/([1-d]\{S[1-v-f] - I\}$
   $= 1 - 5,529/([1-0.3000]\{51,294[1-0.1900$
   $-0.5800] - 513\}$
   $= \underline{30.00\%}$

Law-3481:
If both (v= V/S= Variable Portion= 19.00%), (I= Interest Expense= 513 millions USD), (R= Retained Earnings= 5,529 millions USD), (f= F/S= Fixed Portion= 58.00%), (S= Sales or Revenues= 51,294 millions USD), and (t= T/B= Tax Rate= 30.00%), are known, then it's (d= Dividend Portion or Payout planned), is:

$$d = 1-R/([1-t]\{S[1-v-f]-I\})$$
$$= 1-5,529/([1-0.3000]\{51,294$$
$$[1-0.1900-0.5800]-513\})$$
$$= 30.00\%$$

Law-3482:
If both (v= V/S= Variable Portion= 19.00%), (i= I/S= Interest Portion= 1.00%), (d= D/A= Dividend Portion or Payout= 30.00%), (f= F/S= Fixed Portion= 58.00%), (S= Sales or Revenues= 51,294 millions USD), and (t= T/B= Tax Rate= 30.00%), are known, then it's (R= Retained Earnings planned), is:

$$R = [1-d][1-t]\{S-Sv-Sf-Si\} = [1-d][1-t]\{S[1-v-f-i]\}$$
$$= [1-0.3000][1-0.3000]\{51,294$$
$$[1-0.1900-0.5800-0.0100]\}$$
$$= 5,529 \text{ millions USD}$$

Finance Construction-4, *Tim Asikin, Steve Asikin, Indra Senihardja*

Law-3483:
If both (**v**= V/S= Variable Portion= 19.00%), (**i**= I/S= Interest Portion= 1.00%), (**d**= D/A= Dividend Portion or Payout= 30.00%), (**f**= F/S=Fixed Portion= 58.00%), (**R**= Retained Earnings= 5,529 millions USD),and (**t**= T/B= Tax Rate= 30.00%), are known, then it's (**$**= Sales or Revenues planned), is:
$$\$= (R/\{[1-t][1-d]\})/[1-v-f-i]$$
$$= (5,529/\{[1-0.3000][1-0.3000]\})/[1-0.1900-0.5800-0.0100]$$
$$= 51,294 \text{ millions USD}$$

Law-3484:
If both (**$**= Sales or Revenues= 51,294 millions USD), (**i**= I/S= Interest Portion= 1.00%), (**d**= D/A= Dividend Portion or Payout= 30.00%), (**f**= F/S= Fixed Portion= 58.00%), (**R**= Retained Earnings= 5,529 millions USD),and (**t**= T/B= Tax Rate= 30.00%), are known, then it's (**v**= Variable Portion planned), is:
$$v= 1-f-i-(R/(\$[1-t][1-d]\})$$
$$= 1-0.5800-0.0100-(5,529/\{51,294[1-0.3000][1-0.3000]\})$$
$$= 19.00\%$$

Law-3485:

If both ($= Sales or Revenues= 51,294 millions USD), (i= I/S= Interest Portion= 1.00%), (d= D/A= Dividend Portion or Payout= 30.00%), (v= V/S= Variable Portion= 19.00%), (R= Retained Earnings= 5,529 millions USD), and (t= T/B= Tax Rate= 30.00%), are known, then it's (f= Fixed Portion planned), is:

$$f = 1-v-i-(R/(\$[1-t][1-d]\}))$$
$$= 1-0.1900-0.0100-(5,529/\{51,294\\ [1-0.3000][1-0.3000]\})$$
$$= \underline{58.00\%}$$

Law-3486:

If both ($= Sales or Revenues= 51,294 millions USD), (f= F/S= Fixed Portion= 58.00%), (d= D/A= Dividend Portion or Payout= 30.00%), (v= V/S= Variable Portion= 19.00%), (R= Retained Earnings= 5,529 millions USD), and (t= T/B= Tax Rate= 30.00%), are known, then it's (i= Interest Portion planned), is:

$$i = 1-v-f-(R/(\$[1-t][1-d]\}))$$
$$= 1-0.1900-0.5800-(5,529/\{51,294\\ [1-0.3000][1-0.3000]\})$$
$$= \underline{1.00\%}$$

Law-3487:

If both ($= Sales or Revenues= 51,294 millions USD), (f= F/S= Fixed Portion= 58.00%), (d= D/A= Dividend Portion or Payout= 30.00%), (v= V/S= Variable Portion= 19.00%), (R= Retained Earnings= 5,529 millions USD),and (i= I/S= Interest Portion= 1.00%), are known, then it's (t= Tax Rate planned), is:

$$t = 1-R/([1-d]\{\$[1-v-f-i]\})$$
$$= 1-5,529/([1-0.3000]\{51,294$$
$$[1-0.1900-0.5800-0.0100]\})$$
$$= 30.00\%$$

Law-3488:

If both ($= Sales or Revenues= 51,294 millions USD), (f= F/SFixed Portion= 58.00%), (t= T/B= Tax Rate= 30.00%), (v= V/S= Variable Portion= 19.00%), (R= Retained Earnings= 5,529 millions USD),and (i= I/S= Interest Portion= 1.00%), are known, then it's (d= Dividend Portion or Payout planned), is:

$$d = 1-R/([1-t]\{\$[1-v-f-i]\})$$
$$= 1-5,529/([1-0.3000]\{51,294$$
$$[1-0.1900-0.5800-0.0100]\})$$
$$= 30.00\%$$

Law-3489:

If both ($'$= Sales of Past Year= 48,851 millions USD), (s= [S/S']-1= Sales Growth= 5.00%), ($= Sales or Revenues= 51,294 millions USD), (f= F/S=Fixed Portion= 58.00%), (t= T/B= Tax Rate= 30.00%), (v= V/S= Variable Portion= 19.00%), (d= D/A= Dividend Portion or Payout= 30.00%), and (i= I/S= Interest Portion= 1.00%), are known, then it's (R= Retained Earnings planned), is:

$$R = [1-t][1-d]\{\$-\$v-\$f-\$'i[1+s]\}$$
$$= [1-t][1-d]\{\$-[1-v-f]-\$'i[1+s]\}$$
$$= [1-0.3000][1-0.3000]\{51,294$$
$$[1-0.1900-0.5800]$$
$$-48,851*0.0100[1+0.0500]\}$$
$$= \underline{5,529} \text{ millions USD}$$

Law-3490:

If both ($'$= Sales of Past Year= 48,851 millions USD), (s= [S/S']-1= Sales Growth= 5.00%), (R= Retained Earnings= 5,529 millions USD), (f= F/S=Fixed Portion= 58.00%), (t= T/B= Tax Rate= 30.00%), (v= V/S= Variable Portion= 19.00%), (d= D/A= Dividend Portion or Payout= 30.00%), and (i= I/S= Interest Portion= 1.00%), are known, then it's ($= Retained Earnings planned), is:

$$\$ = (\$'i[1+s]+R/\{[1-t][1-d]\})/[1-v-f]$$
$$= (48,851*0.0100[1+0.0500]+5,529$$
$$/\{[1-0.3000][1-0.3000]\})$$
$$/[1-0.1900-0.5800]$$
$$= \underline{51,294} \text{ millions USD}$$

Law-3491:

If both ($\$'$= Sales of Past Year= <u>48,851</u> millions USD), (s= [S/S']-1= Sales Growth= <u>5.00</u>%), (R= Retained Earnings= <u>5,529</u> millions USD), (f= F/S= Fixed Portion= <u>58.00</u>%), (t= T/B= Tax Rate= <u>30.00</u>%), ($\$$= Sales or Revenues= <u>51,294</u> millions USD), (d= D/A= Dividend Portion or Payout= 30.00%), and (i= I/S= Interest Portion= <u>1.00</u>%), are known, then it's (v= Variable Portion planned), is:

v= 1-f-($\$'i$[1+$s$]+$R$/{[1-$t$][1-$d$]})/$\$$
  = 1-0.5800-(48,851*0.0100[1+0.0500]
    +5,529/{[1-0.3000][1-0.3000]})
    /51,294
  = <u>19.00</u>%

Law-3492:

If both ($\$'$= Sales of Past Year= <u>48,851</u> millions USD), (s= [S/S']-1= Sales Growth= <u>5.00</u>%), (R= Retained Earnings= <u>5,529</u> millions USD), (v= V/S= Variable Portion= <u>19.00</u>%), (t= T/B= Tax Rate= <u>30.00</u>%), ($\$$= Sales or Revenues= <u>51,294</u> millions USD), (d= D/A= Dividend Portion or Payout= 30.00%), and (i= I/S= Interest Portion= <u>1.00</u>%), are known, then it's (f= Fixed Portion planned), is:

f= 1-v-($\$'i$[1+$s$]+$R$/{[1-$t$][1-$d$]})/$\$$
  = 1-0.1900-(48,851*0.0100[1+0.0500]
    +5,529/{[1-0.3000][1-0.3000]})
    /51,294
  = <u>58.00</u>%

Law-3493:

If both (**f**= F/S= Fixed Portion= 58.00%), (**s**= [S/S']-1= Sales Growth= 5.00%), (**R**= Retained Earnings= 5,529 millions USD), (**v**= V/S= Variable Portion= 19.00%), (**t**= T/B= Tax Rate= 30.00%), (**$**= Sales or Revenues= 51,294 millions USD), (**d**= D/A= Dividend Portion or Payout= 30.00%), and (**i**= I/S= Interest Portion= 1.00%), are known, then it's (**$'**= Sales Past), must be:

$$\$' = (\$[1-v-f]-R/\{[1-t][1-d]\})/\{i[1+s]\}$$
$$= (51{,}294[1-0.1900-0.5800]-5{,}529$$
$$/\{[1-0.3000][1-0.3000]\})$$
$$/\{\,0.0100[1+0.0500]\}$$
$$= 48{,}851 \text{ millions USD}$$

Law-3494:

If both (**f**= F/S= Fixed Portion= 58.00%), (**s**= [S/S']-1= Sales Growth= 5.00%), (**R**= Retained Earnings= 5,529 millions USD), (**v**= V/S= Variable Portion= 19.00%), (**t**= T/B= Tax Rate= 30.00%), (**$**= Sales or Revenues= 51,294 millions USD), (**d**= D/A= Dividend Portion or Payout= 30.00%), and (**$'**= Sales of Past Year= 48,851 millions USD), are known, then it's (**i**= Interest Portion planned), is:

$$i = (\$[1-v-f]-R/\{[1-t][1-d]\})/\{\$'[1+s]\}$$
$$= (51{,}294[1-0.1900-0.5800]-5{,}529$$
$$/\{[1-0.3000][1-0.3000]\})$$
$$/\{\,48{,}851[1+0.0500]\}$$
$$= 1.00\%$$

Law-3495:

If both (**f**= F/S= Fixed Portion= 58.00%), (**i**= I/S= Interest Portion= 1.00%), (**R**= Retained Earnings= 5,529 millions USD), (**v**= V/S= Variable Portion= 19.00%), (**t**= T/B= Tax Rate= 30.00%), (**$**= Sales or Revenues= 51,294 millions USD), (**d**= D/A= Dividend Portion or Payout= 30.00%), and (**$'**= Sales of Past Year= 48,851 millions USD), are known, then it's (**s**= Sales Growth planned), is:

$$s = (\$[1-v-f]-R/\{[1-t][1-d]\})/[\$'i]-1$$
$$= (51{,}294[1-0.1900-0.5800]-5{,}529 \; /\{[1-0.3000][1-0.3000]\}) \; /[48{,}851*0.0100]-1$$
$$= 5.00\%$$

Law-3496:

If both (**f**= F/S= Fixed Portion= 58.00%), (**i**= I/S= Interest Portion= 1.00%), (**R**= Retained Earnings= 5,529 millions USD), (**v**= V/S= Variable Portion= 19.00%), (**s**= [S/S']-1= Sales Growth= 5.00%), (**$**= Sales or Revenues= 51,294 millions USD), (**d**= D/A= Dividend Portion or Payout= 30.00%), and (**$'**= Sales of Past Year= 48,851 millions USD), are known, then it's (**t**= Tax Rate planned), is:

$$t = 1-R/([1-d]\{\$[1-v-f]-\$'i[1+s]\})$$
$$= 1-5{,}529/([1-0.3000]\{51{,}294[1-0.1900 -0.5800]- 48{,}851*0.0100 [1+0.0500]\})$$
$$= 30.00\%$$

Law-3497:

If both (**f**= F/S= Fixed Portion= 58.00%), (**i**= I/S= Interest Portion= 1.00%), (**R**= Retained Earnings= 5,529 millions USD), (**v**= V/S= Variable Portion= 19.00%), (**s**= [S/S']-1= Sales Growth= 5.00%), (**$**= Sales or Revenues= 51,294 millions USD), (**t**= T/B= Tax Rate= 30.00%), and (**$'**= Sales of Past Year= 48,851 millions USD), are known, then it's (**d**= Dividend Portion or Payout planned), is:

$$d = 1-R/([1-t]\{\$[1-v-f]-\$'i[1+s]\})$$
$$= 1-5,529/([1-0.3000]\{51,294[1-0.1900-0.5800]-48,851*0.0100[1+0.0500]\})$$
$$= 30.00\%$$

Law-3498:

If both (**f**= F/S= Fixed Portion= 58.00%), (**I**= Interest Expense= 513 millions USD), (**d**= D/A= Dividend Portion or Payout= 30.00%), (**v**= V/S= Variable Portion= 19.00%), (**s**= [S/S']-1= Sales Growth= 5.00%), (**$**= Sales or Revenues= 51,294 millions USD), (**t**= T/B= Tax Rate= 30.00%), and (**$'**= Sales of Past Year= 48,851 millions USD), are known, then it's (**R**= Retained Earnings planned), is:

$$R = [1-t][1-d]\{\$-\$v-\$'f[1+s]-I\}$$
$$= [1-t][1-d]\{\$[1-v]-\$'f[1+s]-I\}$$
$$= [1-0.3000][1-0.3000]\{51,294[1-0.1900]-48,851*0.5800[1+0.0500]-513\}$$
$$= 5,529 \text{ millions USD}$$

Law-3499:

If both (**f**= F/S= Fixed Portion= 58.00%), (**I**= Interest Expense= 513 millions USD), (**d**= D/A= Dividend Portion or Payout= 30.00%), (**v**= V/S= Variable Portion= 19.00%), (**s**= [S/S']-1= Sales Growth= 5.00%), (**R**= Retained Earnings= 5,529 millions USD), (**t**= T/B= Tax Rate= 30.00%), and (**S'**= Sales of Past Year= 48,851 millions USD), are known, then it's (**S**= Sales or Revenues planned), is:

$S = (S'f[1+s]+I+R/\{[1-t][1-d]\})/[1-v]$
$= (48,851*0.5800[1+0.0500]+513$
$\quad +5,529/\{[1-0.3000][1-0.3000]\})$
$\quad /[1-0.1900]$
$= 51,294$ millions USD

Law-3500:

If both (**f**= F/S= Fixed Portion= 58.00%), (**I**= Interest Expense= 513 millions USD), (**d**= D/A= Dividend Portion or Payout= 30.00%), (**S**= Sales or Revenues= 51,294 millions USD), (**s**= [S/S']-1= Sales Growth= 5.00%), (**R**= Retained Earnings= 5,529 millions USD), (**t**= T/B= Tax Rate= 30.00%), and (**S'**= Sales of Past Year= 48,851 millions USD), are known, then it's (**v**= Variable Portion planned), is:

$v = 1-(S'f[1+s]+I+R/\{[1-t][1-d]\})/S$
$= 1-(48,851*0.5800[1+0.0500]+513$
$\quad +5,529/\{[1-0.3000]$
$\quad\quad [1-0.3000]\})/51,294$
$= 19.00\%$

Law-3501:

If both (f= F/S= Fixed Portion= 58.00%), (I= Interest Expense= 513 millions USD), (d= D/A= Dividend Portion or Payout= 30.00%), (S= Sales or Revenues= 51,294 millions USD), (s= [S/S']-1= Sales Growth= 5.00%), (R= Retained Earnings= 5,529 millions USD), (t= T/B= Tax Rate= 30.00%), and (v= V/S= Variable Portion= 19.00%), are known, then it's (S'= Sales Past), must be:

S' = (S[1-v]-I-R/{[1-t][1-d]})/{f[1+s]}
= (51,294[1-0.1900]-513-5,529
/{[1-0.3000][1-0.3000]})
/ {0.5800[1+0.0500]}
= 48,851 millions USD

Law-3502:

If both (S'= Sales of Past Year= 48,851 millions USD), (I= Interest Expense= 513 millions USD), (d= D/A= Dividend Portion or Payout= 30.00%), (S= Sales or Revenues= 51,294 millions USD), (s= [S/S']-1= Sales Growth= 5.00%), (R= Retained Earnings= 5,529 millions USD), (t= T/B= Tax Rate= 30.00%), and (v= V/S= Variable Portion= 19.00%), are known, then it's (f= Fixed Portion planned), is:

f = (S[1-v]-I-R/{[1-t][1-d]})/{S'[1+s]}
= (51,294[1-0.1900]-513-5,529
/{[1-0.3000][1-0.3000]})
/{48,851[1+0.0500]}
= 58.00%

Law-3503:
If both ($'= Sales of Past Year= 48,851 millions USD), (I= Interest Expense= 513 millions USD), (d= D/A= Dividend Portion or Payout= 30.00%), ($= Sales or Revenues= 51,294 millions USD), (f= F/S= Fixed Portion= 58.00%), (R= Retained Earnings= 5,529 millions USD), (t= T/B= Tax Rate= 30.00%), and (v= V/S= Variable Portion= 19.00%), are known, then it's (s= Sales Growth planned), is:

s= ($[1-v]-I-R/{[1-t][1-d]})/[$'f]-1
= (51,294[1-0.1900]-513-5,529
/{[1-0.3000][1-0.3000]})
/[48,851*0.5800]-1
= 5.00%

Law-3504:
If both ($'= Sales of Past Year= 48,851 millions USD), (s= [S/S']-1= Sales Growth= 5.00%), (d= D/A= Dividend Portion or Payout= 30.00%), ($= Sales or Revenues= 51,294 millions USD), (f= F/S= Fixed Portion= 58.00%), (R= Retained Earnings= 5,529 millions USD), (t= T/B= Tax Rate= 30.00%), and (v= V/S= Variable Portion= 19.00%), are known, then it's (I= Interest Expense planned), is:

I= ($[1-v]-$'f[1+s]-R/{[1-t][1-d]}
= (51,294[1-0.1900]- 48,851*0.5800
[1+0.0500]-5,529/{[1-0.3000]
[1-0.3000]}
= 513 millions USD

Law-3505:

If both ($'= Sales of Past Year= 48,851 millions USD), (s= [S/S']-1= Sales Growth= 5.00%), (d= D/A= Dividend Portion or Payout= 30.00%), ($= Sales or Revenues= 51,294 millions USD), (f= F/S= Fixed Portion= 58.00%), (R= Retained Earnings= 5,529 millions USD), (I= Interest Expense= 513 millions USD), and (v= V/S= Variable Portion= 19.00%), are known, then it's (t= Tax Rate planned), is:

$t= 1-R/([1-d]\{\$[1-v]-\$'f[1+s]-I\}$
$= 1-5,529/([1-0.3000]\{51,294[1-0.1900]$
$-48,851*0.5800[1+0.0500]-513\}$

$= 30.00\%$

Law-3506:

If both ($'= Sales of Past Year= 48,851 millions USD), (s= [S/S']-1= Sales Growth= 5.00%), (t= T/B= Tax Rate= 30.00%), ($= Sales or Revenues= 51,294 millions USD), (f= F/S= Fixed Portion= 58.00%), (R= Retained Earnings= 5,529 millions USD), (I= Interest Expense= 513 millions USD), and (v= V/S= Variable Portion= 19.00%), are known, then it's (d= Dividend Portion or Payout planned), is:

$d= 1-R/([1-t]\{\$[1-v]-\$'f[1+s]-I\}$
$= 1-5,529/([1-0.3000]\{51,294[1-0.1900]$
$-48,851*0.5800[1+0.0500]-513\}$

$= 30.00\%$

Law-3507:

If both ($\$'$= Sales of Past Year= 48,851 millions USD), (s= [S/S']-1= Sales Growth= 5.00%), (t= T/B= Tax Rate= 30.00%), ($\$$= Sales or Revenues= 51,294 millions USD), (f= F/S= Fixed Portion= 58.00%), (d= D/A= Dividend Portion or Payout= 30.00%), (i= I/S= Interest Portion= 1.00%), and (v= V/S= Variable Portion= 19.00%), are known, then it's (R= Retained Earnings planned), is:

R = [1-t][1-d]{$\$$-$\$v$-$\$'f$[1+$s$]-$\$i$}
= [1-t][1-d]{$\$$[1-v-i]-$\$'f$[1+$s$]}
= [1-.03000][1-0.3000]{51,294
[1-0.1900-0.0100]- 48,851
*0.5800[1+0.0500]}
= 5,529 millions USD

Law-3508:

If both ($\$'$= Sales of Past Year= 48,851 millions USD), (s= [S/S']-1= Sales Growth= 5.00%), (t= T/B= Tax Rate= 30.00%), (R= Retained Earnings= 5,529 millions USD),(f= F/S= Fixed Portion= 58.00%), (d= D/A= Dividend Portion or Payout= 30.00%), (i= I/S= Interest Portion= 1.00%), and (v= V/S= Variable Portion= 19.00%), are known, then it's ($\$$= Sales or Revenues planned), is:

$\$$ = ($\$'f$[1+$s$]+$R$/{[1-$t$][1-$d$]})/[1-$v$-$i$]
= (48,851*0.5800[1+0.0500]+5,529
/{[1-0.3000][1-0.3000]})
/[1-0.1900-0.0100]
= 51,294 millions USD

Law-3509:

If both ($\$'$= Sales of Past Year= 48,851 millions USD), (s= [S/S']-1= Sales Growth= 5.00%), (t= T/B= Tax Rate= 30.00%), (R= Retained Earnings= 5,529 millions USD), (f= F/S= Fixed Portion= 58.00%), (d= D/A= Dividend Portion or Payout= 30.00%), (i= I/S= Interest Portion= 1.00%), and ($\$$= Sales or Revenues= 51,294 millions USD), are known, then it's (v= Variable Portion planned), is:

v= 1-i-($\$'f$[1+$s$]+$R$/{[1-$t$][1-$d$]})/$\$$
 = 1-0.0100-(48,851*0.5800[1+0.0500]
   +5,529/{[1-0.3000][1-0.3000]})
   / 51,294
 = 19.00%

Law-3510:

If both ($\$'$= Sales of Past Year= 48,851 millions USD), (s= [S/S']-1= Sales Growth= 5.00%), (t= T/B= Tax Rate= 30.00%), (R= Retained Earnings= 5,529 millions USD), (f= F/S= Fixed Portion= 58.00%), (d= D/A= Dividend Portion or Payout= 30.00%), (v= V/S= Variable Portion= 19.00%), and ($\$$= Sales or Revenues= 51,294 millions USD), are known, then it's (i= Interest Portion planned), is:

i= 1-v-($\$'f$[1+$s$]+$R$/{[1-$t$][1-$d$]})/$\$$
 = 1-0.1900-(48,851*0.5800[1+0.0500]
   +5,529/{[1-0.3000][1-0.3000]})
   /51,294
 = 1.00%

Law-3511:

If both (i= I/S= Interest Portion= 1.00%), (s= [S/S']-1= Sales Growth= 5.00%), (t= T/B= Tax Rate= 30.00%), (R= Retained Earnings= 5,529 millions USD),(f= F/S= Fixed Portion= 58.00%), (d= D/A= Dividend Portion or Payout= 30.00%), (v= V/S= Variable Portion= 19.00%), and (S= Sales or Revenues= 51,294 millions USD), are known, then it's (S'= Sales Past), must be:

$$S' = (S[1-v-i]-R/\{[1-t][1-d]\})/\{f[1+s]\}$$
$$= (51,294\,[1-0.1900-0.0100]-5,529$$
$$/\{[1-0.3000][1-0.3000]\})$$
$$/\{0.5800[1+0.0500]\}$$
$$= 48,851 \text{ millions USD}$$

Law-3512:

If both (i= I/S= Interest Portion= 1.00%), (s= [S/S']-1= Sales Growth= 5.00%), (t= T/B= Tax Rate= 30.00%), (R= Retained Earnings= 5,529 millions USD), (S'= Sales of Past Year= 48,851 millions USD), (d= D/A= Dividend Portion or Payout= 30.00%), (v= V/S= Variable Portion= 19.00%), and (S= Sales or Revenues= 51,294 millions USD), are known, then it's (f= Fixed Portion planned), is:

$$f = (S[1-v-i]-R/\{[1-t][1-d]\})/\{S'[1+s]\}$$
$$= (51,294\,[1-0.1900-0.0100]-5,529$$
$$/\{[1-0.3000][1-0.3000]\})$$
$$/\{48,851[1+0.0500]\}$$
$$= 58.00\%$$

Law-3513:
If both (**i**= I/S= Interest Portion= 1.00%), (**f**= F/S= Fixed Portion= 58.00%), (**t**= T/B= Tax Rate= 30.00%), (**R**= Retained Earnings= 5,529 millions USD), (**$'**= Sales of Past Year= 48,851 millions USD), (**d**= D/A= Dividend Portion or Payout= 30.00%), (**v**= V/S= Variable Portion= 19.00%), and (**$**= Sales or Revenues= 51,294 millions USD), are known, then it's (**s**= Sales Growth planned), is:

$$s = (\$[1-v-i]-R/\{[1-t][1-d]\})/[\$'f]-1$$
$$= (51,294\,[1-0.1900-0.0100]-5,529$$
$$/\{[1-0.3000][1-0.3000]\})$$
$$/[48,851*0.5800]-1$$
$$= 5.00\%$$

Law-3514:
If both (**i**= I/S= Interest Portion= 1.00%), (**f**= F/S= Fixed Portion= 58.00%), (**s**= [S/S']-1= Sales Growth= 5.00%), (**R**= Retained Earnings= 5,529 millions USD), (**$'**= Sales of Past Year= 48,851 millions USD), (**d**= D/A= Dividend Portion or Payout= 30.00%), (**v**= V/S= Variable Portion= 19.00%), and (**$**= Sales or Revenues= 51,294 millions USD), are known, then it's (**t**= Tax Rate planned), is:

$$t = 1-R/([1-d][\$[1-v-i]-\$'f[1+s]\})$$
$$= 1-5,529/([1-0.3000]\{51,294[1-0.1900$$
$$-0.0100]- 48,851*0.5800$$
$$[1+0.0500]\}$$
$$= 30.00\%$$

Law-3515:

If both (**i**= I/S= Interest Portion= 1.00%), (**f**= F/S= Fixed Portion= 58.00%), (**s**= [S/S']-1= Sales Growth= 5.00%), (**R**= Retained Earnings= 5,529 millions USD), (**$'**= Sales of Past Year= 48,851 millions USD), (**t**= T/B= Tax Rate= 30.00%), (**v**= V/S= Variable Portion= 19.00%), and (**$**= Sales or Revenues= 51,294 millions USD), are known, then it's (**d**= Dividend Portion or Payout planned), is:

$$d = 1 - R/([1-t][\$[1-v-i] - \$'f[1+s]])$$
$$= 1 - 5,529/([1-0.3000]\{51,294[1-0.1900 - 0.0100] - 48,851 \cdot 0.5800 [1+0.0500]\})$$
$$= 30.00\%$$

Law-3516:

If both (**i**= I/S= Interest Portion= 1.00%), (**f**= F/S= Fixed Portion= 58.00%), (**s**= [S/S']-1= Sales Growth= 5.00%), (**d**= D/A= Dividend Portion or Payout= 30.00%), (**$'**= Sales of Past Year= 48,851 millions USD), (**t**= T/B= Tax Rate= 30.00%), (**v**= V/S= Variable Portion= 19.00%), and (**$**= Sales or Revenues= 51,294 millions USD), are known, then it's (**R**= Retained Earnings planned), is:

$$R = [1-t][1-d]\{\$ - \$v - \$'f[1+s] - \$'i[1+s]\}$$
$$= [1-t][1-d]\{\$[1-v] - \$'[1+s][f+i]\}$$
$$= [1-0.3000][1-0.3000]\{51,294 [1-0.1900] - 48,851[1+0.0500] [0.5800+0.0100]\}$$
$$= 5,529 \text{ millions USD}$$

Finance Construction-4, *Tim Asikin, Steve Asikin, Indra Senihardja*

Law-3517:
If both (**i**= I/S= Interest Portion= 1.00%), (**f**= F/S= Fixed Portion= 58.00%), (**s**= [S/S']-1= Sales Growth= 5.00%), (**d**= D/A= Dividend Portion or Payout= 30.00%), (**$'**= Sales of Past Year= 48,851 millions USD), (**t**= T/B= Tax Rate= 30.00%), (**v**= V/S= Variable Portion= 19.00%), and (**R**= Retained Earnings= 5,529 millions USD), are known, then it's (**$**= Sales or Revenues planned), is:

$$\$ = (\$'[1+s][f+i]+R/\{[1-t][1-d]\})/[1-v]$$
$$= (48,851[1+0.0500][0.5800+0.0100]$$
$$+5,529/\{[1-0.3000][1-0.3000]\})$$
$$/[1-0.1900]$$
$$= 51,294 \text{ millions USD}$$

Law-3518:
If both (**i**= I/S= Interest Portion= 1.00%), (**f**= F/S= Fixed Portion= 58.00%), (**s**= [S/S']-1= Sales Growth= 5.00%), (**d**= D/A= Dividend Portion or Payout= 30.00%), (**$'**= Sales of Past Year= 48,851 millions USD), (**t**= T/B= Tax Rate= 30.00%), (**$**= Sales or Revenues= 51,294 millions USD), and (**R**= Retained Earnings= 5,529 millions USD), are known, then it's (**v**= Variable Portion planned), is:

$$v = 1-(\$'[1+s][f+i]+R/\{[1-t][1-d]\})/\$$$
$$= 1-(48,851[1+0.0500][0.5800+0.0100]$$
$$+5,529/\{[1-0.3000][1-0.3000]\})$$
$$/51,294$$
$$= 19.00\%$$

Law-3519:
If both (**i**= I/S= Interest Portion= 1.00%), (**f**= F/S= Fixed Portion= 58.00%), (**s**= [S/S']-1= Sales Growth= 5.00%), (**d**= D/A= Dividend Portion or Payout= 30.00%), (**v**= V/S= Variable Portion= 19.00%), (**t**= T/B= Tax Rate= 30.00%), (**$**= Sales or Revenues= 51,294 millions USD), and (**R**= Retained Earnings= 5,529 millions USD), are known, then it's (**$'**= Sales Past), must be:

$$\$' = (\$[1-v]-R/\{[1-t][1-d]\})/\{[1+s][f+i]\}$$
$$= (51,294[1-0.1900]-5,529/\{[1-0.3000][1-0.3000]\})/\{[1+0.0500][0.5800+0.0100]\}$$
$$= 48,851 \text{ millions USD}$$

Law-3520:
If both (**i**= I/S= Interest Portion= 1.00%), (**f**= F/S= Fixed Portion= 58.00%), (**$'**= Sales of Past Year= 48,851 millions USD), (**d**= D/A= Dividend Portion or Payout= 30.00%), (**v**= V/S= Variable Portion= 19.00%), (**t**= T/B= Tax Rate= 30.00%), (**$**= Sales or Revenues= 51,294 millions USD), and (**R**= Retained Earnings= 5,529 millions USD), are known, then it's (**s**= Sales Growth planned), is:

$$s = (\$[1-v]-R/\{[1-t][1-d]\})/\{\$'[f+i]\}-1$$
$$= (51,294[1-0.1900]-5,529/\{[1-0.3000][1-0.3000]\})/\{48,851[0.5800+0.0100]\}-1$$
$$= 5.00\%$$

Law-3521:

If both (s= [S/S']-1= Sales Growth= 5.00%), ($\$'$= Sales of Past Year= 48,851 millions USD), (i= I/S= Interest Portion= 1.00%), (d= D/A= Dividend Portion or Payout= 30.00%), (v= V/S= Variable Portion= 19.00%), (t= T/B= Tax Rate= 30.00%), ($\$$= Sales or Revenues= 51,294 millions USD),and (R= Retained Earnings= 5,529 millions USD), are known, then it's (f= Fixed Portion planned), is:

f= ($\$[1-v]-R/\{[1-t][1-d]\})/\{\$'[1+s]\}-i$
= (51,294[1-0.1900]-5,529/{[1-0.3000]
[1-0.3000]})/{ 48,851
[1+0.0500]}-0.0100
= 58.00%

Law-3522:

If both (s= [S/S']-1= Sales Growth= 5.00%), ($\$'$= Sales of Past Year= 48,851 millions USD), (f= F/S= Fixed Portion= 58.00%), (d= D/A= Dividend Portion or Payout= 30.00%), (v= V/S= Variable Portion= 19.00%), (t= T/B= Tax Rate= 30.00%), ($\$$= Sales or Revenues= 51,294 millions USD),and (R= Retained Earnings= 5,529 millions USD), are known, then it's (i= Interest Portion planned), is:

i= ($\$[1-v]-R/\{[1-t][1-d]\})/\{\$'[1+s]\}-f$
= (51,294[1-0.1900]-5,529/{[1-0.3000]
[1-0.3000]})/{ 48,851
[1+0.0500]}-0.5800
= 1.00%

Law-3523:
If both (s= [S/S']-1= Sales Growth= 5.00%), (s'= Sales of Past Year= 48,851 millions USD), (f= F/S= Fixed Portion= 58.00%), (d= D/A= Dividend Portion or Payout= 30.00%), (v= V/S= Variable Portion= 19.00%), (i= I/S= Interest Portion= 1.00%), (S= Sales or Revenues= 51,294 millions USD), and (R= Retained Earnings= 5,529 millions USD), are known, then it's (t= Tax Rate planned), is:

t= 1-R/([1-d]{S[1-v]-S'[1+s][f+i]})
  = 1-5,529/([1-0.3000]{51,294[1-0.1900]
    -48,851[1+0.0500][0.5800
    +0.0100]})
  = 30.00%

Law-3524:
If both (s= [S/S']-1= Sales Growth= 5.00%), (s'= Sales of Past Year= 48,851 millions USD), (f= F/S= Fixed Portion= 58.00%), (t= T/B= Tax Rate= 30.00%), (v= V/S= Variable Portion= 19.00%), (i= I/S= Interest Portion= 1.00%), (S= Sales or Revenues= 51,294 millions USD), and (R= Retained Earnings= 5,529 millions USD), are known, then it's (d= Dividend Portion or Payout planned), is:

d= 1-R/([1-t]{S[1-v]-S'[1+s][f+i]})
  = 1-5,529/([1-0.3000]{51,294[1-0.1900]
    -48,851[1+0.0500][0.5800
    +0.0100]})
  = 30.00%

Law-3525:

If both (s= [S/S']-1= Sales Growth= 5.00%), (S'= Sales of Past Year= 48,851 millions USD), (F= Fixed Cost= 29,750 millions USD), (I= Interest Expense= 513 millions USD), (t= T/B= Tax Rate= 30.00%), (v= V/S= Variable Portion= 19.00%), (S= Sales or Revenues= 51,294 millions USD),and (d= D/A= Dividend Portion or Payout= 30.00%), are known, then it's (R= Retained Earnings planned), is:

$$R = [1-t][1-d]\{S-F-I-S'v[1+s]\}$$
$$= [1-0.3000][1-0.3000]\{51,294-29,750-513-48,851*0.1900[1+0.0500]\}$$
$$= 5,529 \text{ millions USD}$$

Law-3526:

If both (s= [S/S']-1= Sales Growth= 5.00%), (S'= Sales of Past Year= 48,851 millions USD), (F= Fixed Cost= 29,750 millions USD), (I= Interest Expense= 513 millions USD), (t= T/B= Tax Rate= 30.00%), (v= V/S= Variable Portion= 19.00%), (R= Retained Earnings= 5,529 millions USD), and (d= D/A= Dividend Portion or Payout= 30.00%), are known, then it's (S= Sales or Revenues planned), is:

$$S = S'v[1+s]+F+I+R/\{[1-t][1-d]\}$$
$$= 48,851*0.1900[1+0.0500]+29,750+513+5,529/\{[1-0.3000][1-0.3000]\}$$
$$= 51,294 \text{ millions USD}$$

Law-3527:
If both (s= [S/S']-1= Sales Growth= 5.00%), (S= Sales or Revenues= 51,294 millions USD), (F= Fixed Cost= 29,750 millions USD), (I= Interest Expense= 513 millions USD), (t= T/B= Tax Rate= 30.00%), (v= V/S= Variable Portion= 19.00%), (R= Retained Earnings= 5,529 millions USD), and (d= D/A= Dividend Portion or Payout= 30.00%), are known, then it's (S'= Sales Past), must be:

S'= (S-F-I-R/{[1-t][1-d]})/{v[1+s]}
= (51,294-29,750 -513-5,529
/{[1-0.3000][1-0.3000]})
/{0.1900[1+0.0500]}
= 48,851 millions USD

Law-3528:
If both (s= [S/S']-1= Sales Growth= 5.00%), (S= Sales or Revenues= 51,294 millions USD), (F= Fixed Cost= 29,750 millions USD), (I= Interest Expense= 513 millions USD), (t= T/B= Tax Rate= 30.00%), (S'= Sales of Past Year= 48,851 millions USD), (R= Retained Earnings= 5,529 millions USD), and (d= D/A= Dividend Portion or Payout= 30.00%), are known, then it's (v= Variable Portion planned), is:

v= (S-F-I-R/{[1-t][1-d]})/{S'[1+s]}
= (51,294-29,750 -513-5,529
/{[1-0.3000][1-0.3000]})
/{48,851[1+0.0500]}
= 19.00%

Law-3529:

If both (**v**= V/S= Variable Portion= 19.00%), (**$**= Sales or Revenues= 51,294 millions USD), (**F**= Fixed Cost= 29,750 millions USD), (**I**= Interest Expense= 513 millions USD), (**t**= T/B= Tax Rate= 30.00%), (**$'**= Sales of Past Year= 48,851 millions USD), (**R**= Retained Earnings= 5,529 millions USD), and (**d**= D/A= Dividend Portion or Payout= 30.00%), are known, then it's (**s**= Sales Growth planned), is:

$s = (\$-F-I-R/\{[1-t][1-d]\})/[\$'v]-1$
$= (51,294-29,750-513-5,529$
$/\{[1-0.3000][1-0.3000]\})$
$/[48,851*0.1900]-1$
$= 5.00\%$

Law-3530:

If both (**v**= V/S= Variable Portion= 19.00%), (**$**= Sales or Revenues= 51,294 millions USD), (**s**= [S/S']-1= Sales Growth= 5.00%), (**I**= Interest Expense= 513 millions USD), (**t**= T/B= Tax Rate= 30.00%), (**$'**= Sales of Past Year=48,851 millions USD), (**R**= Retained Earnings= 5,529 millions USD), and (**d**= D/A= Dividend Portion or Payout= 30.00%), are known, then it's (**F**= Fixed Cost planned), is:

$F = \$-I-\$'v[1+s]-R/\{[1-t][1-d]\}$
$= 51,294-513-48,851*0.1900$
$[1+0.0500]-5,529$
$/\{[1-0.3000][1-0.3000]\}$
$= 29,750$ millions USD

Law-3531:

If both (v= V/S= Variable Portion= 19.00%), (S= Sales or Revenues= 51,294 millions USD), (s= [S/S']-1= Sales Growth= 5.00%), (F= Fixed Cost= 29,750 millions USD), (t= T/B= Tax Rate= 30.00%), (S'= Sales of Past Year=48,851 millions USD), (R= Retained Earnings= 5,529 millions USD), and (d= D/A= Dividend Portion or Payout= 30.00%), are known, then it's (I= Interest Expense planned), is:

$I = S-F-S'v[1+s]-R/\{[1-t][1-d]\}$
  = 51,294-29,750-48,851*0.1900
    [1 +0.0500]-5,529
    /{[1-0.3000][1-0.3000]}
  = 513 millions USD

Law-3532:

If both (v= V/S= Variable Portion= 19.00%), (S= Sales or Revenues= 51,294 millions USD), (s= [S/S']-1= Sales Growth= 5.00%), (F= Fixed Cost= 29,750 millions USD), (I= Interest Expense= 513 millions USD), (S'= Sales of Past Year=48,851 millions USD), (R= Retained Earnings= 5,529 millions USD), and (d= D/A= Dividend Portion or Payout= 30.00%), are known, then it's (t= Tax Rate planned), is:

$t = 1 - R/([1-d]\{S-F-S'v[1+s]-I\})$
  = 1-5,529/([1-0.3000]{51,294-29,750
    -48,851*0.1900[1 +0.0500]
    -513})
  = 30.00%

Law-3533:

If both (v= V/S= Variable Portion= 19.00%), (S= Sales or Revenues= 51,294 millions USD), (s= [S/S']-1= Sales Growth= 5.00%), (F= Fixed Cost= 29,750 millions USD), (I= Interest Expense= 513 millions USD), (S'= Sales of Past Year=48,851 millions USD), (R= Retained Earnings= 5,529 millions USD), and (t= T/B= Tax Rate= 30.00%), are known, then it's (d= Dividend Portion or Payout planned), is:

$$d = 1 - R/([1-t]\{S-F-S'v[1+s]-I\})$$
$$= 1 - 5{,}529/([1-0.3000]\{51{,}294-29{,}750$$
$$-48{,}851*0.1900[1+0.0500]$$
$$-513\})$$
$$= 30.00\%$$

Law-3534:

If both (v= V/S= Variable Portion= 19.00%), (S= Sales or Revenues= 51,294 millions USD), (s= [S/S']-1= Sales Growth= 5.00%), (F= Fixed Cost= 29,750 millions USD), (i= I/S= Interest Portion= 1.00%), (S'= Sales of Past Year=48,851 millions USD), (d= D/A= Dividend Portion or Payout= 30.00%), and (t= T/B= Tax Rate= 30.00%), are known, then it's (R= Retained Earnings planned), is:

$$R = [1-t][1-d]\{S-S'v[1+s]-F-Si\}$$
$$= [1-0.3000][1-0.3000]\{51{,}294$$
$$-48{,}851*0.1900[1+0.0500]$$
$$-29{,}750-51{,}294*0.0100\}$$
$$= 5{,}529 \text{ millions USD}$$

Law-3535:
If both (**v**= V/S= Variable Portion= 19.00%), (**R**= Retained Earnings= 5,529 millions USD), (**s**= [S/S']-1= Sales Growth= 5.00%), (**F**= Fixed Cost= 29,750 millions USD), (**i**= I/S= Interest Portion= 1.00%), (**$'**= Sales of Past Year=48,851 millions USD), (**d**= D/A= Dividend Portion or Payout= 30.00%), and (**t**= T/B= Tax Rate= 30.00%), are known, then it's (**$**= Sales or Revenues planned), is:

$$\$= (F+\$'v[1+s]+R/\{[1-t][1-d]\})/[1-i]$$
$$= (29{,}750+48{,}851*0.1900[1+0.0500]$$
$$+5{,}529/\{[1-0.3000][1-0.3000]\})$$
$$/[1-0.0100]$$
$$= \underline{51{,}294} \text{ millions USD}$$

Law-3536:
If both (**v**= V/S= Variable Portion= 19.00%), (**R**= Retained Earnings= 5,529 millions USD), (**s**= [S/S']-1= Sales Growth= 5.00%), (**F**= Fixed Cost= 29,750 millions USD), (**$**= Sales or Revenues= 51,294 millions USD), (**$'**= Sales of Past Year=48,851 millions USD), (**d**= D/A= Dividend Portion or Payout= 30.00%), and (**t**= T/B= Tax Rate= 30.00%), are known, then it's (**i**= Interest Portion planned), is:

$$i= 1-(F+\$'v[1+s]+R/\{[1-t][1-d]\})/\$$$
$$= 1-(29{,}750+48{,}851*0.1900[1+0.0500]$$
$$+5{,}529/\{[1-0.3000][1-0.3000]\})$$
$$/51{,}294$$
$$= \underline{1.00\%}$$

Law-3537:
If both (v= V/S= Variable Portion= 19.00%), (R= Retained Earnings= 5,529 millions USD), (s= [S/S']-1= Sales Growth= 5.00%), (F= Fixed Cost= 29,750 millions USD), ($\$$= Sales or Revenues= 51,294 millions USD), (i= I/S= Interest Portion= 1.00%), (d= D/A= Dividend Portion or Payout= 30.00%), and (t= T/B= Tax Rate= 30.00%), are known, then it's ($\$'$= Sales Past), is:

$\$' = (\$[1-i]-F-R/\{[1-t][1-d]\})/\{v[1+s]\}$
$= (51,294[1-0.0100]-29,750-5,529$
$/\{[1-0.3000][1-0.3000]\})$
$/\{0.1900[1+0.0500]\}$
$= 48,851$ millions USD

Law-3538:
If both ($\$'$= Sales of Past Year= 48,851 millions USD), (R= Retained Earnings= 5,529 millions USD), (s= [S/S']-1= Sales Growth= 5.00%), (F= Fixed Cost= 29,750 millions USD), ($\$$= Sales or Revenues= 51,294 millions USD), (i= I/S= Interest Portion= 1.00%), (d= D/A= Dividend Portion or Payout= 30.00%), and (t= T/B= Tax Rate= 30.00%), are known, then it's (v= Variable Portion planned), is:

$v = (\$[1-i]-F-R/\{[1-t][1-d]\})/\{\$'[1+s]\}$
$= (51,294[1-0.0100]-29,750-5,529$
$/\{[1-0.3000][1-0.3000]\})$
$/\{48,851[1+0.0500]\}$
$= 19.00\%$

Law-3539:
If both ($'= Sales of Past Year= 48,851 millions USD), (R= Retained Earnings= 5,529 millions USD), (v= V/S= Variable Portion= 19.00%), (F= Fixed Cost= 29,750 millions USD), ($= Sales or Revenues= 51,294 millions USD), (i= I/S= Interest Portion= 1.00%), (d= D/A= Dividend Portion or Payout= 30.00%), and (t= T/B= Tax Rate= 30.00%), are known, then it's (s= Sales Growth planned), is:
$$s= (\$[1-i]-F-R/\{[1-t][1-d]\})/[\$'v]-1$$
$$= (51,294[1-0.0100]-29,750-5,529$$
$$/\{[1-0.3000][1-0.3000]\})$$
$$/[48,851*0.1900]-1$$
$$= 5.00\%$$

Law-3540:
If both ($'= Sales of Past Year= 48,851 millions USD), (R= Retained Earnings= 5,529 millions USD), (v= V/S= Variable Portion= 19.00%), (s= [S/S']-1= Sales Growth= 5.00%), ($= Sales or Revenues= 51,294 millions USD), (i= I/S= Interest Portion= 1.00%), (d= D/A= Dividend Portion or Payout= 30.00%), and (t= T/B= Tax Rate= 30.00%), are known, then it's (F= Fixed Cost planned), is:
$$F= \$[1-i]-\$'v[1+s]-R/\{[1-t][1-d]\}$$
$$= 51,294[1-0.0100]-48,851*0.1900$$
$$[1+0.0500]-5,529/\{[1-0.3000]$$
$$[1-0.3000]\}$$
$$= 29,750 \text{ millions USD}$$

Law-3541:

If both ($'$= Sales of Past Year= 48,851 millions USD), (R= Retained Earnings= 5,529 millions USD), (v= V/S= Variable Portion= 19.00%), (s= [S/S']-1= Sales Growth= 5.00%), ($\$$= Sales or Revenues= 51,294 millions USD), (i= I/S= Interest Portion= 1.00%), (d= D/A=Dividend Portion or Payout= 30.00%), and (F= Fixed Cost= 29,750 millions USD), are known, then it's (t= Tax Rate planned), is:

t= 1-R/([1-d]{$\$$[1-i]-$\$'v$[1+$s$]-$F$})
= 1-5,529/{[1-0.3000]{51,294[1-0.0100]
-48,851*0.1900 [1+0.0500]
-29,750
= 30.00%

Law-3542:

If both ($'$= Sales of Past Year= 48,851 millions USD), (R= Retained Earnings= 5,529 millions USD), (v= V/S= Variable Portion= 19.00%), (s= [S/S']-1= Sales Growth= 5.00%), ($\$$= Sales or Revenues= 51,294 millions USD), (i= I/S= Interest Portion= 1.00%), (t= T/B= Tax Rate= 30.00%), and (F= Fixed Cost= 29,750 millions USD), are known, then it's (d= Dividend Portion or Payout planned), is:

d= 1-R/([1-t]{$\$$[1-i]-$\$'v$[1+$s$]-$F$})
= 1-5,529/{[1-0.3000]{51,294[1-0.0100]
-48,851*0.1900 [1+0.0500]
-29,750
= 30.00%

Law-3543:
If both ($'$= Sales of Past Year= 48,851 millions USD), (d= D/A= Dividend Portion or Payout= 30.00%), (v= V/S= Variable Portion= 19.00%), (s= [S/S']-1= Sales Growth= 5.00%), ($\$$= Sales or Revenues= 51,294 millions USD), (i= I/S= Interest Portion= 1.00%), (t= T/B= Tax Rate= 30.00%), and (F= Fixed Cost= 29,750 millions USD), are known, then it's (R= Retained Earnings planned), is:

$R = [1-t][1-d]\{\$-\$'v[1+s]-F-\$'i[1+s]\}$
$= [1-t][1-d]\{\$-F-\$'v[1+s][v+i]\}$
$= [1-0.3000][1-0.3000]\{51,294$
$-29,750-48,851[1+0.0500]$
$[0.1900+0.0100]\}$
$= \underline{5,529}$ millions USD

Law-3544:
If both ($'$= Sales of Past Year= 48,851 millions USD), (d= D/A= Dividend Portion or Payout= 30.00%), (v= V/S= Variable Portion= 19.00%), (s= [S/S']-1= Sales Growth= 5.00%), (R= Retained Earnings= 5,529 millions USD), (i= I/S= Interest Portion= 1.00%), (t= T/B= Tax Rate= 30.00%), and (F= Fixed Cost= 29,750 millions USD), are known, then it's ($\$$= Sales or Revenues planned), is:

$\$ = F+\$'[1+s][v+i]+R/[1-t][1-d]\}$
$= 29,750+48,851[1+0.0500][0.1900$
$+0.0100] +5,529/\{[1-0.3000]$
$[1-0.3000]\}$
$= \underline{51,294}$ millions USD

Law-3545:
If both ($= Sales or Revenues= 51,294 millions USD), (d= D/A= Dividend Portion or Payout= 30.00%), (v= V/S= Variable Portion= 19.00%), (s= [S/S']-1= Sales Growth= 5.00%), (R= Retained Earnings= 5,529 millions USD), (i= I/S= Interest Portion= 1.00%), (t= T/B= Tax Rate= 30.00%), and (F= Fixed Cost= 29,750 millions USD), are known, then it's ($'= Sales Past), must be:

$$\$' = (\$ - F - R/[1-t][1-d])/\{[1+s][v+i]\}$$
$$= (51,294 - 29,750] - 5,529/\{[1-0.3000][1-0.3000]\})/\{[1+0.0500][0.1900+0.0100]\}$$
$$= \underline{48,851} \text{ millions USD}$$

Law-3546:
If both ($= Sales or Revenues= 51,294 millions USD), (d= D/A= Dividend Portion or Payout= 30.00%), ($'= Sales of Past Year= 48,851 millions USD), (s= [S/S']-1= Sales Growth= 5.00%), (R= Retained Earnings= 5,529 millions USD), (i= I/S= Interest Portion= 1.00%), (t= T/B= Tax Rate= 30.00%), and (F= Fixed Cost= 29,750 millions USD), are known, then it's (v= Variable Portion planned), is:

$$v = (\$ - F - R/[1-t][1-d])/\{\$'[1+s]\} - i$$
$$= (51,294 - 29,750] - 5,529/\{[1-0.3000][1-0.3000]\})/\{48,851[1+0.0500]\} - 0.0100$$
$$= \underline{19.00}\%$$

Law-3547:
If both ($= Sales or Revenues= 51,294 millions USD), (d= D/A= Dividend Portion or Payout= 30.00%), ($'= Sales of Past Year= 48,851 millions USD), (s= [S/S']-1= Sales Growth= 5.00%), (R= Retained Earnings= 5,529 millions USD), (v= V/S= Variable Portion= 19.00%), (t= T/B= Tax Rate= 30.00%), and (F= Fixed Cost= 29,750 millions USD), are known, then it's (i= Interest Portion planned), is:

$$i = (\$-F-R/[1-t][1-d]\})/\{\$'[1+s]\}-v$$
$$= (51,294-29,750]-5,529/\{[1-0.3000][1-0.3000]\})/\{48,851[1+0.0500]\}-0.1900$$
$$= \underline{1.00\%}$$

Law-3548:
If both ($= Sales or Revenues= 51,294 millions USD), (d= D/A= Dividend Portion or Payout= 30.00%), ($'= Sales of Past Year= 48,851 millions USD), (i= I/S= Interest Portion= 1.00%), (R= Retained Earnings= 5,529 millions USD), (v= V/S= Variable Portion= 19.00%), (t= T/B= Tax Rate= 30.00%), and (F= Fixed Cost= 29,750 millions USD), are known, then it's (s= Sales Growth planned), is:

$$s = (\$-F-R/[1-t][1-d]\})/\{\$'[v+i]\}-1$$
$$= (51,294-29,750]-5,529/\{[1-0.3000][1-0.3000]\})/\{48,851[0.1900+0.0100]\}-1$$
$$= \underline{5.00\%}$$

Law-3549:

If both ($= Sales or Revenues= 51,294 millions USD), (**d**= D/A= Dividend Portion or Payout= 30.00%), ($'= Sales of Past Year= 48,851 millions USD), (**i**= I/S= Interest Portion= 1.00%), (**R**= Retained Earnings= 5,529 millions USD), (**v**= V/S= Variable Portion= 19.00%), (**t**= T/B= Tax Rate= 30.00%), and (**s**= [S/S']-1= Sales Growth= 5.00%), are known, then it's (**F**= Fixed Cost planned), is:

$$F = \$-\$'[v+i][1+s]-R/\{[1-t][1-d]\}$$
$$= 51{,}294-48{,}851\,[0.1900+0.0100]$$
$$[1+0.0500]\,-5{,}529/\{[1-0.3000]$$
$$[1-0.3000]\}$$
$$= 29{,}750 \text{ millions USD}$$

Law-3550:

If both ($= Sales or Revenues= 51,294 millions USD), (**d**= D/A= Dividend Portion or Payout= 30.00%), ($'= Sales of Past Year= 48,851 millions USD), (**i**= I/S= Interest Portion= 1.00%), (**R**= Retained Earnings= 5,529 millions USD), (**v**= V/S= Variable Portion= 19.00%), (**F**= Fixed Cost= 29,750 millions USD), and (**s**= [S/S']-1= Sales Growth= 5.00%), are known, then it's (**t**= Tax Rate planned), is:

$$t = 1-R/([1-d]\{\$-\$'[v+i][1+s]-F\}$$
$$= 1-5{,}529/([1-0.3000]\{51{,}294-48{,}851$$
$$[0.1900+0.0100][1+0.0500]$$
$$-29{,}750\})$$
$$= 30.00\%$$

Law-3551:

If both ($= Sales or Revenues= 5l,294 millions USD), (t= T/B= Tax Rate= 30.00%), ($'= Sales of Past Year= 48,851 millions USD), (i= I/S= Interest Portion= 1.00%), (R= Retained Earnings= 5,529 millions USD), (v= V/S= Variable Portion= 19.00%), (F= Fixed Cost= 29,750 millions USD), and (s= [S/S']-1= Sales Growth= 5.00%), are known, then it's (d= Dividend Portion or Payout planned), is:

$$d = 1-R/([1-t]\{\$-\$'[v+i][1+s]-F\}$$
$$= 1-5,529/([1-0.3000]\{51,294-48,851$$
$$[0.1900+0.0100][1+0.0500]$$
$$-29,750\})$$
$$= 30.00\%$$

Law-3552:

If both ($= Sales or Revenues= 51,294 millions USD), (t= T/B= Tax Rate= 30.00%), ($'= Sales of Past Year= 48,851 millions USD), (d= D/A= Dividend Portion or Payout= 30.00%), (I= Interest Expense= 513 millions USD), (v= V/S= Variable Portion= 19.00%), (f= F/S=Fixed Portion= 58.00%), and (s= [S/S']-1= Sales Growth= 5.00%), are known, then it's (R= Retained Earnings planned), is:

$$R = [1-t][1-d]\{\$-\$'v[1+s]-\$'f[1+s]-I\}$$
$$= [1-t][1-d]\{\$[1-f]-\$'v[1+s]-I\}$$
$$= [1-0.3000][1-0.3000]\{51,294$$
$$[1-0.5800]-48,851*0.1900$$
$$[1+0.0500] -513\}$$
$$= 5,529 \text{ millions USD}$$

Law-3553:
If both (**R**= Retained Earnings= 5,529 millions USD), (**t**= T/B= Tax Rate= 30.00%), (**$'**= Sales of Past Year= 48,851 millions USD), (**d**= D/A= Dividend Portion or Payout= 30.00%), (**I**= Interest Expense= 513 millions USD), (**v**= V/S= Variable Portion= 19.00%), (**f**= F/S= Fixed Portion= 58.00%), and (**s**= [S/S']-1= Sales Growth= 5.00%), are known, then it's (**$**= Sales or Revenues planned), is:

$$\$= (\$'v[1+s]+I+R/\{[1-t][1-d]\})/[1-f]$$
$$= (48,851*0.1900\ [1+0.0500]+ 513$$
$$+5,529/\{[1-0.3000][1-0.3000]\})$$
$$/[1-0.5800]$$
$$= \underline{51,294}\ \text{millions USD}$$

Law-3554:
If both (**R**= Retained Earnings= 5,529 millions USD), (**t**= T/B= Tax Rate= 30.00%), (**$'**= Sales of Past Year= 48,851 millions USD), (**d**= D/A= Dividend Portion or Payout= 30.00%), (**I**= Interest Expense= 513 millions USD), (**v**= V/S= Variable Portion= 19.00%), (**$**= Sales or Revenues= 51,294 millions USD), and (**s**= [S/S']-1= Sales Growth= 5.00%), are known, then it's (**f**= Fixed Portion planned), is:

$$f= 1-(\$'v[1+s]+I+R/\{[1-t][1-d]\})/\$$$
$$= 1-(48,851*0.1900\ [1+0.0500]+ 513$$
$$+5,529/\{[1-0.3000][1-0.3000]\})$$
$$/51,294$$
$$= \underline{58.00\%}$$

Law-3555:

If both (**R**= Retained Earnings= 5,529 millions USD), (**t**= T/B= Tax Rate= 30.00%), (**f**= F/S= Fixed Portion= 58.00%), (**d**= D/A= Dividend Portion or Payout= 30.00%), (**l**= Interest Expense= 513 millions USD), (**v**= V/S= Variable Portion= 19.00%), (**$**= Sales or Revenues= 51,294 millions USD), and (**s**= [S/S']-1= Sales Growth= 5.00%), are known, then it's (**$'**= Sales Past), must be:

$$\$' = (\$[1-f]-I-R/\{[1-t][1-d]\})/\{v[1+s]\}$$
$$= (51,294[1-0.5800]- 513-5,529$$
$$/\{[1-0.3000][1-0.3000]\})$$
$$/\{0.1900 [1+0.0500]\}$$
$$= \underline{4,8851} \text{ millions USD}$$

Law-3556:

If both (**R**= Retained Earnings= 5,529 millions USD), (**t**= T/B= Tax Rate= 30.00%), (**f**= F/S=Fixed Portion= 58.00%), (**d**= D/A= Dividend Portion or Payout= 30.00%), (**l**= Interest Expense= 513 millions USD), (**$'**= Sales of Past Year= 48,851 millions USD), (**$**= Sales or Revenues= 51,294 millions USD), and (**s**= [S/S']-1= Sales Growth= 5.00%), are known, then it's (**v**= Variable Portion planned), is:

$$v = (\$[1-f]-I-R/\{[1-t][1-d]\})/\{\$'[1+s]\}$$
$$= (51,294[1-0.5800]- 513-5,529$$
$$/\{[1-0.3000][1-0.3000]\})$$
$$/\{48,851 [1+0.0500]\}$$
$$= \underline{19.00\%}$$

Finance Construction-4, *Tim Asikin, Steve Asikin, Indra Senihardja*

Law-3557:
If both (**R**= Retained Earnings= 5,529 millions USD), (**t**= T/B= Tax Rate= 30.00%), (**f**= F/S= Fixed Portion= 58.00%), (**d**= D/A= Dividend Portion or Payout= 30.00%), (**I**= Interest Expense= 513 millions USD), (**$'**= Sales of Past Year= 48,851 millions USD), (**$**= Sales or Revenues= 51,294 millions USD), and (**v**= V/S= Variable Portion= 19.00%), are known, then it's (**s**= Sales Growth planned), is:

$$s = (\$[1-f]-I-R/\{[1-t][1-d]\})/[\$'v]-1$$
$$= (51,294[1-0.5800]- 513-5,529$$
$$/\{[1-0.3000][1-0.3000]\})$$
$$/\{48,851*0.1900\}-1$$
$$= 5.00\%$$

Law-3558:
If both (**R**= Retained Earnings= 5,529 millions USD), (**t**= T/B= Tax Rate= 30.00%), (**f**= F/S= Fixed Portion= 58.00%), (**d**= D/A= Dividend Portion or Payout= 30.00%), (**s**= [S/S']-1= Sales Growth= 5.00%), (**$'**= Sales of Past Year= 48,851 millions USD), (**$**= Sales or Revenues= 51,294 millions USD), and (**v**= V/S= Variable Portion= 19.00%), are known, then it's (**I**= Interest Expense planned), is:

$$I = \$[1-f]-\$'v[1+s]-R/\{[1-t][1-d]\}$$
$$= 51,294[1-0.5800]- 48,851*0.1900$$
$$[1+0.0500]-5,529/\{[1-0.3000]$$
$$[1-0.3000]\}$$
$$= 513 \text{ millions USD}$$

Law-3559:

If both (**R**= Retained Earnings= 5,529 millions USD), (**I**= Interest Expense= 513 millions USD), (**f**= F/S= Fixed Portion= 58.00%), (**d**= D/A= Dividend Portion or Payout= 30.00%), (**s**= [S/S']-1= Sales Growth= 5.00%), (**S'**= Sales of Past Year= 48,851 millions USD), (**S**= Sales or Revenues= 51,294 millions USD), and (**v**= V/S= Variable Portion= 19.00%), are known, then it's (**t**= Tax Rate planned), is:

$$\mathbf{t}= 1-\mathbf{R}/([1-\mathbf{d}]\{\mathbf{S}[1-\mathbf{f}]-\mathbf{S'v}[1+\mathbf{s}]-\mathbf{R}/\{[1-\mathbf{t}][1-\mathbf{d}]\}$$
$$= 1-5,529/([1-0.3000]\{51,294$$
$$[1-0.5800]- 48,851*0.1900$$
$$[1+0.0500]-513\})$$
$$= \underline{30.00\%}$$

Law-3560:

If both (**R**= Retained Earnings= 5,529 millions USD), (**I**= Interest Expense= 513 millions USD), (**f**= F/S= Fixed Portion= 58.00%), (**t**= T/B= Tax Rate= 30.00%), (**s**= [S/S']-1= Sales Growth= 5.00%), (**S'**= Sales of Past Year= 48,851 millions USD), (**S**= Sales or Revenues= 51,294 millions USD), and (**v**= V/S= Variable Portion= 19.00%), are known, then it's (**d**= Dividend Portion or Payout planned), is:

$$\mathbf{d}= 1-\mathbf{R}/([1-\mathbf{t}]\{\mathbf{S}[1-\mathbf{f}]-\mathbf{S'v}[1+\mathbf{s}]-\mathbf{R}/\{[1-\mathbf{t}][1-\mathbf{d}]\}$$
$$= 1-5,529/([1-0.3000]\{51,294$$
$$[1-0.5800]- 48,851*0.1900$$
$$[1+0.0500]-513\})$$
$$= \underline{30.00\%}$$

Law-3561:
If both (**d**= D/A= Dividend Portion or Payout= 30.00%), (**i**= I/S= Interest Portion= 1.00%), (**f**= F/S= Fixed Portion= 58.00%), (**t**= T/B= Tax Rate= 30.00%), (**s**= [S/S']-1= Sales Growth= 5.00%), (**$'**= Sales of Past Year= 48,851 millions USD), (**$**= Sales or Revenues= 51,294 millions USD), and (**v**= V/S= Variable Portion= 19.00%), are known, then it's (**R**= Retained Earnings planned), is:

$$R = [1-t][1-d]\{\$-\$'v[1+s]\}-\$f-\$i\}$$
$$= [1-t][1-d]\{\$[1-f-i]-\$'v[1+s]\}$$
$$= [1-0.3000][1-0.3000]\{51,294$$
$$[1-0.5800-0.0100]- 48,851$$
$$*0.1900[1+0.0500]\}$$
$$= 5,529 \text{ millions USD}$$

Law-3562:
If both (**d**= D/A= Dividend Portion or Payout= 30.00%), (**i**= I/S= Interest Portion= 1.00%), (**f**= F/S= Fixed Portion= 58.00%), (**t**= T/B= Tax Rate= 30.00%), (**s**= [S/S']-1= Sales Growth= 5.00%), (**$'**= Sales of Past Year= 48,851 millions USD), (**R**= Retained Earnings= 5,529 millions USD), and (**v**= V/S= Variable Portion= 19.00%), are known, then it's (**$**= Sales or Revenues planned), is:

$$\$ = (R\{[1-t][1-d]\}+\$'v[1+s])/[1-f-i]$$
$$= (5,529/\{[1-0.3000][1-0.3000]\}$$
$$+ 48,851*0.1900[1+0.0500])$$
$$/[1-0.5800-0.0100]$$
$$= 51,294 \text{ millions USD}$$

Law-3563:
If both (**d**= D/A= Dividend Portion or Payout= 30.00%), (**i**= I/S= Interest Portion= 1.00%), (**$**= Sales or Revenues= 51,294 millions USD), (**t**= T/B= Tax Rate= 30.00%), (**s**= [S/S']-1= Sales Growth= 5.00%), (**$'**= Sales of Past Year= 48,851 millions USD), (**R**= Retained Earnings= 5,529 millions USD), and (**v**= V/S= Variable Portion= 19.00%), are known, then it's (**f**= Fixed Portion planned), is:

$$f = 1-i-(R\{[1-t][1-d]\}+\$'v[1+s])/\$$$
$$= 1-0.0100-(5,529/\{[1-0.3000]$$
$$[1-0.3000]\}+ 48,851*0.1900$$
$$[1+0.0500])/ 51,294$$
$$= \underline{58.00\%}$$

Law-3564:
If both (**d**= D/A= Dividend Portion or Payout= 30.00%), (**f**= F/S= Fixed Portion= 58.00%), (**$**= Sales or Revenues= 51,294 millions USD), (**t**= T/B= Tax Rate= 30.00%), (**s**= [S/S']-1= Sales Growth= 5.00%), (**$'**= Sales of Past Year= 48,851 millions USD), (**R**= Retained Earnings= 5,529 millions USD), and (**v**= V/S= Variable Portion= 19.00%), are known, then it's (**i**= Interest Portion planned), is:

$$i = 1-f-(R\{[1-t][1-d]\}+\$'v[1+s])/\$$$
$$= 1-0.5800-(5,529/\{[1-0.3000]$$
$$[1-0.3000]\}+ 48,851*0.1900$$
$$[1+0.0500])/ 51,294$$
$$= \underline{1.00\%}$$

Finance Construction-4, *Tim Asikin, Steve Asikin, Indra Senihardja*

Law-3565:
If both (**d**= D/A= Dividend Portion or Payout= 30.00%), (**f**= F/S= Fixed Portion= 58.00%), (**$**= Sales or Revenues= 51,294 millions USD), (**t**= T/B= Tax Rate= 30.00%), (**s**= [S/S']-1= Sales Growth= 5.00%), (**i**= I/S= Interest Portion= 1.00%), (**R**= Retained Earnings= 5,529 millions USD), and (**v**= V/S= Variable Portion= 19.00%), are known, then it's (**$'**= Sales Past), must be:

$'= ($[1-**f**-**i**]-**R**{[1-**t**][1-**d**]})/{**v**[1+**s**]}
  = (51,294[1-0.5800-0.0100]-5,529
    /{[1-0.3000][1-0.3000]})
    /{0.1900[1+0.0500]}
  = 48,851 millions USD

Law-3566:
If both (**d**= D/A= Dividend Portion or Payout= 30.00%), (**f**= F/S= Fixed Portion= 58.00%), (**$**= Sales or Revenues= 51,294 millions USD), (**t**= T/B= Tax Rate= 30.00%), (**s**= [S/S']-1= Sales Growth= 5.00%), (**i**= I/S= Interest Portion= 1.00%), (**R**= Retained Earnings= 5,529 millions USD), and (**$'**= Sales of Past Year= 48,851 millions USD), are known, then it's (**v**= Variable Portion planned), is:

v= ($[1-**f**-**i**]-**R**{[1-**t**][1-**d**]})/{$'[1+**s**]}
  = (51,294[1-0.5800-0.0100]-5,529
    /{[1-0.3000][1-0.3000]})
    /{48,851[1+0.0500]}
  = 19.00%

Law-3567:

If both (**d**= D/A= Dividend Portion or Payout= 30.00%), (**f**= F/S= Fixed Portion= 58.00%), (**$**= Sales or Revenues= 51,294 millions USD), (**t**= T/B= Tax Rate= 30.00%), (**v**= V/S= Variable Portion= 19.00%), (**i**= I/S= Interest Portion= 1.00%), (**R**= Retained Earnings= 5,529 millions USD), and (**$'**= Sales of Past Year= 48,851 millions USD), are known, then it's (**s**= Sales Growth planned), is:

$$s = (\$[1\text{-}f\text{-}i]\text{-}R\{[1\text{-}t][1\text{-}d]\})/[\$'v]\text{-}1$$
$$= (51{,}294[1\text{-}0.5800\text{-}0.0100]\text{-}5{,}529$$
$$/\{[1\text{-}0.3000][1\text{-}0.3000]\})$$
$$/[48{,}851*0.1900]\text{-}1$$
$$= 5.00\%$$

Law-3568:

If both (**d**= D/A= Dividend Portion or Payout= 30.00%), (**f**= F/S= Fixed Portion= 58.00%), (**$**= Sales or Revenues= 51,294 millions USD), (**s**= [S/S']-1= Sales Growth= 5.00%), (**v**= V/S= Variable Portion= 19.00%), (**i**= I/S= Interest Portion= 1.00%), (**R**= Retained Earnings= 5,529 millions USD), and (**$'**= Sales of Past Year= 48,851 millions USD), are known, then it's (**t**= Tax Rate planned), is:

$$t = 1\text{-}R/([1\text{-}d]\{\$[1\text{-}f\text{-}i]\text{-}\$'v[1+s]\})$$
$$= (51{,}294[1\text{-}0.5800\text{-}0.0100]$$
$$- 48{,}851*0.1900[1+0.0500]\})$$
$$= 30.00\%$$

Law-3569:

If both (**t**= T/B= Tax Rate= 30.00%), (**f**= F/S= Fixed Portion= 58.00%), (**$**= Sales or Revenues= 51,294 millions USD), (**s**= [S/S']-1= Sales Growth= 5.00%), (**v**= V/S= Variable Portion= 19.00%), (**i**= I/S= Interest Portion= 1.00%), (**R**= Retained Earnings= 5,529 millions USD), and (**$'**= Sales of Past Year= 48,851 millions USD), are known, then it's (**d**= Dividend Portion or Payout planned), is:

$$\mathbf{d}= 1-\mathbf{R}/([1-\mathbf{t}]\{\$[1-\mathbf{f}-\mathbf{i}]-\$'\mathbf{v}[1+\mathbf{s}]\})$$
$$= (51,294[1-0.5800-0.0100]-48,851$$
$$*0.1900[1+0.0500]\})$$
$$= 30.00\%$$

Law-3570:

If both (**t**= T/B= Tax Rate= 30.00%), (**f**= F/S= Fixed Portion= 58.00%), (**$**= Sales or Revenues= 51,294 millions USD), (**s**= [S/S']-1= Sales Growth= 5.00%), (**v**= V/S= Variable Portion= 19.00%), (**i**= I/S= Interest Portion= 1.00%), (**d**= D/A= Dividend Portion or Payout= 30.00%), and (**$'**= Sales of Past Year= 48,851 millions USD), are known, then it's (**R**= Retained Earnings planned), is:

$$\mathbf{R}= [1-\mathbf{t}][1-\mathbf{d}]\{\$-\$'\mathbf{v}[1+\mathbf{s}]-\$\mathbf{f}-\$'\mathbf{i}[1+\mathbf{s}]\}$$
$$= [1-\mathbf{t}][1-\mathbf{d}]\{\$[1-\mathbf{f}]-\$'[1+\mathbf{s}][\mathbf{v}+\mathbf{i}]\}$$
$$= 51,294[1-0.5800]- 48,851 [1+0.0500]$$
$$[0.1900+0.0100]\}$$
$$= 5,529 \text{ millions USD}$$

Law-3571:

If both (**t**= T/B= Tax Rate= 30.00%), (**f**= F/S= Fixed Portion= 58.00%), (**R**= Retained Earnings= 5,529 millions USD), (**s**= [S/S']-1= Sales Growth= 5.00%), (**v**= V/S= Variable Portion= 19.00%), (**i**= I/S= Interest Portion= 1.00%), (**d**= D/A= Dividend Portion or Payout= 30.00%), and (**S'**= Sales of Past Year= 48,851 millions USD), are known, then it's (**S**= Sales or Revenues planned), is:

$$S = (S'[1+s][v+i] + R/\{[1-t][1-d]\})/[1-f]$$
$$= (48,851\ [1+0.0500][0.1900+0.0100]$$
$$+5,529/\{1-0.3000][1-0.3000]\})$$
$$/[1-0.5800]$$
$$= \underline{51,294}\ \text{millions USD}$$

Law-3572:

If both (**t**= T/B= Tax Rate= 30.00%), (**R**= Retained Earnings= 5,529 millions USD), (**s**= [S/S']-1= Sales Growth= 5.00%), (**S**= Sales or Revenues= 51,294 millions USD), (**v**= V/S= Variable Portion= 19.00%), (**i**= I/S= Interest Portion= 1.00%), (**d**= D/A= Dividend Portion or Payout= 30.00%), and (**S'**= Sales of Past Year= 48,851 millions USD), are known, then it's (**f**= Fix Portion planned), is:

$$f = 1 - (S'[1+s][v+i] + R/\{[1-t][1-d]\})/S$$
$$= 1 - (48,851\ [1+0.0500][0.1900+0.0100]$$
$$+5,529/\{1-0.3000][1-0.3000]\})$$
$$/51,294$$
$$= \underline{58.00\%}$$

Law-3573:
If both (**t**= T/B= Tax Rate= 30.00%), (**R**= Retained Earnings= 5,529 millions USD), (**s**= [S/S']-1= Sales Growth= 5.00%), (**$**= Sales or Revenues= 51,294 millions USD), (**v**= V/S= Variable Portion= 19.00%), (**i**= I/S= Interest Portion= 1.00%), (**d**= D/A= Dividend Portion or Payout= 30.00%), and (**f**= F/S= Fixed Portion= 58.00%), are known, then it's (**$'**= Sales Past), must be:

$'= ($[1-f]1-R/{[1-t][1-d]})/{[1+s][v+i]}
= (51,294[1-0.5800]-5,529/{1-0.3000]
   [1-0.3000]})/{[1+0.0500]
   [0.1900+0.0100]}
= 48,851 millions USD

Law-3574:
If both (**t**= T/B= Tax Rate= 30.00%), (**R**= Retained Earnings= 5,529 millions USD), (**$'**= Sales of Past Year= 48,851 millions USD), (**$**= Sales or Revenues= 51,294 millions USD), (**v**= V/S= Variable Portion= 19.00%), (**i**= I/S= Interest Portion= 1.00%), (**d**= D/A= Dividend Portion or Payout= 30.00%), and (**f**= F/S= Fixed Portion= 58.00%), are known, then it's (**s**= Sales Growth planned), is:

s= ($[1-f]1-R/{[1-t][1-d]})/{$'[v+i]}-1
= (51,294[1-0.5800]-5,529/{1-0.3000]
   [1-0.3000]})/{[48,851[0.1900
   +0.0100]}-1
= 5.00%

Law-3575:

If both (**t**= T/B= Tax Rate= 30.00%), (**R**= Retained Earnings= 5,529 millions USD), (**$'**= Sales of Past Year= 48,851 millions USD), (**$**= Sales or Revenues= 51,294 millions USD), (**s**= [S/S']-1= Sales Growth= 5.00%), (**i**= I/S= Interest Portion= 1.00%), (**d**= D/A= Dividend Portion or Payout= 30.00%), and (**f**= F/S= Fixed Portion= 58.00%), are known, then it's (**s**= Sales Growth planned), is:

$$v = (\$[1-f]1-R/\{[1-t][1-d]\})/\{\$'[1+s]\}-i$$
$$= (51{,}294[1-0.5800]-5{,}529/\{1-0.3000\}$$
$$[1-0.3000]\})/\{[48{,}851$$
$$[1+0.0500]\}-0.0100$$
$$= 19.00\%$$

Law-3576:

If both (**t**= T/B= Tax Rate= 30.00%), (**R**= Retained Earnings= 5,529 millions USD), (**$'**= Sales of Past Year= 48,851 millions USD), (**$**= Sales or Revenues= 51,294 millions USD), (**s**= [S/S']-1= Sales Growth= 5.00%), (**v**= V/S= Variable Portion= 19.00%), (**d**= D/A= Dividend Portion or Payout= 30.00%), and (**f**= F/S= Fixed Portion= 58.00%), are known, then it's (**i**= Interest Portion planned), is:

$$i = (\$[1-f]-R/\{[1-t][1-d]\})/\{\$'[1+s]\}-v$$
$$= (51{,}294[1-0.5800]-5{,}529/\{1-0.3000\}$$
$$[1-0.3000]\})/\{48{,}851$$
$$[1+0.0500]\}-0.1900$$
$$= 1.00\%$$

Law-3577:
If both (**i**= I/S= Interest Portion= 1.00%), (**R**= Retained Earnings= 5,529 millions USD), (**$'**= Sales of Past Year= 48,851 millions USD), (**$**= Sales or Revenues= 51,294 millions USD), (**s**= [S/S']-1= Sales Growth= 5.00%), (**v**= V/S= Variable Portion= 19.00%), (**d**= D/A= Dividend Portion or Payout= 30.00%), and (**f**= F/S= Fixed Portion= 58.00%), are known, then it's (**t**= Tax Rate planned), is:

$$\mathbf{t} = 1 - \mathbf{R}/([1-\mathbf{d}]\{\$[1-\mathbf{f}] - \$'[1+\mathbf{s}][\mathbf{v}+\mathbf{i}]\})$$
$$= 1 - 5{,}529/([1-0.3000]\{51{,}294$$
$$[1-0.5800] - 48{,}851[1+0.0500]$$
$$[0.1900+0.0100]\})$$
$$= 30.00\%$$

Law-3578:
If both (**i**= I/S= Interest Portion= 1.00%), (**R**= Retained Earnings= 5,529 millions USD), (**$'**= Sales of Past Year= 48,851 millions USD), (**$**= Sales or Revenues= 51,294 millions USD), (**s**= [S/S']-1= Sales Growth= 5.00%), (**v**= V/S= Variable Portion= 19.00%), (**t**= T/B= Tax Rate= 30.00%), and (**f**= F/S= Fixed Portion= 58.00%), are known, then it's (**d**= Dividend Portion or Payout planned), is:

$$\mathbf{d} = 1 - \mathbf{R}/([1-\mathbf{t}]\{\$[1-\mathbf{f}] - \$'[1+\mathbf{s}][\mathbf{v}+\mathbf{i}]\})$$
$$= 1 - 5{,}529/([1-0.3000]\{51{,}294$$
$$[1-0.5800] - 48{,}851[1+0.0500]$$
$$[0.1900+0.0100]\})$$
$$= 30.00\%$$

Law-3579:

If both (**I**= Interest Expense= 513 millions USD), (**d**= D/A= Dividend Portion or Payout= 30.00%), (**$'**= Sales of Past Year= 48,851 millions USD), (**$**= Sales or Revenues= 51,294 millions USD), (**s**= [S/S']-1= Sales Growth= 5.00%), (**v**= V/S= Variable Portion= 19.00%), (**t**= T/B= Tax Rate= 30.00%), and (**f**= F/S= Fixed Portion= 58.00%), are known, then it's (**R**= Retained Earnings planned), is:

$$R = [1-t][1-d]\{\$-\$'v[1+s]-\$'f[1+s]-I\}$$
$$= [1-t][1-d]\{\$-\$[1+s][v+f]-I\}$$
$$= [1-0.3000][1-0.3000]\{51,294$$
$$-48,851[1+0.0500][0.1900$$
$$+0.5800]-513\}$$
$$= \underline{5,529} \text{ millions USD}$$

Law-3580:

If both (**I**= Interest Expense= 513 millions USD), (**d**= D/A= Dividend Portion or Payout= 30.00%), (**$'**= Sales of Past Year= 48,851 millions USD), (**R**= Retained Earnings= 5,529 millions USD), (**s**= [S/S']-1= Sales Growth= 5.00%), (**v**= V/S= Variable Portion= 19.00%), (**t**= T/B= Tax Rate= 30.00%), and (**f**= F/S=Fixed Portion= 58.00%), are known, then it's (**$**= Sales or Revenues planned), is:

$$\$ = \$'[1+s][v+f]+I+R/\{[1-t][1-d]\}$$
$$= 48,851[1+0.0500][0.1900+0.5800]$$
$$+513+5,529/\{[1-0.3000]$$
$$[1-0.3000]\}$$
$$= \underline{51,294} \text{ millions USD}$$

Law-3581:
If both (**I**= Interest Expense= 513 millions USD), (**d**= D/A= Dividend Portion or Payout= 30.00%), (**$**= Sales or Revenues= 51,294 millions USD), (**R**= Retained Earnings= 5,529 millions USD), (**s**= [S/S']-1= Sales Growth= 5.00%), (**v**= V/S= Variable Portion= 19.00%), (**t**= T/B= Tax Rate= 30.00%), and (**f**= F/S= Fixed Portion= 58.00%), are known, then it's (**$'**= Sales Past), must be:

$\$' = (\$-I-R/\{[1-t][1-d]\})/\{[1+s][v+f]\}$
$= (51,294-513-5,529/\{[1-0.3000]$
$[1-0.3000]\})/\{[1+0.0500]$
$[0.1900+0.5800]\}$
$= 48,851$ millions USD

Law-3582:
If both (**I**= Interest Expense= 513 millions USD), (**d**= D/A= Dividend Portion or Payout= 30.00%), (**$**= Sales or Revenues= 51,294 millions USD), (**R**= Retained Earnings= 5,529 millions USD), (**$'**= Sales of Past Year= 48,851 millions USD), (**v**= V/S= Variable Portion= 19.00%), (**t**= T/B= Tax Rate= 30.00%), and (**f**= F/S= Fixed Portion= 58.00%), are known, then it's (**s**= Sales Growth planned), is:

$s = (\$-I-R/\{[1-t][1-d]\})/\{\$'[v+f]\}-1$
$= (51,294-513-5,529/\{[1-0.3000]$
$[1-0.3000]\})/\{48,851[0.1900$
$+0.5800]\}-1$
$= 5.00\%$

Law-3583:
If both (**l**= Interest Expense= 513 millions USD), (**d**= D/A= Dividend Portion or Payout= 30.00%), (**$**= Sales or Revenues= 51,294 millions USD), (**R**= Retained Earnings= 5,529 millions USD), (**$'**= Sales of Past Year= 48,851 millions USD), (**s**= [S/S']-1= Sales Growth= 5.00%), (**t**= T/B= Tax Rate= 30.00%), and (**f**= F/S=Fixed Portion= 58.00%), are known, then it's (**v**= Variable Portion planned), is:
$$v = (\$\text{-}l\text{-}R/\{[1\text{-}t][1\text{-}d]\})/\{\$'[1+s]\}\text{-}f$$
$$= (51{,}294\text{-}513\text{-}5{,}529/\{[1\text{-}0.3000]$$
$$[1\text{-}0.3000]\})/\{48{,}851$$
$$[1+0.0500]\})\text{-}0.5800$$
$$= 19.00\%$$

Law-3584:
If both (**l**= Interest Expense= 513 millions USD), (**d**= D/A= Dividend Portion or Payout= 30.00%), (**$**= Sales or Revenues= 51,294 millions USD), (**R**= Retained Earnings= 5,529 millions USD), (**$'**= Sales of Past Year= 48,851 millions USD), (**s**= [S/S']-1= Sales Growth= 5.00%), (**t**= T/B= Tax Rate= 30.00%), and (**v**= V/S= Variable Portion= 19.00%), are known, then it's (**f**= Fixed Portion planned), is:
$$f = (\$\text{-}l\text{-}R/\{[1\text{-}t][1\text{-}d]\})/\{\$'[1+s]\}\text{-}v$$
$$= (51{,}294\text{-}513\text{-}5{,}529/\{[1\text{-}0.3000]$$
$$[1\text{-}0.3000]\})/\{48{,}851$$
$$[1+0.0500]\})\text{-}0.1900$$
$$= 58.00\%$$

Law-3585:

If both (f= F/S= Fixed Portion= 58.00%), (d= D/A= Dividend Portion or Payout= 30.00%), ($\$$= Sales or Revenues= 51,294 millions USD), (R= Retained Earnings= 5,529 millions USD), ($\$'$= Sales of Past Year= 48,851 millions USD), (s= [S/S']-1= Sales Growth= 5.00%), (t= T/B= Tax Rate= 30.00%), and (v= V/S= Variable Portion= 19.00%), are known, then it's (I= Interest Expense planned), is:

$$I = \$-\$'[1+s][v+f]-R/\{[1-t][1-d]\}$$
$$= 51{,}294-48{,}851[1+0.0500]$$
$$[0.1900+0.5800]-5{,}529$$
$$/\{[1-0.3000][1-0.3000]\}$$
$$= 513 \text{ millions USD}$$

Law-3586:

If both (f= F/S= Fixed Portion= 58.00%), (d= D/A= Dividend Portion or Payout= 30.00%), ($\$$= Sales or Revenues= 51,294 millions USD), (R= Retained Earnings= 5,529 millions USD), ($\$'$= Sales of Past Year= 48,851 millions USD), (s= [S/S']-1= Sales Growth= 5.00%), (I= Interest Expense= 513 millions USD), and (v= V/S= Variable Portion= 19.00%), are known, then it's (t= Tax Rate planned), is:

$$t = 1 - R/([1-d]\{\$-\$'[1+s][v+f]-I\})$$
$$= 1-5{,}529/([1-0.3000]\{51{,}294-48{,}851$$
$$[1+0.0500][0.1900+0.5800]$$
$$-513\})$$
$$= 30.00\%$$

Law-3587:

If both (**f**= F/S= Fixed Portion= 58.00%), (**t**= T/B= Tax Rate= 30.00%), (**$**= Sales or Revenues= 51,294 millions USD), (**R**= Retained Earnings= 5,529 millions USD), (**$'**= Sales of Past Year= 48,851 millions USD), (**s**= [S/S']-1= Sales Growth= 5.00%), (**I**= Interest Expense= 513 millions USD), and (**v**= V/S= Variable Portion= 19.00%), are known, then it's (**d**= Dividend Portion or Payout planned), is:

$$\mathbf{d} = 1 - \mathbf{R}/([1-\mathbf{t}]\{\$-\$'[1+\mathbf{s}][\mathbf{v}+\mathbf{f}]-\mathbf{I}\})$$
$$= 1 - 5{,}529/([1-0.3000]\{51{,}294$$
$$-48{,}851[1+0.0500]$$
$$[0.1900+0.5800]-513\})$$
$$= 30.00\%$$

Law-3588:

If both (**f**= F/S= Fixed Portion= 58.00%), (**t**= T/B= Tax Rate= 30.00%), (**$**= Sales or Revenues= 51,294 millions USD), (**d**= D/A= Dividend Portion or Payout= 30.00%), (**$'**= Sales of Past Year= 48,851 millions USD), (**s**= [S/S']-1= Sales Growth= 5.00%), (**i**= I/S= Interest Portion= 1.00%), and (**v**= V/S= Variable Portion= 19.00%), are known, then it's (**R**= Retained Earnings planned), is:

$$\mathbf{R} = [1-\mathbf{t}][1-\mathbf{d}]\{\$-\$'\mathbf{v}[1+\mathbf{s}]-\$'\mathbf{f}[1+\mathbf{s}]-\$\mathbf{i}\}$$
$$= [1-\mathbf{t}][1-\mathbf{d}]\{\$[1-\mathbf{i}]-\$'[1+\mathbf{s}][\mathbf{v}+\mathbf{f}]\}$$
$$= [1-0.3000][1-0.3000]\{51{,}294$$
$$[1-0.0100]-48{,}851[1+0.0500]$$
$$[0.1900+0.5800]\}$$
$$= 5{,}529 \text{ millions USD}$$

Law-3589:
If both (**f**= F/S= Fixed Portion= 58.00%), (**t**= T/B= Tax Rate= 30.00%), (**R**= Retained Earnings= 5,529 millions USD), (**d**= D/A= Dividend Portion or Payout= 30.00%), (**$'**= Sales of Past Year= 48,851 millions USD), (**s**= [S/S']-1= Sales Growth= 5.00%), (**i**= I/S= Interest Portion= 1.00%), and (**v**= V/S= Variable Portion= 19.00%), are known, then it's (**$**= Sales or Revenues planned), is:

$= ($'[1+s][v+f]+R/{[1-t][1-d]})/[1-i]
 = (48,851[1+0.0500][0.1900+0.5800]
   +5,529/{[1-0.3000][1-0.3000]})
   /[1-0.0100]
 = 51,294 millions USD

Law-3590:
If both (**f**= F/S= Fixed Portion= 58.00%), (**t**= T/B= Tax Rate= 30.00%), (**R**= Retained Earnings= 5,529 millions USD), (**d**= D/A= Dividend Portion or Payout= 30.00%), (**$'**= Sales of Past Year= 48,851 millions USD), (**s**= [S/S']-1= Sales Growth= 5.00%), (**$**= Sales or Revenues= 51,294 millions USD), and (**v**= V/S= Variable Portion= 19.00%), are known, then it's (**i**= Interest Portion planned), is:

i= 1-($'[1+s][v+f]+R/{[1-t][1-d]})/$
 = 1-(48,851[1+0.0500][0.1900+0.5800]
   +5,529/{[1-0.3000][1-0.3000]})
   /51,294
 = 1.00%

Law-3591:
If both (**f**= F/S= Fixed Portion= 58.00%), (**t**= T/B= Tax Rate= 30.00%), (**R**= Retained Earnings= 5,529 millions USD), (**d**= D/A= Dividend Portion or Payout= 30.00%), (**i**= I/S= Interest Portion= 1.00%), (**s**= [S/S']-1= Sales Growth= 5.00%), (**$**= Sales or Revenues= 51,294 millions USD), and (**v**= V/S= Variable Portion= 19.00%), are known, then it's (**$'**= Sales Past), must be:

$'= ($[1-**i**]-[**R**/{[1-**t**][1-**d**]})/{[1+**s**][**v**+**f**]}
 = (51,294[1-0.0100]-5,529
   /{[1-0.3000][1-0.3000]})
   /{[1+0.0500][0.1900+0.5800]}
 = 48,851 millions USD

Law-3592:
If both (**f**= F/S= Fixed Portion= 58.00%), (**t**= T/B= Tax Rate= 30.00%), (**R**= Retained Earnings= 5,529 millions USD), (**d**= D/A= Dividend Portion or Payout= 30.00%), (**i**= I/S= Interest Portion= 1.00%), (**$'**= Sales of Past Year= 48,851 millions USD), (**$**= Sales or Revenues= 51,294 millions USD), and (**v**= V/S= Variable Portion= 19.00%), are known, then it's (**s**= Sales Growth planned), is:

s= ($[1-**i**]-[**R**/{[1-**t**][1-**d**]})/{$'[**v**+**f**]}-1
 = (51,294[1-0.0100]- 5,529
   /{[1-0.3000][1-0.3000]})
   /{48,851[0.1900+0.5800]}-1
 = 5.00%

Finance Construction-4, *Tim Asikin, Steve Asikin, Indra Senihardja*

Law-3593:
If both (**f**= F/S= Fixed Portion= 58.00%), (**t**= T/B= Tax Rate= 30.00%), (**R**= Retained Earnings= 5,529 millions USD), (**d**= D/A= Dividend Portion or Payout= 30.00%), (**i**= I/S= Interest Portion= 1.00%), (**$'**= Sales of Past Year= 48,851 millions USD), (**$**= Sales or Revenues= 51,294 millions USD), and (**s**= [S/S']-1= Sales Growth= 5.00%), are known, then it's (**v**= Variable Portion planned), is:

$$v = (\$[1-i]-R/\{[1-t][1-d]\})/\{\$'[1+s]\} - f$$
$$= (51,294[1-0.0100] - 5,529/\{[1-0.3000][1-0.3000]\})/\{48,851[1+0.0500]\} - 0.5800$$
$$= 19.00\%$$

Law-3594:
If both (**v**= V/S= Variable Portion= 19.00%), (**t**= T/B= Tax Rate= 30.00%), (**R**= Retained Earnings= 5,529 millions USD), (**d**= D/A= Dividend Portion or Payout= 30.00%), (**i**= I/S= Interest Portion= 1.00%), (**$'**= Sales of Past Year= 48,851 millions USD), (**$**= Sales or Revenues= 51,294 millions USD), and (**s**= [S/S']-1= Sales Growth= 5.00%), are known, then it's (**f**= Fixed Portion planned), is:

$$f = (\$[1-i]-R/\{[1-t][1-d]\})/\{\$'[1+s]\} - v$$
$$= (51,294[1-0.0100] - 5,529/\{[1-0.3000][1-0.3000]\})/\{48,851[1+0.0500]\} - 0.1900$$
$$= 58.00\%$$

Law-3595:

If both (**v**= V/S= Variable Portion= 19.00%), (**f**= F/S= Fixed Portion= 58.00%), (**R**= Retained Earnings= 5,529 millions USD), (**d**= D/A= Dividend Portion or Payout= 30.00%), (**i**= I/S= Interest Portion= 1.00%), (**$'**= Sales of Past Year= 48,851 millions USD), (**$**= Sales or Revenues= 51,294 millions USD), and (**s**= [S/S']-1= Sales Growth= 5.00%), are known, then it's (**t**= Tax Rate planned), is:

$$\mathbf{t}= 1-\mathbf{R}/([1-\mathbf{d}]\{\mathbf{\$}[1-\mathbf{i}]-\mathbf{\$'}[1+\mathbf{s}][\mathbf{v}+\mathbf{f}]\})$$
$$= 5,529/([1-0.3000]\{51,294[1-0.0100]$$
$$- 48,851 [1+0.0500][0.1900$$
$$+0.5800]\})$$
$$= 30.00\%$$

Law-3596:

If both (**v**= V/S= Variable Portion= 19.00%), (**f**= F/S= Fixed Portion= 58.00%), (**R**= Retained Earnings= 5,529 millions USD), (**t**= T/B= Tax Rate= 30.00%), (**i**= I/S= Interest Portion= 1.00%), (**$'**= Sales of Past Year= 48,851 millions USD), (**$**= Sales or Revenues= 51,294 millions USD), and (**s**= [S/S']-1= Sales Growth= 5.00%), are known, then it's (**d**= Dividend Portion or Payout planned), is:

$$\mathbf{d}= 1-\mathbf{R}/([1-\mathbf{t}]\{\mathbf{\$}[1-\mathbf{i}]-\mathbf{\$'}[1+\mathbf{s}][\mathbf{v}+\mathbf{f}]\})$$
$$= 5,529/([1-0.3000]\{51,294[1-0.0100]$$
$$- 48,851 [1+0.0500][0.1900$$
$$+0.5800]\})$$
$$= 30.00\%$$

Law-3597:

If both (v= V/S= Variable Portion= 19.00%), (f= F/S= Fixed Portion= 58.00%), (d= D/A= Dividend Portion or Payout= 30.00%), (t= T/B= Tax Rate= 30.00%), (i= I/S= Interest Portion= 1.00%), (S'= Sales of Past Year= 48,851 millions USD), (S= Sales or Revenues= 51,294 millions USD), and (s= [S/S']-1= Sales Growth= 5.00%), are known, then it's (R= Retained Earnings planned), is:

$$R = [1-t][1-d]\{S-S'v[1+s]-S'f[1+s]-S'i[1+s]\}$$
$$= [1-t][1-d]\{S-S'[1+s][v+f+i]\}$$
$$= [1-0.3000][1-0.3000]\{51,294-48,851$$
$$[1+0.0500][0.1900+0.5800$$
$$+0.0100]\})$$
$$= 5,529 \text{ millions USD}$$

Law-3598:

If both (v= V/S= Variable Portion= 19.00%), (f= F/S= Fixed Portion= 58.00%), (d= D/A= Dividend Portion or Payout= 30.00%), (t= T/B= Tax Rate= 30.00%), (i= I/S= Interest Portion= 1.00%), (S'= Sales of Past Year= 48,851 millions USD), (R= Retained Earnings= 5,529 millions USD), and (s= [S/S']-1= Sales Growth= 5.00%), are known, then it's (S= Sales or Revenues planned), is:

$$S = S'[1+s][v+f+i]+R/\{[1-t][1-d]\}$$
$$= 48,851 [1+0.0500][0.1900+0.5800$$
$$+0.0100]+5,529/\{[1-0.3000]$$
$$[1-0.3000]\}$$
$$= 51,294 \text{ millions USD}$$

Law-3599:

If both (v= V/S= Variable Portion= 19.00%), (f= F/S= Fixed Portion= 58.00%), (d= D/A= Dividend Portion or Payout= 30.00%), (t= T/B= Tax Rate= 30.00%), (i= I/S= Interest Portion= 1.00%), (S= Sales or Revenues= 51,294 millions USD), (R= Retained Earnings= 5,529 millions USD), and (s= [S/S']-1= Sales Growth= 5.00%), are known, then it's (S'= Sales Past), must be:

$$S' = (S-R/\{[1-t][1-d]\})/\{[1+s][v+f+i]\}$$
$$= (51,294-5,529/\{[1-0.3000][1-0.3000]\})/\{[1+0.0500][0.1900+0.5800+0.0100]\}$$
$$= 48,851 \text{ millions USD}$$

Law-3600:

If both (v= V/S= Variable Portion= 19.00%), (f= F/S= Fixed Portion= 58.00%), (d= D/A= Dividend Portion or Payout= 30.00%), (t= T/B= Tax Rate= 30.00%), (i= I/S= Interest Portion= 1.00%), (S= Sales or Revenues= 51,294 millions USD), (R= Retained Earnings= 5,529 millions USD), and (S'= Sales of Past Year= 48,851 millions USD), are known, then it's (s= Sales Growth planned), is:

$$s = (S-R/\{[1-t][1-d]\})/\{S'[v+f+i]\}-1$$
$$= (51,294-5,529/\{[1-0.3000][1-0.3000]\})/\{48,851[0.1900+0.5800+0.0100]\}-1$$
$$= 5.00\%$$

Law-3601:

If both (s= [S/S']-1= Sales Growth= 5.00%), (f= F/S= Fixed Portion= 58.00%), (d= D/A= Dividend Portion or Payout= 30.00%), (t= T/B= Tax Rate= 30.00%), (i= I/S= Interest Portion= 1.00%), ($\$$= Sales or Revenues= 51,294 millions USD), (R= Retained Earnings= 5,529 millions USD), and ($\$'$= Sales of Past Year= 48,851 millions USD), are known, then it's (v= Variable Portion planned), is:

v= ($\$$-R/{[1-t][1-d]})/{$\$'$[1+$s$]}-$f$-$i$
= (51,294-5,529/{[1-0.3000]
[1-0.3000]})/{48,851
[1+0.0500]}-0.5800-0.0100
= 19.00%

Law-3602:

If both (s= [S/S']-1= Sales Growth= 5.00%), (v= V/S= Variable Portion= 19.00%), (d= D/A= Dividend Portion or Payout= 30.00%), (t= T/B= Tax Rate= 30.00%), (i= I/S= Interest Portion= 1.00%), ($\$$= Sales or Revenues= 51,294 millions USD), (R= Retained Earnings= 5,529 millions USD), and ($\$'$= Sales of Past Year= 48,851 millions USD), are known, then it's (f= Fixed Portion planned), is:

f= ($\$$-R/{[1-t][1-d]})/{$\$'$[1+$s$]}-$v$-$i$
= (51,294-5,529/{[1-0.3000]
[1-0.3000]})/{48,851
[1+0.0500]}- 0.1900-0.0100
= 58.00%

Law-3603:
If both (s= [S/S']-1= Sales Growth= 5.00%), (v= V/S= Variable Portion= 19.00%), (d= D/A= Dividend Portion or Payout= 30.00%), (t= T/B= Tax Rate= 30.00%), (f= F/S= Fixed Portion= 58.00%), ($\$$= Sales or Revenues= 51,294 millions USD), (R= Retained Earnings= 5,529 millions USD), and ($\$'$= Sales of Past Year= 48,851 millions USD), are known, then it's (i= Interest Portion planned), is:
i= ($\$$-R/{[1-t][1-d]})/{$\$'$[1+$s$]}-$v$-$f$
= (51,294-5,529/{[1-0.3000]
[1-0.3000]})/{48,851
[1+0.0500]}-0.1900-0.5800
= 1.00%

Law-3604:
If both (s= [S/S']-1= Sales Growth= 5.00%), (v= V/S= Variable Portion= 19.00%), (d= D/A= Dividend Portion or Payout= 30.00%), (i= I/S= Interest Portion= 1.00%), (f= F/S= Fixed Portion= 58.00%), ($\$$= Sales or Revenues= 51,294 millions USD), (R= Retained Earnings= 5,529 millions USD), and ($\$'$= Sales of Past Year= 48,851 millions USD), are known, then it's (t= Tax Rate planned), is:
t= 1-R/([1-d]{$\$$-$\$'$[1+$s$][$v$+$f$+$i$]}
=1-5,529/([1-0.3000]{51,294-48,851
[1+0.0500][0.1900+0.5800
+0.0100]})
= 30.00%

Law-3605:
If both (s= [S/S']-1= Sales Growth= 5.00%), (v= V/S= Variable Portion= 19.00%), (t= T/B= Tax Rate= 30.00%), (i= I/S= Interest Portion= 1.00%), (f= F/S=Fixed Portion= 58.00%), (S= Sales or Revenues= 51,294 millions USD), (R= Retained Earnings= 5,529 millions USD), and (S'= Sales of Past Year= 48,851 millions USD), are known, then it's (d= Dividend Portion or Payout planned), is:
d= 1-R/([1-t]{S-S'[1+s][v+f+i]})
= 1-5,529/([1-0.3000]{51,294
-48,851[1+0.0500][0.1900
+0.5800+0.0100]})
= 30.00%

Law-3606:
If both (s= [S/S']-1= Sales Growth= 5.00%), (F= Fixed Cost= 29,750 millions USD), (t= T/B= Tax Rate= 30.00%), (I= Interest Expense= 513 millions USD), (V= Variable Cost= 9,746 millions USD), (d= D/A= Dividend Portion or Payout= 30.00%), and (S'= Sales of Past Year= 48,851 millions USD), are known, then it's (R= Retained Earnings planned), is:
R= [1-t][1-d]{S'[1+s]-V-F-I}
= [1-0.3000][1-0.3000]{48,851
[1+0.0500]-9,746-29,750-513]}
= 5,529 millions USD

Law-3607:
If both (**s**= [S/S']-1= Sales Growth= 5.00%), (**F**= Fixed Cost= 29,750 millions USD), (**t**= T/B= Tax Rate= 30.00%), (**I**= Interest Expense= 513 millions USD), (**V**= Variable Cost= 9,746 millions USD), (**d**= D/A= Dividend Portion or Payout= 30.00%), and (**R**= Retained Earnings= 5,529 millions USD), are known, then it's (**S'**= Sales Past), must be:

$$S' = (V+F+I+R/\{[1-t][1-d]\})/[1+s]$$
$$= (9{,}746+29{,}750+513+5{,}529 \\ /\{[1-0.3000][1-0.3000]\}) \\ /[1+0.0500]$$
$$= \underline{48{,}851} \text{ millions USD}$$

Law-3608:
If both (**S'**= Sales of Past Year= 48,851 millions USD), (**F**= Fixed Cost= 29,750 millions USD), (**t**= T/B= Tax Rate= 30.00%), (**I**= Interest Expense= 513 millions USD), (**V**= Variable Cost= 9,746 millions USD), (**d**= D/A= Dividend Portion or Payout= 30.00%), and (**R**= Retained Earnings= 5,529 millions USD), are known, then it's (**s**= Sales Growth planned), is:

$$s = (V+F+I+R/\{[1-t][1-d]\})/S' - 1$$
$$= (9{,}746+29{,}750+513+5{,}529 \\ /\{[1-0.3000][1-0.3000]\}) \\ /48{,}851 - 1$$
$$= \underline{5.00\%}$$

Law-3609:

If both ($'=$ Sales of Past Year= 48,851 millions USD), (F= Fixed Cost= 29,750 millions USD), (t= T/B= Tax Rate= 30.00%), (I= Interest Expense= 513 millions USD), (s= [S/S']-1= Sales Growth= 5.00%), (d= D/A= Dividend Portion or Payout= 30.00%), and (R= Retained Earnings= 5,529 millions USD), are known, then it's (V= Variable Cost planned), is:

$$V = \$'[1+s]-F-I-R/\{[1-t][1-d]\}$$
$$= 48,851[1+0.0500]-29,750-513-5,529$$
$$/\{[1-0.3000][1-0.3000]\}$$
$$= 9,746 \text{ millions USD}$$

Law-3610:

If both ($'=$ Sales of Past Year= 48,851 millions USD), (V= Variable Cost= 9,746 millions USD), (t= T/B= Tax Rate= 30.00%), (I= Interest Expense= 513 millions USD), (s= [S/S']-1= Sales Growth= 5.00%), (d= D/A= Dividend Portion or Payout= 30.00%), and (R= Retained Earnings= 5,529 millions USD), are known, then it's (F= Fixed Cost planned), is:

$$F = \$'[1+s]-V-I-R/\{[1-t][1-d]\}$$
$$= 48,851[1+0.0500]-9,746-513-5,529$$
$$/\{[1-0.3000][1-0.3000]\}$$
$$= 29,750 \text{ millions USD}$$

Law-3611:
If both ($**'**= Sales of Past Year= 48,851 millions USD), (**V**= Variable Cost= 9,746 millions USD), (**t**= T/B= Tax Rate= 30.00%), (**F**= Fixed Cost= 29,750 millions USD), (**s**= [S/S']-1= Sales Growth= 5.00%), (**d**= D/A= Dividend Portion or Payout= 30.00%), and (**R**= Retained Earnings= 5,529 millions USD), are known, then it's (**I**= Interest Expense planned), is:

$$I = S'[1+s]-V-F-R/\{[1-t][1-d]\}$$
$$= 48,851[1+0.0500]-9,746-29,750$$
$$-5,529/\{[1-0.3000][1-0.3000]\}$$
$$= 513 \text{ millions USD}$$

Law-3612:
If both ($**'**= Sales of Past Year= 48,851 millions USD), (**V**= Variable Cost= 9,746 millions USD), (**I**= Interest Expense= 513 millions USD), (**F**= Fixed Cost= 29,750 millions USD), (**s**= [S/S']-1= Sales Growth= 5.00%), (**d**= D/A= Dividend Portion or Payout= 30.00%), and (**R**= Retained Earnings= 5,529 millions USD), are known, then it's (**t**= Tax Rate planned), is:

$$t = 1-R/([1-d]\{S'[1+s]-V-F-I\})$$
$$= 1-5,529/([1-0.3000]\{48,851$$
$$[1+0.0500]-9,746-29,750-513\})$$
$$= 30.00\%$$

Law-3613:

If both ($\textbf{S'}$= Sales of Past Year= 48,851 millions USD), ($\textbf{V}$= Variable Cost= 9,746 millions USD), ($\textbf{I}$= Interest Expense= 513 millions USD), ($\textbf{F}$= Fixed Cost= 29,750 millions USD), ($\textbf{s}$= [S/S']-1= Sales Growth= 5.00%), ($\textbf{t}$= T/B= Tax Rate= 30.00%), and ($\textbf{R}$= Retained Earnings= 5,529 millions USD), are known, then it's ($\textbf{d}$= Dividend Portion or Payout planned), is:

$\textbf{d}$= 1-$\textbf{R}$/([1-$\textbf{t}$]{$\textbf{S'}$[1+$\textbf{s}$]-$\textbf{V}$-$\textbf{F}$-$\textbf{I}$})
= 1-5,529/([1-0.3000]{48,851
[1+0.0500]-9,746-29,750-513})
= 30.00%

Law-3614:

If both ($\textbf{S'}$= Sales of Past Year= 48,851 millions USD), ($\textbf{V}$= Variable Cost= 9,746 millions USD), ($\textbf{S}$= Sales or Revenues= 51,294 millions USD), ($\textbf{i}$= I/S= Interest Portion= 1.00%), ($\textbf{F}$= Fixed Cost= 29,750 millions USD), ($\textbf{s}$= [S/S']-1= Sales Growth= 5.00%), ($\textbf{t}$= T/B= Tax Rate= 30.00%), and ($\textbf{d}$= D/A= Dividend Portion or Payout= 30.00%), are known, then it's ($\textbf{R}$= Retained Earnings planned), is:

$\textbf{R}$= [1-$\textbf{t}$][1-$\textbf{d}$]{$\textbf{S'}$[1+$\textbf{s}$]-$\textbf{V}$-$\textbf{F}$-$\textbf{Si}$}
= [1-0.3000][1-0.3000]{48,851
[1+0.0500]-9,746-29,750
-51,294*0.0100}
= 5,529 millions USD

Law-3615:
If both (**R**= Retained Earnings= 5,529 millions USD), (**V**= Variable Cost= 9,746 millions USD), (**$**= Sales or Revenues= 51,294 millions USD), (**i**= I/S= Interest Portion= 1.00%), (**F**= Fixed Cost= 29,750 millions USD), (**s**= [S/S']-1= Sales Growth= 5.00%), (**t**= T/B= Tax Rate= 30.00%), and (**d**= D/A= Dividend Portion or Payout= 30.00%), are known, then it's (**$'**= Sales Past), must be:

$$\$' = (V+F+\$i+R/\{[1-t][1-d]\})/[1+s]$$
$$= (9,746+29,750+51,294*0.0100$$
$$+5,529/\{[1-0.3000][1-0.3000]\})$$
$$/[1+0.0500]$$
$$= 48,851 \text{ millions USD}$$

Law-3616:
If both (**R**= Retained Earnings= 5,529 millions USD), (**V**= Variable Cost= 9,746 millions USD), (**$**= Sales or Revenues= 51,294 millions USD), (**i**= I/S= Interest Portion= 1.00%), (**F**= Fixed Cost= 29,750 millions USD), (**$'**= Sales of Past Year= 48,851 millions USD), (**t**= T/B= Tax Rate= 30.00%), and (**d**= D/A= Dividend Portion or Payout= 30.00%), are known, then it's (**s**= Sales Growth planned), is:

$$s = (V+F+\$i+R/\{[1-t][1-d]\})/\$'-1$$
$$= (9,746+29,750+51,294*0.0100+5,529$$
$$/\{[1-0.3000][1-0.3000]\})$$
$$/48,851-1$$
$$= 5.00\%$$

Law-3617:

If both (**R**= Retained Earnings= 5,529 millions USD), (**s**= [S/S']-1= Sales Growth= 5.00%), (**S**= Sales or Revenues= 51,294 millions USD), (**i**= I/S= Interest Portion= 1.00%), (**F**= Fixed Cost= 29,750 millions USD), (**S'**= Sales of Past Year= 48,851 millions USD), (**t**= T/B= Tax Rate= 30.00%), and (**d**= D/A= Dividend Portion or Payout= 30.00%), are known, then it's (**V**= Variable Cost planned), is:

$$V = S'[1+s] - F - Si - R/\{[1-t][1-d]\}$$
$$= 48,851[1+0.0500] - 29,750 - 51,294$$
$$*0.0100 - 5,529/\{[1-0.3000][1-0.3000]\}$$
$$= 9,746 \text{ millions USD}$$

Law-3618:

If both (**R**= Retained Earnings= 5,529 millions USD), (**s**= [S/S']-1= Sales Growth= 5.00%), (**S**= Sales or Revenues= 51,294 millions USD), (**i**= I/S= Interest Portion= 1.00%), (**V**= Variable Cost= 9,746 millions USD), (**S'**= Sales of Past Year= 48,851 millions USD), (**t**= T/B= Tax Rate= 30.00%), and (**d**= D/A= Dividend Portion or Payout= 30.00%), are known, then it's (**F**= Fixed Cost planned), is:

$$F = S'[1+s] - V - Si - R/\{[1-t][1-d]\}$$
$$= 48,851[1+0.0500] - 9,746 - 51,294$$
$$*0.0100 - 5,529/\{[1-0.3000][1-0.3000]\}$$
$$= 29,750 \text{ millions USD}$$

Law-3619:

If both (**R**= Retained Earnings= 5,529 millions USD), (**s**= [S/S']-1= Sales Growth= 5.00%), (**F**= Fixed Cost= 29,750 millions USD), (**i**= I/S= Interest Portion= 1.00%), (**V**= Variable Cost= 9,746 millions USD), (**S'**= Sales of Past Year= 48,851 millions USD), (**t**= T/B= Tax Rate= 30.00%), and (**d**= D/A= Dividend Portion or Payout= 30.00%), are known, then it's (**S**= Sales or Revenues planned), is:

$$S = (S'[1+s]-V-F-R/\{[1-t][1-d]\})/i$$
$$= (48,851[1+0.0500]-9,746-29,750$$
$$-5,529/\{[1-0.3000][1-0.3000]\})$$
$$/0.0100$$
$$= \underline{51,294} \text{ millions USD}$$

Law-3620:

If both (**R**= Retained Earnings= 5,529 millions USD), (**s**= [S/S']-1= Sales Growth= 5.00%), (**F**= Fixed Cost= 29,750 millions USD), (**S**= Sales or Revenues= 51,294 millions USD), (**V**= Variable Cost= 9,746 millions USD), (**S'**= Sales of Past Year= 48,851 millions USD), (**t**= T/B= Tax Rate= 30.00%), and (**d**= D/A= Dividend Portion or Payout= 30.00%), are known, then it's (**i**= Interest Portion planned), is:

$$i = (S'[1+s]-V-F-R/\{[1-t][1-d]\})/S$$
$$= (48,851[1+0.0500]-9,746-29,750$$
$$-5,529/\{[1-0.3000][1-0.3000]\})$$
$$/51,294$$
$$= \underline{1.00}\%$$

Law-3621:
If both (**R**= Retained Earnings= 5,529 millions USD), (**s**= [S/S']-1= Sales Growth= 5.00%), (**F**= Fixed Cost= 29,750 millions USD), (**S**= Sales or Revenues= 51,294 millions USD), (**V**= Variable Cost= 9,746 millions USD), (**S'**= Sales of Past Year= 48,851 millions USD), (**i**= I/S= Interest Portion= 1.00%), and (**d**= D/A= Dividend Portion or Payout= 30.00%), are known, then it's (**t**= Tax Rate planned), is:
 t= 1-**R**/([1-**d**]{**S'**[1+**s**]-**V**-**F**-**Si**})
  = 1-5,529/([1-0.3000]{48,851
   [1+0.0500]-9,746-29,750
   -51,294*0.0100})
  = 30.00%

Law-3622:
If both (**R**= Retained Earnings= 5,529 millions USD), (**s**= [S/S']-1= Sales Growth= 5.00%), (**F**= Fixed Cost= 29,750 millions USD), (**S**= Sales or Revenues= 51,294 millions USD), (**V**= Variable Cost= 9,746 millions USD), (**S'**= Sales of Past Year= 48,851 millions USD), (**i**= I/S= Interest Portion= 1.00%), and (**t**= T/B= Tax Rate= 30.00%), are known, then it's (**d**= Dividend Portion or Payout planned), is:
 d= 1-**R**/([1-**t**]{**S'**[1+**s**]-**V**-**F**-**Si**})
  = 1-5,529/([1-0.3000]{48,851
   [1+0.0500]-9,746-29,750
   -51,294*0.0100})
  = 30.00%

Law-3623:
If both (**d**= D/A= Dividend Portion or Payout= 30.00%), (**s**= [S/S']-1= Sales Growth= 5.00%), (**F**= Fixed Cost= 29,750 millions USD), (**V**= Variable Cost= 9,746 millions USD), (**S'**= Sales of Past Year= 48,851 millions USD), (**i**= I/S= Interest Portion= 1.00%), and (**t**= T/B= Tax Rate= 30.00%), are known, then it's (**R**= Retained Earnings planned), is:

$$R = [1-t][1-d]\{S'[1+s]-V-F-S'i[1+s]\}$$
$$= [1-t][1-d]\{S'[1+s][1-i]-V-F\}$$
$$= [1-0.3000][1-0.3000]\{48,851$$
$$[1+0.0500][1-0.0100]-9,746$$
$$-29,750\}$$
$$= \underline{5,529} \text{ millions USD}$$

Law-3624:
If both (**d**= D/A= Dividend Portion or Payout= 30.00%), (**s**= [S/S']-1= Sales Growth= 5.00%), (**F**= Fixed Cost= 29,750 millions USD), (**V**= Variable Cost= 9,746 millions USD), (**R**= Retained Earnings= 5,529 millions USD), (**i**= I/S= Interest Portion= 1.00%), and (**t**= T/B= Tax Rate= 30.00%), are known, then it's (**S'**= Sales Past), must be:

$$S' = (V+F+R/\{[1-t][1-d]\})/\{[1+s][1-i]\}$$
$$= (9,746+29,750+5,529/\{[1-0.3000]$$
$$[1-0.3000]\})/\{[1+0.0500]$$
$$[1-0.0100]\}$$
$$= \underline{48,851} \text{ millions USD}$$

Law-3625:
If both (**d**= D/A=Dividend Portion or Payout= 30.00%), (**$'**= Sales of Past Year= 48,851 millions USD), (**F**= Fixed Cost= 29,750 millions USD), (**V**= Variable Cost= 9,746 millions USD), (**R**= Retained Earnings= 5,529 millions USD), (**i**= I/S= Interest Portion= 1.00%), and (**t**= T/B= Tax Rate= 30.00%), are known, then it's (**s**= Sales Growth planned), is:

$s = (V+F+R/\{[1-t][1-d]\})/\{S'[1-i]\}-1$
$= (9,746+29,750+5,529$
$/\{[1-0.3000][1-0.3000]\})$
$/\{48,851[1-0.0100]\}-1$
$= 5.00\%$

Law-3626:
If both (**d**= D/A= Dividend Portion or Payout= 30.00%), (**$'**= Sales of Past Year= 48,851 millions USD), (**F**= Fixed Cost= 29,750 millions USD), (**V**= Variable Cost= 9,746 millions USD), (**R**= Retained Earnings= 5,529 millions USD), (**s**= [S/S']-1= Sales Growth= 5.00%), and (**t**= T/B= Tax Rate= 30.00%), are known, then it's (**i**= Interest Portion planned), is:

$i = 1-(V+F+R/\{[1-t][1-d]\})/\{S'[1+s]\}$
$= 1-(9,746+29,750+5,529$
$/\{[1-0.3000][1-0.3000]\})$
$/\{48,851[1+0.0500]\}$
$= 1.00\%$

Law-3627:

If both (**d**= D/A= Dividend Portion or Payout= 30.00%), (**$'**= Sales of Past Year= 48,851 millions USD), (**F**= Fixed Cost= 29,750 millions USD), (**i**= I/S= Interest Portion= 1.00%), (**R**= Retained Earnings= 5,529 millions USD), (**s**= [S/S']-1= Sales Growth= 5.00%), and (**t**= T/B= Tax Rate= 30.00%), are known, then it's (**V**= Variable Cost planned), is:

$V = \$'[1+s][1-i] - F - R/\{[1-t][1-d]\}$
$= 48,851[1+0.0500][1-0.0100] - 29,750$
$\quad -5,529/\{[1-0.3000][1-0.3000]\}$
$= \underline{9,746}$ millions USD

Law-3628:

If both (**d**= D/A= Dividend Portion or Payout= 30.00%), (**$'**= Sales of Past Year= 48,851 millions USD), (**V**= Variable Cost= 9,746 millions USD), (**i**= I/S= Interest Portion= 1.00%), (**R**= Retained Earnings= 5,529 millions USD), (**s**= [S/S']-1= Sales Growth= 5.00%), and (**t**= T/B= Tax Rate= 30.00%), are known, then it's (**F**= Fixed Cost planned), is:

$F = \$'[1+s][1-i] - V - R/\{[1-t][1-d]\}$
$= 48,851[1+0.0500][1-0.0100] - 9,746$
$\quad -5,529/\{[1-0.3000][1-0.3000]\}$
$= \underline{29,750}$ millions USD

Law-3629:
If both (**d**= D/A= Dividend Portion or Payout= 30.00%), (**$'**= Sales of Past Year= 48,851 millions USD), (**V**= Variable Cost= 9,746 millions USD), (**i**= I/S= Interest Portion= 1.00%), (**R**= Retained Earnings= 5,529 millions USD), (**s**= [S/S']-1= Sales Growth= 5.00%), and (**F**= Fixed Cost= 29,750 millions USD), are known, then it's (**t**= Tax Rate planned), is:

t= 1-**R**/([1-**d**]{**$'**[1+**s**][1-**i**]-**V**-**F**}
  = 1-5,529/ ([1-0.3000]{48,851
         [1+0.0500][1-0.0100]-9,746
         -29,750}
  = 30.00%

Law-3630:
If both (**t**= T/B= Tax Rate= 30.00%), (**$'**= Sales of Past Year= 48,851 millions USD), (**V**= Variable Cost= 9,746 millions USD), (**i**= I/S= Interest Portion= 1.00%), (**R**= Retained Earnings= 5,529 millions USD), (**s**= [S/S']-1= Sales Growth= 5.00%), and (**F**= Fixed Cost= 29,750 millions USD), are known, then it's (**d**= Dividend Portion or Payout planned), is:

d= 1-**R**/([1-**t**]{**$'**[1+**s**][1-**i**]-**V**-**F**}
  = 1-5,529/ ([1-0.3000]{48,851
         [1+0.0500][1-0.0100]-9,746
         -29,750}
  = 30.00%

Law-3631:

If both (**t**= T/B= Tax Rate= 30.00%), (**$'**= Sales of Past Year= 48,851 millions USD), (**V**= Variable Cost= 9,746 millions USD), (**I**= Interest Expense= 513 millions USD), (**$**= Sales or Revenues= 51,294 millions USD), (**d**= D/A= Dividend Portion or Payout= 30.00%), (**s**= [S/S']-1= Sales Growth= 5.00%), and (**f**= F/S= Fixed Portion= 58.00%), are known, then it's (**R**= Retained Earnings planned), is:

$$R = [1-t][1-d]\{\$'[1+s]-V-\$f-I\}$$
$$= [1-0.3000][1-0.3000]\{48,851$$
$$[1+0.0500]- 9,746-51,294$$
$$*0.5800-513\}$$
$$= 5,529 \text{ millions USD}$$

Law-3632:

If both (**t**= T/B= Tax Rate= 30.00%), (**R**= Retained Earnings= 5,529 millions USD), (**V**= Variable Cost= 9,746 millions USD), (**I**= Interest Expense= 513 millions USD), (**$**= Sales or Revenues= 51,294 millions USD), (**d**= D/A= Dividend Portion or Payout= 30.00%), (**s**= [S/S']-1= Sales Growth= 5.00%), and (**f**= F/S= Fixed Portion= 58.00%), are known, then it's (**$'**= Sales Past), must be:

$$\$' = (V+\$f+I+R/\{[1-t][1-d]\})/[1+s]$$
$$= (9,746+51,294*0.5800+513+5,529$$
$$/\{ [1-0.3000][1-0.3000]\})$$
$$/ [1+0.0500]$$
$$= 48,851 \text{ millions USD}$$

Law-3633:

If both (**t**= T/B= Tax Rate= 30.00%), (**R**= Retained Earnings= 5,529 millions USD), (**V**= Variable Cost= 9,746 millions USD), (**I**= Interest Expense= 513 millions USD), (**$**= Sales or Revenues= 51,294 millions USD), (**d**= D/A= Dividend Portion or Payout= 30.00%), (**$'**= Sales of Past Year= 48,851 millions USD), and (**f**= F/S= Fixed Portion= 58.00%), are known, then it's (**s**= Sales Growth planned), is:

$$s = (V+\$f+I+R/\{[1-t][1-d]\})/\$' - 1$$
$$= (9,746+51,294*0.5800+513+5,529$$
$$/\{[1-0.3000][1-0.3000]\})$$
$$/48,851 - 1$$
$$= 5.00\%$$

Law-3634:

If both (**t**= T/B= Tax Rate= 30.00%), (**R**= Retained Earnings= 5,529 millions USD), (**s**= [S/S']-1= Sales Growth= 5.00%), (**I**= Interest Expense= 513 millions USD), (**$**= Sales or Revenues= 51,294 millions USD), (**d**= D/A= Dividend Portion or Payout= 30.00%), (**$'**= Sales of Past Year= 48,851 millions USD), and (**f**= F/S= Fixed Portion= 58.00%), are known, then it's (**V**= Variable Cost planned), is:

$$V = \$'[1+s] - \$f - I - R/\{[1-t][1-d]\}$$
$$= 48,851[1+0.0500] - 51,294*0.5800$$
$$-513 - 5,529/\{[1-0.3000]$$
$$[1-0.3000]\}$$
$$= 9,746 \text{ millions USD}$$

Law-3635:
If both (**t**= T/B= Tax Rate= 30.00%), (**R**= Retained Earnings= 5,529 millions USD), (**s**= [S/S']-1= Sales Growth= 5.00%), (**I**= Interest Expense= 513 millions USD), (**V**= Variable Cost= 9,746 millions USD), (**d**= D/A= Dividend Portion or Payout= 30.00%), (**S'**= Sales of Past Year= 48,851 millions USD), and (**f**= F/S= Fixed Portion= 58.00%), are known, then it's (**S**= Sales or Revenues planned), is:

$S = (S'[1+s]-V-I-R/\{[1-t][1-d]\})/f$
$= (48,851[1+0.0500]-9,746-513-5,529$
$/\{[1-0.3000][1-0.3000]\})$
$/0.5800$
$= 51,294$ millions USD

Law-3636:
If both (**t**= T/B= Tax Rate= 30.00%), (**R**= Retained Earnings= 5,529 millions USD), (**s**= [S/S']-1= Sales Growth= 5.00%), (**I**= Interest Expense= 513 millions USD), (**V**= Variable Cost= 9,746 millions USD), (**d**= D/A= Dividend Portion or Payout= 30.00%), (**S'**= Sales of Past Year= 48,851 millions USD), and (**S**= Sales or Revenues= 51,294 millions USD), are known, then it's (**f**= Fixed Portion planned), is:

$f = (S'[1+s]-V-I-R/\{[1-t][1-d]\})/S$
$= (48,851[1+0.0500]-9,746-513-5,529$
$/\{[1-0.3000][1-0.3000]\})$
$/51,294$
$= 58.00\%$

Law-3637:
If both (**t**= T/B= Tax Rate= 30.00%), (**R**= Retained Earnings= 5,529 millions USD), (**s**= [S/S']-1= Sales Growth= 5.00%), (**f**= F/S= Fixed Portion= 58.00%), (**V**= Variable Cost= 9,746 millions USD), (**d**= D/A= Dividend Portion or Payout= 30.00%), (**S'**= Sales of Past Year= 48,851 millions USD), and (**S**= Sales or Revenues= 51,294 millions USD), are known, then it's (**I**= Interest Expense planned), is:

I= **S'**[1+**s**]-**V**-**Sf**-**R**/{[1-**t**][1-**d**]}
= 48,851[1+0.0500]-9,746-51,294
*0.5800-5,529/{ [1-0.3000]
[1-0.3000]}
= 513 millions USD

Law-3638:
If both (**I**= Interest Expense= 513 millions USD), (**R**= Retained Earnings= 5,529 millions USD), (**s**= [S/S']-1= Sales Growth= 5.00%), (**f**= F/S= Fixed Portion= 58.00%), (**V**= Variable Cost= 9,746 millions USD), (**d**= D/A= Dividend Portion or Payout= 30.00%), (**S'**= Sales of Past Year= 48,851 millions USD), and (**S**= Sales or Revenues= 51,294 millions USD), are known, then it's (**t**= Tax Rate planned), is:

t= 1-**R**/([1-**d**]{**S'**[1+**s**]-**V**-**Sf**-**I**})
= 1-5,529/([1-0.3000]{48,851
[1+0.0500]-9,746-51,294
*0.5800-513}
= 30.00%

Law-3639:

If both (**I**= Interest Expense= 513 millions USD), (**R**= Retained Earnings= 5,529 millions USD), (**s**= [S/S']-1= Sales Growth= 5.00%), (**f**= F/S= Fixed Portion= 58.00%), (**V**= Variable Cost= 9,746 millions USD), (**t**= T/B= Tax Rate= 30.00%), (**$'**= Sales of Past Year= 48,851 millions USD), and (**$**= Sales or Revenues= 51,294 millions USD), are known, then it's (**d**= Dividend Portion or Payout planned), is:

$$d = 1 - R/([1-t]\{\$'[1+s]-V-\$f-I\})$$
$$= 1-5,529/([1-0.3000]\{48,851$$
$$[1+0.0500]-9,746-51,294$$
$$*0.5800-513\}$$
$$= 30.00\%$$

Law-3640:

If both (**i**= I/S= Interest Portion= 1.00%), (**d**= D/A= Dividend Portion or Payout= 30.00%), (**s**= [S/S']-1= Sales Growth= 5.00%), (**f**= F/S= Fixed Portion= 58.00%), (**V**= Variable Cost= 9,746 millions USD), (**t**= T/B= Tax Rate= 30.00%), (**$'**= Sales of Past Year= 48,851 millions USD), and (**$**= Sales or Revenues= 51,294 millions USD), are known, then it's (**R**= Retained Earnings planned), is:

$$R = [1-t][1-d]\{\$'[1+s]-V-\$f-\$i\}$$
$$= [1-t][1-d]\{\$'[1+s]-V-\$[f+i]\}$$
$$= [1-0.3000][1-0.3000]\{48,851$$
$$[1+0.0500]-9,746-51,294$$
$$[0.5800+0.0100]\}$$
$$= 5,529 \text{ millions USD}$$

Law-3641:
If both (**i**= I/S= Interest Portion= 1.00%), (**d**= D/A= Dividend Portion or Payout= 30.00%), (**s**= [S/S']-1= Sales Growth= 5.00%), (**f**= F/S= Fixed Portion= 58.00%), (**V**= Variable Cost= 9,746 millions USD), (**t**= T/B= Tax Rate= 30.00%), (**R**= Retained Earnings= 5,529 millions USD), and (**S**= Sales or Revenues= 51,294 millions USD), are known, then it's (**S'**= Sales Past), must be:

$S' = (V+S[f+i]+R/\{[1-t][1-d]\})/[1+s]$
  = (9,746+51,294[0.5800+0.0100]
      +5,529/{[1-0.3000][1-0.3000]})
      /[1+0.0500]
  = 48,851 millions USD

Law-3642:
If both (**i**= I/S= Interest Portion= 1.00%), (**d**= D/A= Dividend Portion or Payout= 30.00%), (**S'**= Sales of Past Year= 48,851 millions USD), (**f**= F/S= Fixed Portion= 58.00%), (**V**= Variable Cost= 9,746 millions USD), (**t**= T/B= Tax Rate= 30.00%), (**R**= Retained Earnings= 5,529 millions USD), and (**S**= Sales or Revenues= 51,294 millions USD), are known, then it's (**s**= Sales Growth planned), is:

$s = (V+S[f+i]+R/\{[1-t][1-d]\})/S' - 1$
  = (9,746+51,294[0.5800+0.0100]
      +5,529/{[1-0.3000][1-0.3000]})
      /48,851-1
  = 5.00%

Law-3643:
If both (**i**= I/S= Interest Portion= 1.00%), (**d**= D/A= Dividend Portion or Payout= 30.00%), (**$'**= Sales of Past Year= 48,851 millions USD), (**f**= F/S= Fixed Portion= 58.00%), (**s**= [S/S']-1= Sales Growth= 5.00%), (**t**= T/B= Tax Rate= 30.00%), (**R**= Retained Earnings= 5,529 millions USD), and (**$**= Sales or Revenues= 51,294 millions USD), are known, then it's (**V**= Variable Cost planned), is:

V= **$'**[1+**s**]-**$**[**f**+**i**]-**R**/{[1-**t**][1-**d**]}
 = 48,851[1+0.0500]-51,294[0.5800
  +0.0100]-5,529/{[1-0.3000]
  [1-0.3000]}
 = 9,746 millions USD

Law-3644:
If both (**i**= I/S= Interest Portion= 1.00%), (**d**= D/A= Dividend Portion or Payout= 30.00%), (**$'**= Sales of Past Year= 48,851 millions USD), (**f**= F/S= Fixed Portion= 58.00%), (**s**= [S/S']-1= Sales Growth= 5.00%), (**t**= T/B= Tax Rate= 30.00%), (**R**= Retained Earnings= 5,529 millions USD), and (**V**= Variable Cost= 9,746 millions USD), are known, then it's (**$**= Sales or Revenues planned), is:

$**= **$'[1+**s**]-**V**-**R**/{[1-**t**][1-**d**]})/[**f**+**i**]
 = 48,851[1+0.0500]-9,746-5,529
  /{[1-0.3000][1-0.3000]})
  /[0.5800+0.0100]
 = 51,294 millions USD

Law-3645:
If both (**i**= I/S= Interest Portion= 1.00%), (**d**= D/A= Dividend Portion or Payout= 30.00%), (**$'**= Sales of Past Year= 48,851 millions USD), (**$**= Sales or Revenues= 51,294 millions USD), (**s**= [S/S']-1= Sales Growth= 5.00%), (**t**= T/B= Tax Rate= 30.00%), (**R**= Retained Earnings= 5,529 millions USD), and (**V**= Variable Cost= 9,746 millions USD), are known, then it's (**f**= Fixed Portion planned), is:

$$f = (\$'[1+s]-V-R/\{[1-t][1-d]\})/\$-i$$
$$= (48,851[1+0.0500]-9,746-5,529$$
$$/\{[1-0.3000][1-0.3000]\})$$
$$/51,294-0.0100$$
$$= 58.00\%$$

Law-3646:
If both (**f**= F/S= Fixed Portion= 58.00%), (**d**= D/A= Dividend Portion or Payout= 30.00%), (**$'**= Sales of Past Year= 48,851 millions USD), (**$**= Sales or Revenues= 51,294 millions USD), (**s**= [S/S']-1= Sales Growth= 5.00%), (**t**= T/B= Tax Rate= 30.00%), (**R**= Retained Earnings= 5,529 millions USD), and (**V**= Variable Cost= 9,746 millions USD), are known, then it's (**i**= Interest Portion planned), is:

$$i = (\$'[1+s]-V-R/\{[1-t][1-d]\})/\$-f$$
$$= (48,851[1+0.0500]-9,746-5,529$$
$$/\{[1-0.3000][1-0.3000]\})$$
$$/51,294-0.5800$$
$$= 1.00\%$$

Law-3647:
If both (**f**= F/S= Fixed Portion= 58.00%), (**d**= D/A= Dividend Portion or Payout= 30.00%), (**$'**= Sales of Past Year= 48,851 millions USD), (**$**= Sales or Revenues= 51,294 millions USD), (**s**= [S/S']-1= Sales Growth= 5.00%), (**i**= I/S= Interest Portion= 1.00%), (**R**= Retained Earnings= 5,529 millions USD), and (**V**= Variable Cost= 9,746 millions USD), are known, then it's (**t**= Tax Rate planned), is:

t= 1-**R**/([1-**d**]{**$'**[1+**s**]-**V**-**$**[**f**+**i**]})
  = 1-5,529/ ([1-0.3000]{48,851
     [1+0.0500]-9,746-51,294
     [0.5800+0.0100]})
  = 30.00%

Law-3648:
If both (**f**= F/S= Fixed Portion= 58.00%), (**t**= T/B= Tax Rate= 30.00%), (**$'**= Sales of Past Year= 48,851 millions USD), (**$**= Sales or Revenues= 51,294 millions USD), (**s**= [S/S']-1= Sales Growth= 5.00%), (**i**= I/S= Interest Portion= 1.00%), (**R**= Retained Earnings= 5,529 millions USD), and (**V**= Variable Cost= 9,746 millions USD), are known, then it's (**d**= Dividend Portion or Payout planned), is:

d= 1-**R**/([1-**t**]{**$'**[1+**s**]-**V**-**$**[**f**+**i**]})
  = 1-5,529/([1-0.3000]{48,851
     [1+0.0500]-9,746-51,294
     [0.5800+0.0100]})
  = 30.00%

Law-3649:
If both (**f**= F/S= Fixed Portion= 58.00%), (**t**= T/B= Tax Rate= 30.00%), (**$'**= Sales of Past Year= 48,851 millions USD), (**$**= Sales or Revenues= 51,294 millions USD), (**s**= [S/S']-1= Sales Growth= 5.00%), (**i**= I/S= Interest Portion= 1.00%), (**d**= D/A= Dividend Portion or Payout= 30.00%), and (**V**= Variable Cost= 9,746 millions USD), are known, then it's (**R**= Retained Earnings planned), is:
$$R = [1-t][1-d]\{\$'[1+s]-V-\$f-\$'i[1+s]\}$$
$$= [1-t][1-d]\{\$'[1+s][1-i]-V-\$f\}$$
$$= [1-0.3000][1-0.3000]\{48,851$$
$$[1+0.0500][1-0.0100]-9,746$$
$$-51,294*0.5800\}$$
$$= 5,529 \text{ millions USD}$$

Law-3650:
If both (**f**= F/S= Fixed Portion= 58.00%), (**t**= T/B= Tax Rate= 30.00%), (**R**= Retained Earnings= 5,529 millions USD), (**$**= Sales or Revenues= 51,294 millions USD), (**s**= [S/S']-1= Sales Growth= 5.00%), (**i**= I/S= Interest Portion= 1.00%), (**d**= D/A= Dividend Portion or Payout= 30.00%), and (**V**= Variable Cost= 9,746 millions USD), are known, then it's (**$'**= Sales Past), must be:
$$\$' = (V+\$f+R/\{[1-t][1-d]\})/\{[1+s][1-i]\}$$
$$= (9,746+51,294*0.5800+5,529$$
$$/\{[1-0.3000][1-0.3000]\})$$
$$/\{[1+0.0500][1-0.0100]\}$$
$$= 48,851 \text{ millions USD}$$

Law-3651:
If both (**f**= F/S= Fixed Portion= 58.00%), (**t**= T/B= Tax Rate= 30.00%), (**R**= Retained Earnings= 5,529 millions USD), (**$**= Sales or Revenues= 51,294 millions USD), (**$'**= Sales of Past Year= 48,851 millions USD), (**i**= I/S= Interest Portion= 1.00%), (**d**= D/A= Dividend Portion or Payout= 30.00%), and (**V**= Variable Cost= 9,746 millions USD), are known, then it's (**s**= Sales Growth planned), is:

$s = (V+\$f+R/\{[1-t][1-d]\})/\{\$'[1-i]\} - 1$
$= (9{,}746 + 51{,}294 \ast 0.5800 + 5{,}529$
$/\{[1-0.3000][1-0.3000]\})$
$/\{48{,}851[1-0.0100]\} - 1$
$= 5.00\%$

Law-3652:
If both (**f**= F/S= Fixed Portion= 58.00%), (**t**= T/B= Tax Rate= 30.00%), (**R**= Retained Earnings= 5,529 millions USD), (**$**= Sales or Revenues= 51,294 millions USD), (**$'**= Sales of Past Year= 48,851 millions USD), (**s**= [S/S']-1= Sales Growth= 5.00%), (**d**= D/A= Dividend Portion or Payout= 30.00%), and (**V**= Variable Cost= 9,746 millions USD), are known, then it's (**i**= Interest Portion planned), is:

$i = 1 - (V+\$f+R/\{[1-t][1-d]\})/\{\$'[1+s]\}$
$= 1 - (9{,}746 + 51{,}294 \ast 0.5800 + 5{,}529$
$/\{[1-0.3000][1-0.3000]\})$
$/\{48{,}851[1+0.0500]\}$
$= 1.00\%$

Law-3653:
If both (**f**= F/S= Fixed Portion= 58.00%), (**t**= T/B= Tax Rate= 30.00%), (**R**= Retained Earnings= 5,529 millions USD), (**$**= Sales or Revenues= 51,294 millions USD), (**$'**= Sales of Past Year= 48,851 millions USD), (**s**= [S/S']-1= Sales Growth= 5.00%), (**d**= D/A= Dividend Portion or Payout= 30.00%), and (**i**= I/S= Interest Portion= 1.00%), are known, then it's (**V**= Variable Cost planned), is:

$$V = \$'[1+s][1-i]-\$f-R/\{[1-t][1-d]\}$$
$$= 48,851[1+0.0500][1-0.0100]-51,294$$
$$*0.5800-5,529/\{[1-0.3000]$$
$$[1-0.3000]\}$$
$$= 9,746 \text{ millions USD}$$

Law-3654:
If both (**f**= F/S= Fixed Portion= 58.00%), (**t**= T/B= Tax Rate= 30.00%), (**R**= Retained Earnings= 5,529 millions USD), (**V**= Variable Cost= 9,746 millions USD), (**$'**= Sales of Past Year= 48,851 millions USD), (**s**= [S/S']-1= Sales Growth= 5.00%), (**d**= D/A= Dividend Portion or Payout= 30.00%), and (**i**= I/S= Interest Portion= 1.00%), are known, then it's (**$**= Sales or Revenues planned), is:

$$\$ = (\$'[1+s][1-i]-V-R/\{[1-t][1-d]\})/f$$
$$= (48,851[1+0.0500][1-0.0100]-9,746$$
$$-5,529/\{[1-0.3000][1-0.3000]\})$$
$$/0.5800$$
$$= 51,294 \text{ millions USD}$$

Law-3655:
If both ($= Sales or Revenues= 51,294 millions USD), (t= T/B= Tax Rate= 30.00%), (R= Retained Earnings= 5,529 millions USD), (V= Variable Cost= 9,746 millions USD), ($'= Sales of Past Year= 48,851 millions USD), (s= [S/S']-1= Sales Growth= 5.00%), (d= D/A= Dividend Portion or Payout= 30.00%), and (i= I/S= Interest Portion= 1.00%), are known, then it's (f= Fixed Portion planned), is:

$$f = (\$'[1+s][1-i]-V-R/\{[1-t][1-d]\})/\$$$
$$= (48,851[1+0.0500][1-0.0100]-9,746$$
$$\quad -5,529/\{[1-0.3000][1-0.3000]\})$$
$$\quad /51,294$$
$$= 58.00\%$$

Law-3656:
If both ($= Sales or Revenues= 51,294 millions USD), (f= F/S= Fixed Portion= 58.00%), (R= Retained Earnings= 5,529 millions USD), (V= Variable Cost= 9,746 millions USD), ($'= Sales of Past Year= 48,851 millions USD), (s= [S/S']-1= Sales Growth= 5.00%), (d= D/A= Dividend Portion or Payout= 30.00%), and (i= I/S= Interest Portion= 1.00%), are known, then it's (t= Tax Rate planned), is:

$$t = 1-R/\{[1-d]\{\$'[1+s][1-i]-V-\$f\}$$
$$= 1-5,529/([1-0.3000]\{48,851$$
$$\quad [1+0.0500]\,[1-0.0100]-9,746$$
$$\quad -51,294*0.5800\})$$
$$= 30.00\%$$

Law-3657:

If both ($= Sales or Revenues= 51,294 millions USD), (f= F/S= Fixed Portion= 58.00%), (R= Retained Earnings= 5,529 millions USD), (V= Variable Cost= 9,746 millions USD), ($'= Sales of Past Year= 48,851 millions USD), (s= [S/S']-1= Sales Growth= 5.00%), (t= T/B= Tax Rate= 30.00%), and (i= I/S= Interest Portion= 1.00%), are known, then it's (d= Dividend Portion or Payout planned), is:

d= 1-R/{[1-t]{$'[1+s][1-i]-V-$f}
  = 1-5,529/ ([1-0.3000]{48,851
      [1+0.0500][1-0.0100]-9,746
      -51,294*0.5800})
  = 30.00%

Law-3658:

If both (f= F/S= Fixed Portion= 58.00%), (d= D/A= Dividend Portion or Payout= 30.00%), (V= Variable Cost= 9,746 millions USD), ($'= Sales of Past Year= 48,851 millions USD), (s= [S/S']-1= Sales Growth= 5.00%), (t= T/B= Tax Rate= 30.00%), and (i= I/S= Interest Portion= 1.00%), are known, then it's (R= Retained Earnings planned), is:

R= [1-t][1-d]{$'[1+s]-V-$'f[1+s]-$'i[1+s]}
  = [1-t][1-d]{$'[1+s][1-f-i]-V}
  = [1-0.3000][1-0.3000]{48,851
      [1+0.0500][1-0.5800-0.0100]
      -9,746}
  = 5,529 millions USD

Law-3659:

If both (**f**= F/S= Fixed Portion= 58.00%), (**d**= D/A= Dividend Portion or Payout= 30.00%), (**V**= Variable Cost= 9,746 millions USD), (**R**= Retained Earnings= 5,529 millions USD), (**s**= [S/S']-1= Sales Growth= 5.00%), (**t**= T/B= Tax Rate= 30.00%), and (**i**= I/S= Interest Portion= 1.00%), are known, then it's (**S'**= Sales Past), must be:

$$S' = (V+R/\{[1-t][1-d]\})/\{[1+s][1-f-i]\}$$
$$= (9,746+5,529/\{[1-0.3000][1-0.3000]\})$$
$$/\{[1+0.0500][1-0.5800-0.0100]\}$$
$$= 48,851 \text{ millions USD}$$

Law-3660:

If both (**f**= F/S= Fixed Portion= 58.00%), (**d**= D/A= Dividend Portion or Payout= 30.00%), (**V**= Variable Cost= 9,746 millions USD), (**R**= Retained Earnings= 5,529 millions USD), (**S'**= Sales of Past Year= 48,851 millions USD), (**t**= T/B= Tax Rate= 30.00%), and (**i**= I/S= Interest Portion= 1.00%), are known, then it's (**s**= Sales Growth planned), is:

$$s = (V+R/\{[1-t][1-d]\})/\{S'[1-f-i]\}-1$$
$$= (9,746+5,529/\{[1-0.3000][1-0.3000]\})$$
$$/\{48,851[1-0.5800-0.0100]\}-1$$
$$= 5.00\%$$

Law-3661:
If both (**s**= [S/S']-1= Sales Growth= 5.00%), (**d**= D/A= Dividend Portion or Payout= 30.00%), (**V**= Variable Cost= 9,746 millions USD), (**R**= Retained Earnings= 5,529 millions USD), (**S'**= Sales of Past Year= 48,851 millions USD), (**t**= T/B= Tax Rate= 30.00%), and (**i**= I/S= Interest Portion= 1.00%), are known, then it's (**f**= Fixed Portion planned), is:

$$f = 1 - i - (V+R/\{[1-t][1-d]\})/\{S'[1+s]\}$$
$$= 1 - 0.0100 - (9,746+5,529/\{[1-0.3000][1-0.3000]\})/\{48,851[1+0.0500]\}$$
$$= 58.00\%$$

Law-3662:
If both (**s**= [S/S']-1= Sales Growth= 5.00%), (**d**= D/A= Dividend Portion or Payout= 30.00%), (**V**= Variable Cost= 9,746 millions USD), (**R**= Retained Earnings= 5,529 millions USD), (**S'**= Sales of Past Year= 48,851 millions USD), (**t**= T/B= Tax Rate= 30.00%), and (**f**= F/S= Fixed Portion= 58.00%), are known, then it's (**i**= Interest Portion planned), is:

$$i = 1 - f - (V+R/\{[1-t][1-d]\})/\{S'[1+s]\}$$
$$= 1 - 0.5800 - (9,746+5,529/\{[1-0.3000][1-0.3000]\})/\{48,851[1+0.0500]\}$$
$$= 1.00\%$$

Law-3663:

If both (**s**= [S/S']-1= Sales Growth= 5.00%), (**d**= D/A= Dividend Portion or Payout= 30.00%), (**i**= I/S= Interest Portion= 1.00%), (**R**= Retained Earnings= 5,529 millions USD), (**S'**= Sales of Past Year= 48,851 millions USD), (**t**= T/B= Tax Rate= 30.00%), and (**f**= F/S= Fixed Portion= 58.00%), are known, then it's (**V**= Variable Cost planned), is:

$$V = S'[1+s][1-f-i] - R/\{[1-t][1-d]\}$$
$$= 48,851[1+0.0500][1-0.5800-0.0100]$$
$$\quad -5,529/\{[1-0.3000][1-0.3000]\}\}$$
$$= 9,746 \text{ millions USD}$$

Law-3664:

If both (**s**= [S/S']-1= Sales Growth= 5.00%), (**d**= D/A= Dividend Portion or Payout= 30.00%), (**i**= I/S= Interest Portion= 1.00%), (**R**= Retained Earnings= 5,529 millions USD), (**S'**= Sales of Past Year= 48,851 millions USD), (**V**= Variable Cost= 9,746 millions USD), and (**f**= F/S= Fixed Portion= 58.00%), are known, then it's (**t**= Tax Rate planned), is:

$$t = 1 - R/([1-d]\{S'[1+s][1-f-i] - V\}$$
$$= 1 - 5,529/([1-0.3000]\{48,851$$
$$\quad [1+0.0500][1-0.5800-0.0100]$$
$$\quad -9,746\})$$
$$= 30.00\%$$

Law-3665:
If both (s= [S/S']-1= Sales Growth= 5.00%), (t= T/B= Tax Rate= 30.00%), (i= I/S= Interest Portion= 1.00%), (R= Retained Earnings= 5,529 millions USD), (S'= Sales of Past Year= 48,851 millions USD), (V= Variable Cost= 9,746 millions USD), and (f= F/S=Fixed Portion= 58.00%), are known, then it's (d= Dividend Portion or Payout planned), is:

d= 1-R/([1-t]{S'[1+s][1-f-i]-V}
 = 1-5,529/([1-0.3000]{48,851
  [1+0.0500][1-0.5800-0.0100]
  -9,746})
 = 30.00%

Law-3666:
If both (s= [S/S']-1= Sales Growth= 5.00%), (t= T/B= Tax Rate= 30.00%), (S= Sales or Revenues= 51,294 millions USD), (I= Interest Expense= 513 millions USD), (d= D/A= Dividend Portion or Payout= 30.00%), (S'= Sales of Past Year= 48,851 millions USD), (v= V/S= Variable Portion= 19.00%), and (F= Fixed Cost= 29,750 millions USD), are known, then it's (R= Retained Earnings planned), is:

R= [1-t][1-d]{S'[1+s]-Sv-F-I}
 = [1-0.3000][1-0.3000]{48,851
  [1+0.0500]-51,294*0.1900
  -29,750-513}
 = 5,529 millions USD

Law-3667:
If both (s= [S/S']-1= Sales Growth= 5.00%), (t= T/B= Tax Rate= 30.00%), ($\$$= Sales or Revenues= 51,294 millions USD), (I= Interest Expense= 513 millions USD), (d= D/A= Dividend Portion or Payout= 30.00%), (R= Retained Earnings= 5,529 millions USD), (v= V/S= Variable Portion= 19.00%), and (F= Fixed Cost= 29,750 millions USD), are known, then it's ($\$$'= Sales Past), must be:

$$\$' = (\$v + F + I + R/\{[1-t][1-d]\})/[1+s]$$
$$= (51,294*0.1900 + 29,750 + 513$$
$$+ 5,529/\{[1-0.3000]$$
$$[1-0.3000]\})/[1+0.0500]$$
$$= \underline{48,851} \text{ millions USD}$$

Law-3668:
If both ($\$$'= Sales of Past Year= 48,851 millions USD), (t= T/B= Tax Rate= 30.00%), ($\$$= Sales or Revenues= 51,294 millions USD), (I= Interest Expense= 513 millions USD), (d= D/A= Dividend Portion or Payout= 30.00%), (R= Retained Earnings= 5,529 millions USD), (v= V/S= Variable Portion= 19.00%), and (F= Fixed Cost= 29,750 millions USD), are known, then it's (s= Sales Growth planned), is:

$$s = (\$v + F + I + R/\{[1-t][1-d]\})/\$' - 1$$
$$= (51,294*0.1900 + 29,750 + 513 + 5,529$$
$$/\{[1-0.3000][1-0.3000]\})$$
$$/48,851 - 1$$
$$= \underline{5.00\%}$$

Law-3669:

If both ($'= Sales of Past Year= 48,851 millions USD), (t= T/B= Tax Rate= 30.00%), (s= [S/S']-1= Sales Growth= 5.00%), (I= Interest Expense= 513 millions USD), (d= D/A= Dividend Portion or Payout= 30.00%), (R= Retained Earnings= 5,529 millions USD), (v= V/S= Variable Portion= 19.00%), and (F= Fixed Cost= 29,750 millions USD), are known, then it's ($= Sales or Revenues planned), is:

$= ($'[1+s]-F-I-R/{[1-t][1-d]})/v
= (48,851[1+0.0500]- 29,750-513
-5,529/{[1-0.3000][1-0.3000]})
/ 0.1900
= 51,294 millions USD

Law-3670:

If both ($'= Sales of Past Year= 48,851 millions USD), (t= T/B= Tax Rate= 30.00%), (s= [S/S']-1= Sales Growth= 5.00%), (I= Interest Expense= 513 millions USD), (d= D/A= Dividend Portion or Payout= 30.00%), (R= Retained Earnings= 5,529 millions USD), ($= Sales or Revenues= 51,294 millions USD), and (F= Fixed Cost= 29,750 millions USD), are known, then it's (v= Variable Portion planned), is:

v= ($'[1+s]-F-I-R/{[1-t][1-d]})/$
= (48,851[1+0.0500]- 29,750-513
-5,529/{[1-0.3000][1-0.3000]})
/51,294
= 19.00%

Law-3671:

If both ($'= Sales of Past Year= 48,851 millions USD), (t= T/B= Tax Rate= 30.00%), (s= [S/S']-1= Sales Growth= 5.00%), (I= Interest Expense= 513 millions USD), (d= D/A=Dividend Portion or Payout= 30.00%), (R= Retained Earnings= 5,529 millions USD), ($= Sales or Revenues= 51,294 millions USD), and (v= V/S= Variable Portion= 19.00%), are known, then it's (F= Fixed Cost planned), is:

$$F= \$'[1+s]-\$v-I-R/\{[1-t][1-d]\}$$
$$= 48,851[1+0.0500]-51,294*0.1900$$
$$-513-5,529/\{[1-0.3000]$$
$$[1-0.3000]\}$$
$$= 29,750 \text{ millions USD}$$

Law-3672:

If both ($'= Sales of Past Year= 48,851 millions USD), (t= T/B= Tax Rate= 30.00%), (s= [S/S']-1= Sales Growth= 5.00%), (F= Fixed Cost= 29,750 millions USD), (d= D/A= Dividend Portion or Payout= 30.00%), (R= Retained Earnings= 5,529 millions USD), ($= Sales or Revenues= 51,294 millions USD), and (v= V/S= Variable Portion= 19.00%), are known, then it's (I= Interest Expense planned), is:

$$I= \$'[1+s]-\$v-F-R/\{[1-t][1-d]\}$$
$$= 48,851[1+0.0500]- 51,294*0.1900$$
$$-29,750-5,529/\{[1-0.3000]$$
$$[1-0.3000]\}$$
$$= 513 \text{ millions USD}$$

Law-3673:
If both ($'= Sales of Past Year= 48,851 millions USD), (I= Interest Expense= 513 millions USD), (s= [S/S']-1= Sales Growth= 5.00%), (F= Fixed Cost= 29,750 millions USD), (d= D/A=Dividend Portion or Payout= 30.00%), (R= Retained Earnings= 5,529 millions USD), ($= Sales or Revenues= 51,294 millions USD), and (v= V/S= Variable Portion= 19.00%), are known, then it's (t= Tax Rate planned), is:

$$t = 1 - R/([1-d]\{\$'[1+s]-\$v-F-I\})$$
$$= 1-5,529/([1-0.3000]\{48,851$$
$$[1+0.0500]-51,294*0.1900$$
$$-29,750-513\}$$
$$= 30.00\%$$

Law-3674:
If both ($'= Sales of Past Year= 48,851 millions USD), (I= Interest Expense= 513 millions USD), (s= [S/S']-1= Sales Growth= 5.00%), (F= Fixed Cost= 29,750 millions USD), (t= T/B= Tax Rate= 30.00%), (R= Retained Earnings= 5,529 millions USD), ($= Sales or Revenues= 51,294 millions USD), and (v= V/S= Variable Portion= 19.00%), are known, then it's (d= Dividend Portion or Payout planned), is:

$$d = 1 - R/([1-t]\{\$'[1+s]-\$v-F-I\})$$
$$= 1-5,529/([1-0.3000]\{48,851$$
$$[1+0.0500]-51,294*0.1900$$
$$-29,750-513\}$$
$$= 30.00\%$$

Law-3675:

If both ($'= Sales of Past Year= 48,851 millions USD), (i= I/S= Interest Portion= 1.00%), (s= [S/S']-1= Sales Growth= 5.00%), (F= Fixed Cost= 29,750 millions USD), (t= T/B= Tax Rate= 30.00%), (d= D/A= Dividend Portion or Payout= 30.00%), ($= Sales or Revenues= 51,294 millions USD), and (v= V/S= Variable Portion= 19.00%), are known, then it's (R= Retained Earnings planned), is:

$$R = [1-t][1-d]\{\$'[1+s]-\$v-F-\$i\}$$
$$= [1-t][1-d]\{\$'[1+s]-\$[v+i]-F\}$$
$$= [1-0.3000][1-0.3000]\{48,851$$
$$[1+0.0500]-51,294$$
$$[0.1900+0.0100]-29,750\}$$
$$= 5,529 \text{ millions USD}$$

Law-3676:

If both (R= Retained Earnings= 5,529 millions USD), (i= I/S= Interest Portion= 1.00%), (s= [S/S']-1= Sales Growth= 5.00%), (F= Fixed Cost= 29,750 millions USD), (t= T/B= Tax Rate= 30.00%), (d= D/A= Dividend Portion or Payout= 30.00%), ($= Sales or Revenues= 51,294 millions USD), and (v= V/S= Variable Portion= 19.00%), are known, then it's ($'= Sales Past), must be:

$$\$' = (F+\$[v+i]+R/\{[1-t][1-d]\}/[1+s]$$
$$= (29,750+51,294[0.1900+0.0100]$$
$$+5,529/\{[1-0.3000][1-0.3000]\})$$
$$/[1+0.0500]$$
$$= 48,851 \text{ millions USD}$$

Law-3677:
If both (**R**= Retained Earnings= 5,529 millions USD), (**i**= I/S= Interest Portion= 1.00%), (**$'**= Sales of Past Year= 48,851 millions USD), (**F**= Fixed Cost= 29,750 millions USD), (**t**= T/B= Tax Rate= 30.00%), (**d**= D/A= Dividend Portion or Payout= 30.00%), (**$**= Sales or Revenues= 51,294 millions USD), and (**v**= V/S= Variable Portion= 19.00%), are known, then it's (**s**= Sales Growth planned), is:

$$s = (F+\$[v+i]+R/\{[1-t][1-d]\})/\$'-1$$
$$= (29{,}750+51{,}294[0.1900+0.0100]$$
$$+5{,}529/\{[1-0.3000][1-0.3000]\})$$
$$/48{,}851-1$$
$$= 5.00\%$$

Law-3678:
If both (**R**= Retained Earnings= 5,529 millions USD), (**i**= I/S= Interest Portion= 1.00%), (**$'**= Sales of Past Year= 48,851 millions USD), (**F**= Fixed Cost= 29,750 millions USD), (**t**= T/B= Tax Rate= 30.00%), (**d**= D/A= Dividend Portion or Payout= 30.00%), (**s**= [S/S']-1= Sales Growth= 5.00%), and (**v**= V/S= Variable Portion= 19.00%), are known, then it's (**$**= Sales or Revenues planned), is:

$$\$ = (\$'[1+s\text{-}F\text{-}R/\{[1-t][1-d]\}/[v+i]$$
$$= (48{,}851[1+0.0500]\text{-}29{,}750\text{-}5{,}529$$
$$/\{[1-0.3000][1-0.3000]\})$$
$$/[0.1900+0.0100]$$
$$= 51{,}294 \text{ millions USD}$$

Law-3679:

If both (**R**= Retained Earnings= 5,529 millions USD), (**i**= I/S= Interest Portion= 1.00%), (**$'**= Sales of Past Year= 48,851 millions USD), (**F**= Fixed Cost= 29,750 millions USD), (**t**= T/B= Tax Rate= 30.00%), (**d**= D/A= Dividend Portion or Payout= 30.00%), (**s**= [S/S']-1= Sales Growth= 5.00%), and (**$**= Sales or Revenues= 51,294 millions USD), are known, then it's (**v**= Variable Portion planned), is:

$$v= (\$'[1+s\text{-}F\text{-}R/\{[1\text{-}t][1\text{-}d]\}/\$\text{-}i$$
$$= (48,851[1+0.0500]\text{-} 29,750\text{-}5,529$$
$$/\{[1\text{-}0.3000][1\text{-}0.3000]\})$$
$$/51,294\text{-}0.0100$$
$$= 19.00\%$$

Law-3680:

If both (**R**= Retained Earnings= 5,529 millions USD), (**v**= V/S= Variable Portion= 19.00%), (**$'**= Sales of Past Year= 48,851 millions USD), (**F**= Fixed Cost= 29,750 millions USD), (**t**= T/B= Tax Rate= 30.00%), (**d**= D/A= Dividend Portion or Payout= 30.00%), (**s**= [S/S']-1= Sales Growth= 5.00%), and (**$**= Sales or Revenues= 51,294 millions USD), are known, then it's (**i**= Interest Portion planned), is:

$$i= (\$'[1+s]\text{-}F\text{-}R/\{[1\text{-}t][1\text{-}d]\}/\$\text{-}v$$
$$= (48,851[1+0.0500]\text{-}29,750\text{-}5,529$$
$$/\{[1\text{-}0.3000][1\text{-}0.3000]\})$$
$$/51,294\text{-}0.1900$$
$$= 1.00\%$$

Law-3681:
If both (**R**= Retained Earnings= 5,529 millions USD), (**v**= V/S= Variable Portion= 19.00%), (**$'**= Sales of Past Year= 48,851 millions USD), (**i**= I/S= Interest Portion= 1.00%), (**t**= T/B= Tax Rate= 30.00%), (**d**= D/A= Dividend Portion or Payout= 30.00%), (**s**= [S/S']-1= Sales Growth= 5.00%), and (**$**= Sales or Revenues= 51,294 millions USD), are known, then it's (**F**= Fixed Cost planned), is:

$$F = \$'[1+s] - \$[v+i] - R/\{[1-t][1-d]\}$$
$$= 48,851[1+0.0500] - 51,294[0.1900 + 0.0100] - 5,529/\{[1-0.3000][1-0.3000]\}$$
$$= 29,750 \text{ millions USD}$$

Law-3682:
If both (**R**= Retained Earnings= 5,529 millions USD), (**v**= V/S= Variable Portion= 19.00%), (**$'**= Sales of Past Year= 48,851 millions USD), (**i**= I/S= Interest Portion= 1.00%), (**F**= Fixed Cost= 29,750 millions USD), (**d**= D/A= Dividend Portion or Payout= 30.00%), (**s**= [S/S']-1= Sales Growth= 5.00%), and (**$**= Sales or Revenues= 51,294 millions USD), are known, then it's (**t**= Tax Rate planned), is:

$$t = 1 - R/([1-d]\{\$'[1+s] - \$[v+i] - F\})$$
$$= 1 - 5,529/([1-0.3000]\{48,851[1+0.0500] - 51,294[0.1900 + 0.0100] - 29,750\})$$
$$= 30.00\%$$

Law-3683:

If both (**R**= Retained Earnings= 5,529 millions USD), (**v**= V/S= Variable Portion= 19.00%), (**$'**= Sales of Past Year= 48,851 millions USD), (**i**= I/S= Interest Portion= 1.00%), (**F**= Fixed Cost= 29,750 millions USD), (**t**= T/B= Tax Rate= 30.00%), (**s**= [S/S']-1= Sales Growth= 5.00%), and (**$**= Sales or Revenues= 51,294 millions USD), are known, then it's (**d**= Dividend Portion or Payout planned), is:

$$\mathbf{d} = 1 - \mathbf{R}/([1-\mathbf{t}]\{\mathbf{\$'}[1+\mathbf{s}] - \mathbf{\$}[\mathbf{v}+\mathbf{i}] - \mathbf{F}\})$$
$$= 1 - 5{,}529/([1-0.3000]\{48{,}851[1+0.0500] - 51{,}294[0.1900 + 0.0100] - 29{,}750\})$$
$$= 30.00\%$$

Law-3684:

If both (**d**= D/A= Dividend Portion or Payout= 30.00%), (**v**= V/S= Variable Portion= 19.00%), (**$'**= Sales of Past Year= 48,851 millions USD), (**i**= I/S= Interest Portion= 1.00%), (**F**= Fixed Cost= 29,750 millions USD), (**t**= T/B= Tax Rate= 30.00%), (**s**= [S/S']-1= Sales Growth= 5.00%), and (**$**= Sales or Revenues= 51,294 millions USD), are known, then it's (**R**= Retained Earnings planned), is:

$$\mathbf{R} = [1-\mathbf{t}][1-\mathbf{d}]\{\mathbf{\$'}[1+\mathbf{s}] - \mathbf{\$v} - \mathbf{F} - \mathbf{\$'i}[1+\mathbf{s}]\}$$
$$= [1-\mathbf{t}][1-\mathbf{d}]\{\mathbf{\$'}[1+\mathbf{s}][1-\mathbf{i}] - \mathbf{\$v} - \mathbf{F}\}$$
$$= [1-0.3000][1-0.3000]\{48{,}851[1+0.0500][1-0.0100] - 51{,}294*0.1900 - 29{,}750\}$$
$$= 5{,}529 \text{ millions USD}$$

Law-3685:

If both (d= D/A= Dividend Portion or Payout= 30.00%), (v= V/S= Variable Portion= 19.00%), (R= Retained Earnings= 5,529 millions USD), (i= I/S= Interest Portion= 1.00%), (F= Fixed Cost= 29,750 millions USD), (t= T/B= Tax Rate= 30.00%), (s= [S/S']-1= Sales Growth= 5.00%), and (S= Sales or Revenues= 51,294 millions USD), are known, then it's (S'= Sales Past), must be:

$S' = (Sv+F+R/\{[1-t][1-d]\})/\{[1+s][1-i]\}$
$= (51,294*0.1900 +29,750+5,529$
$\qquad /\{[1-0.3000][1-0.3000]\})$
$\qquad /\{[1+0.0500][1-0.0100]\}$
$= \underline{48,851}$ millions USD

Law-3686:

If both (d= D/A= Dividend Portion or Payout= 30.00%), (v= V/S= Variable Portion= 19.00%), (R= Retained Earnings= 5,529 millions USD), (i= I/S= Interest Portion= 1.00%), (S'= Sales of Past Year= 48,851 millions USD), (F= Fixed Cost= 29,750 millions USD), (t= T/B= Tax Rate= 30.00%), and (S= Sales or Revenues= 51,294 millions USD), are known, then it's (s= Sales Growth planned), is:

$s = (Sv+F+R/\{[1-t][1-d]\})/\{S'[1-i]\}-1$
$= (51,294*0.1900 +29,750+5,529$
$\qquad /\{[1-0.3000][1-0.3000]\})$
$\qquad /\{48,851[1-0.0100]\}-1$
$= \underline{5.00}\%$

Law-3687:

If both (**d**= D/A= Dividend Portion or Payout= 30.00%), (**v**= V/S= Variable Portion= 19.00%), (**R**= Retained Earnings= 5,529 millions USD), (**s**= [S/S']-1= Sales Growth= 5.00%), (**S'**= Sales of Past Year= 48,851 millions USD), (**F**= Fixed Cost= 29,750 millions USD), (**t**= T/B= Tax Rate= 30.00%), and (**S**= Sales or Revenues= 51,294 millions USD), are known, then it's (**i**= Interest Portion planned), is:

$$i = 1-(Sv+F+R/\{[1-t][1-d]\})/\{S'[1+s]\}$$
$$= 1-(51,294*0.1900+29,750+5,529$$
$$/\{[1-0.3000][1-0.3000]\})$$
$$/\{48,851[1+0.0500]\}$$
$$= 1.00\%$$

Law-3688:

If both (**d**= D/A= Dividend Portion or Payout= 30.00%), (**v**= V/S= Variable Portion= 19.00%), (**R**= Retained Earnings= 5,529 millions USD), (**s**= [S/S']-1= Sales Growth= 5.00%), (**S'**= Sales of Past Year= 48,851 millions USD), (**F**= Fixed Cost= 29,750 millions USD), (**t**= T/B= Tax Rate= 30.00%), and (**i**= I/S= Interest Portion= 1.00%), are known, then it's (**S**= Sales or Revenues planned), is:

$$S = (S'[1+s][1-i]-F-R/\{[1-t][1-d]\})/v$$
$$= (48,851[1+0.0500][1-0.0100]-29,750$$
$$-5,529/\{[1-0.3000][1-0.3000]\})$$
$$/0.1900$$
$$= 51,294 \text{ millions USD}$$

Law-3689:
If both (d= D/A= Dividend Portion or Payout= 30.00%), ($\$$= Sales or Revenues= 51,294 millions USD), (R= Retained Earnings= 5,529 millions USD), (s= [S/S']-1= Sales Growth= 5.00%), ($\$'$= Sales of Past Year= 48,851 millions USD), (F= Fixed Cost= 29,750 millions USD), (t= T/B= Tax Rate= 30.00%), and (i= I/S= Interest Portion= 1.00%), are known, then it's (v= Variable Portion planned), is:

$v = (\$'[1+s][1-i]-F-R/\{[1-t][1-d]\})/\$$
$= (48,851[1+0.0500][1-0.0100]-29,750$
$\quad -5,529/\{[1-0.3000][1-0.3000]\})$
$\quad /51,294$
$= \underline{19.00\%}$

Law-3690:
If both (d= D/A= Dividend Portion or Payout= 30.00%), ($\$$= Sales or Revenues= 51,294 millions USD), (R= Retained Earnings= 5,529 millions USD), (s= [S/S']-1= Sales Growth= 5.00%), ($\$'$= Sales of Past Year= 48,851 millions USD), (v= V/S= Variable Portion= 19.00%), (t= T/B= Tax Rate= 30.00%), and (i= I/S= Interest Portion= 1.00%), are known, then it's (F= Fixed Cost planned), is:

$F = (\$'[1+s][1-i]-\$v-R/\{[1-t][1-d]\}$
$= (48,851[1+0.0500][1-0.0100]$
$\quad -51,294*0.1900-5,529$
$\quad /\{[1-0.3000][1-0.3000]\}$
$= \underline{29,750}$ millions USD

Law-3691:
If both (**d**= D/A= Dividend Portion or Payout= 30.00%), (**$**= Sales or Revenues= 51,294 millions USD), (**R**= Retained Earnings= 5,529 millions USD), (**s**= [S/S']-1= Sales Growth= 5.00%), (**$'**= Sales of Past Year= 48,851 millions USD), (**v**= V/S= Variable Portion= 19.00%), (**F**= Fixed Cost= 29,750 millions USD), and (**i**= I/S= Interest Portion= 1.00%), are known, then it's (**t**= Tax rate planned), is:

$$t = 1 - R/([1-d]\{\$'[1+s][1-i] - \$v - F\})$$
$$= 1 - 5{,}529/([1-0.3000]\{48{,}851[1+0.0500][1-0.0100] - 51{,}294 * 0.1900 - 29{,}750\})$$
$$= 30.00\%$$

Law-3692:
If both (**t**= T/B= Tax Rate= 30.00%), (**$**= Sales or Revenues= 51,294 millions USD), (**R**= Retained Earnings= 5,529 millions USD), (**s**= [S/S']-1= Sales Growth= 5.00%), (**$'**= Sales of Past Year= 48,851 millions USD), (**v**= V/S= Variable Portion= 19.00%), (**F**= Fixed Cost= 29,750 millions USD), and (**i**= I/S= Interest Portion= 1.00%), are known, then it's (**d**= Dividend Portion or Payout planned), is:

$$d = 1 - R/([1-t]\{\$'[1+s][1-i] - \$v - F\})$$
$$= 1 - 5{,}529/([1-0.3000]\{48{,}851[1+0.0500][1-0.0100] - 51{,}294 * 0.1900 - 29{,}750\})$$
$$= 30.00\%$$

Law-3693:
If both (t= T/B= Tax Rate= 30.00%), ($$= Sales or Revenues= 51,294 millions USD), (d= D/A= Dividend Portion or Payout= 30.00%), (s= [S/S']-1= Sales Growth= 5.00%), ($$'= Sales of Past Year= 48,851 millions USD), (v= V/S= Variable Portion= 19.00%), (f= F/S= Fixed Portion= 58.00%), and (I= Interest Expense= 513 millions USD), are known, then it's (R= Retained Earnings planned), is:
R= [1-t][1-d]{$$'[1+s]-$$v$-$$f$-$I$}
 = [1-t][1-d]{$$'[1+s]-$$[v+f]-I})
 = [1-0.3000][1-0.3000]{48,851
   [1+0.0500]-51,294[0.1900
   +0.5800]-513})
 = 5,529 millions USD

Law-3694:
If both (t= T/B= Tax Rate= 30.00%), ($$= Sales or Revenues= 51,294 millions USD), (d= D/A= Dividend Portion or Payout= 30.00%), (s= [S/S']-1= Sales Growth= 5.00%), (R= Retained Earnings= 5,529 millions USD), (v= V/S= Variable Portion= 19.00%), (f= F/S= Fixed Portion= 58.00%), and (I= Interest Expense= 513 millions USD), are known, then it's ($$'= Sales Past), must be:
$$'= ($$[v+f]+I+R/{[1-t][1-d]})/[1+s]
 = (51,294[0.1900+0.5800]+513+5,529
   /{[1-0.3000][1-0.3000]})
   /[1+0.0500]
 = 48,851 millions USD

Law-3695:
If both (**t**= T/B= Tax Rate= 30.00%), (**$**= Sales or Revenues= 51,294 millions USD), (**d**= D/A= Dividend Portion or Payout= 30.00%), (**$'**= Sales of Past Year= 48,851 millions USD), (**R**= Retained Earnings= 5,529 millions USD), (**v**= V/S= Variable Portion= 19.00%), (**f**= F/S=Fixed Portion= 58.00%), and (**I**= Interest Expense= 513 millions USD), are known, then it's (**s**= Sales Growth planned), is:

$s = (\$[v+f]+I+R/\{[1-t][1-d]\})/\$' - 1$
$= (51,294[0.1900+0.5800]+513+5,529$
$\qquad /\{[1-0.3000][1-0.3000]\})$
$\qquad /48,851 - 1$
$= \underline{5.00}\%$

Law-3696:
If both (**t**= T/B= Tax Rate= 30.00%), (**s**= [S/S']-1= Sales Growth= 5.00%), (**d**= D/A= Dividend Portion or Payout= 30.00%), (**$'**= Sales of Past Year= 48,851 millions USD), (**R**= Retained Earnings= 5,529 millions USD), (**v**= V/S= Variable Portion= 19.00%), (**f**= F/S= Fixed Portion= 58.00%), and (**I**= Interest Expense= 513 millions USD), are known, then it's (**$**= Sales or Revenues planned), is:

$\$ = (\$'[1+s]-I-R/\{[1-t][1-d]\})/[v+f]$
$= (48,851[1+0.0500]-513-5,529$
$\qquad /\{[1-0.3000][1-0.3000]\})$
$\qquad /[0.1900+0.5800]$
$= \underline{51,294}$ millions USD

Law-3697:
If both (**t**= T/B= Tax Rate= 30.00%), (**s**= [S/S']-1= Sales Growth= 5.00%), (**d**= D/A= Dividend Portion or Payout= 30.00%), (**$'**= Sales of Past Year= 48,851 millions USD), (**R**= Retained Earnings= 5,529 millions USD), (**$**= Sales or Revenues= 51,294 millions USD), (**f**= F/S=Fixed Portion= 58.00%), and (**I**= Interest Expense= 513 millions USD), are known, then it's (**v**= Variable Portion planned), is:
$$v= (\$'[1+s]-I-R/\{[1-t][1-d]\})/\$-f$$
$$= (48,851[1+0.0500]-513-5,529$$
$$/\{[1-0.3000][1-0.3000]\})$$
$$/51,294-0.5800$$
$$= 19.00\%$$

Law-3698:
If both (**t**= T/B= Tax Rate= 30.00%), (**s**= [S/S']-1= Sales Growth= 5.00%), (**d**= D/A= Dividend Portion or Payout= 30.00%), (**$'**= Sales of Past Year= 48,851 millions USD), (**R**= Retained Earnings= 5,529 millions USD), (**$**= Sales or Revenues= 51,294 millions USD), (**v**= V/S= Variable Portion= 19.00%), and (**I**= Interest Expense= 513 millions USD), are known, then it's (**f**= Fixed Portion planned), is:
$$f= (\$'[1+s]-I-R/\{[1-t][1-d]\})/\$-v$$
$$= (48,851[1+0.0500]-513-5,529$$
$$/\{[1-0.3000][1-0.3000]\})$$
$$/51,294-0.1900$$
$$= 58.00\%$$

Law-3699:

If both (**t**= T/B= Tax Rate= 30.00%), (**s**= [S/S']-1= Sales Growth= 5.00%), (**d**= D/A= Dividend Portion or Payout= 30.00%), (**$'**= Sales of Past Year= 48,851 millions USD), (**R**= Retained Earnings= 5,529 millions USD), (**$**= Sales or Revenues= 51,294 millions USD), (**v**= V/S= Variable Portion= 19.00%), and (**f**= F/S= Fixed Portion= 58.00%), are known, then it's (**I**= Interest Expense planned), is:

$$I = \$'[1+s] - \$[v+f] - R/\{[1-t][1-d]\}$$
$$= 48,851[1+0.0500] - 51,294[0.1900$$
$$+ 0.5800] - 5,529/\{[1-0.3000]$$
$$[1-0.3000]\}$$
$$= 513 \text{ millions USD}$$

Law-3700:

If both (**I**= Interest Expense= 513 millions USD), (**s**= [S/S']-1= Sales Growth= 5.00%), (**d**= D/A= Dividend Portion or Payout= 30.00%), (**$'**= Sales of Past Year= 48,851 millions USD), (**R**= Retained Earnings= 5,529 millions USD), (**$**= Sales or Revenues= 51,294 millions USD), (**v**= V/S= Variable Portion= 19.00%), and (**f**= F/S= Fixed Portion= 58.00%), are known, then it's (**t**= Tax Rate planned), is:

$$t = 1 - R/([1-d]\{\$'[1+s] - \$[v+f] - I\})$$
$$= 1 - 5,529/([1-0.3000]\{48,851$$
$$[1+0.0500] - 51,294[0.1900$$
$$+ 0.5800] - 513]\})$$
$$= 30.00\%$$

Law-3701:

If both (**I**= Interest Expense= 513 millions USD), (**s**= [S/S']-1= Sales Growth= 5.00%), (**t**= T/B= Tax Rate= 30.00%), (**S'**= Sales of Past Year= 48,851 millions USD), (**R**= Retained Earnings= 5,529 millions USD), (**S**= Sales or Revenues= 51,294 millions USD), (**v**= V/S= Variable Portion= 19.00%), and (**f**= F/S= Fixed Portion= 58.00%), are known, then it's (**d**= Dividend Portion or Payout planned), is:

$$d= 1-R/([1-t]\{S'[1+s]-S[v+f]-I\})$$
$$= 1-5,529/([1-0.3000]\{48,851$$
$$[1+0.0500]- 51,294[0.1900$$
$$+0.5800]-513]\})$$
$$= 30.00\%$$

Law-3702:

If both (**i**= I/S= Interest Portion= 1.00%), (**s**= [S/S']-1= Sales Growth= 5.00%), (**t**= T/B= Tax Rate= 30.00%), (**S'**= Sales of Past Year= 48,851 millions USD), (**d**= D/A= Dividend Portion or Payout= 30.00%), (**S**= Sales or Revenues= 51,294 millions USD), (**v**= V/S= Variable Portion= 19.00%), and (**f**= F/S= Fixed Portion= 58.00%), are known, then it's (**R**= Retained Earnings planned), is:

$$R= [1-t][1-d]\{S'[1+s]-Sv-Sf-Si\}$$
$$= [1-t][1-d]\{S'[1+s]-S[v+f+i]\}$$
$$= [1-0.3000][1-0.3000]\{48,851$$
$$[1+0.0500]- 51,294[0.1900$$
$$+0.5800+0.0100]]\})$$
$$= 5,529 \text{ millions USD}$$

Law-3703:
If both (**i**= I/S= Interest Portion= 1.00%), (**s**= [S/S']-1= Sales Growth= 5.00%), (**t**= T/B= Tax Rate= 30.00%), (**R**= Retained Earnings= 5,529 millions USD), (**d**= D/A= Dividend Portion or Payout= 30.00%), (**$**= Sales or Revenues= 51,294 millions USD), (**v**= V/S= Variable Portion= 19.00%), and (**f**= F/S= Fixed Portion= 58.00%), are known, then it's (**$'**= Sales Past), must be:
$$\$' = (\$[v+f+i]+R/\{[1-t][1-d]\})/[1+s]$$
$$= (51{,}294[0.1900+0.5800+0.0100]$$
$$+5{,}529/\{[1-0.3000][1-0.3000]\})$$
$$/[1+0.0500]$$
$$= 48{,}851 \text{ millions USD}$$

Law-3704:
If both (**i**= I/S= Interest Portion= 1.00%), (**$'**= Sales of Past Year= 48,851 millions USD), (**t**= T/B= Tax Rate= 30.00%), (**R**= Retained Earnings= 5,529 millions USD), (**d**= D/A=Dividend Portion or Payout= 30.00%), (**$**= Sales or Revenues= 51,294 millions USD), (**v**= V/S= Variable Portion= 19.00%), and (**f**= F/S= Fixed Portion= 58.00%), are known, then it's (**s**= Sales Growth planned), is:
$$s = (\$[v+f+i]+R/\{[1-t][1-d]\})/\$'-1$$
$$= (51{,}294[0.1900+0.5800+0.0100]$$
$$+5{,}529/\{[1-0.3000][1-0.3000]\})$$
$$/48{,}851-1$$
$$= 5.00\%$$

Law-3705:

If both (i= I/S= Interest Portion= 1.00%), ($\$'$= Sales of Past Year= 48,851 millions USD), (t= T/B= Tax Rate= 30.00%), (R= Retained Earnings= 5,529 millions USD), (d= D/A= Dividend Portion or Payout= 30.00%), (s= [S/S']-1= Sales Growth= 5.00%), (v= V/S= Variable Portion= 19.00%), and (f= F/S= Fixed Portion= 58.00%), are known, then it's ($\$$= Sales or Revenues planned), is:

$$\$ = (\$'[1+s]-R/\{[1-t][1-d]\})/[v+f+i]$$
$$= (48,851[1+0.0500]-5,529$$
$$/\{[1-0.3000][1-0.3000]\})$$
$$/[0.1900+0.5800+0.0100]$$
$$= 51,294 \text{ millions USD}$$

Law-3706:

If both (i= I/S= Interest Portion= 1.00%), ($\$'$= Sales of Past Year= 48,851 millions USD), (t= T/B= Tax Rate= 30.00%), (R= Retained Earnings= 5,529 millions USD), (d= D/A= Dividend Portion or Payout= 30.00%), (s= [S/S']-1= Sales Growth= 5.00%), ($\$$= Sales or Revenues= 51,294 millions USD), and (f= F/S=Fixed Portion= 58.00%), are known, then it's (v= Variable Portion planned), is:

$$v = (\$'[1+s]-R/\{[1-t][1-d]\})/\$-f-i$$
$$= (48,851[1+0.0500]-5,529$$
$$/\{[1-0.3000][1-0.3000]\})$$
$$/51,294-0.5800-0.0100$$
$$= 19.00\%$$

Law-3707:

If both (**i**= I/S= Interest Portion= 1.00%), (**$'**= Sales of Past Year= 48,851 millions USD), (**t**= T/B= Tax Rate= 30.00%), (**R**= Retained Earnings= 5,529 millions USD), (**d**= D/A= Dividend Portion or Payout= 30.00%), (**s**= [S/S']-1= Sales Growth= 5.00%), (**$**= Sales or Revenues= 51,294 millions USD), and (**v**= V/S= Variable Portion= 19.00%), are known, then it's (**f**= Fixed Portion planned), is:

$$v = (\$'[1+s]-R/\{[1-t][1-d]\})/\$-v-i$$
$$= (48,851[1+0.0500]-5,529/\{[1-0.3000][1-0.3000]\})/51,294-0.1900-0.0100$$
$$= 58.00\%$$

Law-3708:

If both (**f**= F/S= Fixed Portion= 58.00%), (**$'**= Sales of Past Year= 48,851 millions USD), (**t**= T/B= Tax Rate= 30.00%), (**R**= Retained Earnings= 5,529 millions USD), (**d**= D/A= Dividend Portion or Payout= 30.00%), (**s**= [S/S']-1= Sales Growth= 5.00%), (**$**= Sales or Revenues= 51,294 millions USD), and (**v**= V/S= Variable Portion= 19.00%), are known, then it's (**i**= Interest Portion planned), is:

$$i = (\$'[1+s]-R/\{[1-t][1-d]\})/\$-v-f$$
$$= (48,851[1+0.0500]-5,529/\{[1-0.3000][1-0.3000]\})/51,294-0.1900-0.5800$$
$$= 1.00\%$$

Law-3709:
If both (**f**= F/S= Fixed Portion= 58.00%), (**$'**= Sales of Past Year= 48,851 millions USD), (**i**= I/S= Interest Portion= 1.00%), (**R**= Retained Earnings= 5,529 millions USD), (**d**= D/A= Dividend Portion or Payout= 30.00%), (**s**= [S/S']-1= Sales Growth= 5.00%), (**$**= Sales or Revenues= 51,294 millions USD), and (**v**= V/S= Variable Portion= 19.00%), are known, then it's (**t**= Tax Rate planned), is:
    t= 1-**R**/([1-**d**]{**$'**[1+**s**]-**$**[**v**+**f** +**i**]})
      = 1-5,529/([1-0.3000]{48,851
          [1+0.0500]-51,294[0.1900
          +0.5800+0.0100]})
      = 30.00%

Law-3710:
If both (**f**= F/S= Fixed Portion= 58.00%), (**$'**= Sales of Past Year= 48,851 millions USD), (**i**= I/S= Interest Portion= 1.00%), (**R**= Retained Earnings= 5,529 millions USD), (**t**= T/B= Tax Rate= 30.00%), (**s**= [S/S']-1= Sales Growth= 5.00%), (**$**= Sales or Revenues= 51,294 millions USD), and (**v**= V/S= Variable Portion= 19.00%), are known, then it's (**d**= Dividend Portion or Payout planned), is:
    d= 1-**R**/([1-**t**]{**$'**[1+**s**]-**$**[**v**+**f**+**i**]})
      = 1-5,529/([1-0.3000]{48,851
          [1+0.0500]-51,294[0.1900
          +0.5800+0.0100]})
      = 30.00%

Law-3711:
If both (f= F/S= Fixed Portion= 58.00%), ($\$'$= Sales of Past Year= 48,851 millions USD), (i= I/S= Interest Portion= 1.00%), (d= D/A= Dividend Portion or Payout= 30.00%), (t= T/B= Tax Rate= 30.00%), (s= [S/S']-1= Sales Growth= 5.00%), ($\$$= Sales or Revenues= 51,294 millions USD), and (v= V/S= Variable Portion= 19.00%), are known, then it's (R= Retained Earnings planned), is:

R = [1-t][1-d]{$\$'$[1+$s$]-$\$v$-$\$f$-$\$'i$[1+$s$]}
 = [1-t][1-d]{$\$'$[1+$s$][1-$i$]-$\$$[v+f]}
 = [1-0.3000][1-0.3000]{48,851
   [1+0.0500][1-0.0100]
   -51,294[0.1900+0.5800]}
 = 5,529 millions USD

Law-3712:
If both (f= F/S= Fixed Portion= 58.00%), (R= Retained Earnings= 5,529 millions USD), (i= I/S= Interest Portion= 1.00%), (d= D/A= Dividend Portion or Payout= 30.00%), (t= T/B= Tax Rate= 30.00%), (s= [S/S']-1= Sales Growth= 5.00%), ($\$$= Sales or Revenues= 51,294 millions USD), and (v= V/S= Variable Portion= 19.00%), are known, then it's ($\$'$= Sales Past), must be:

$\$'$ = ($\$$[v+f]+R/{[1-t][1-d]})/{[1+s][1-i]}
 = (51,294[0.1900+0.5800]+5,529
   /{[1-0.3000][1-0.3000]})
   /{[1+0.0500][1-0.0100]}
 = 48,851 millions USD

Law-3713:

If both (**f**= F/S= Fixed Portion= 58.00%), (**R**= Retained Earnings= 5,529 millions USD), (**i**= I/S= Interest Portion= 1.00%), (**d**= D/A= Dividend Portion or Payout= 30.00%), (**t**= T/B= Tax Rate= 30.00%), (**$'**= Sales of Past Year= 48,851 millions USD), (**$**= Sales or Revenues= 51,294 millions USD), and (**v**= V/S= Variable Portion= 19.00%), are known, then it's (**s**= Sales Growth planned), is:

$s = (\$[v+f]+R/\{[1-t][1-d]\})/\{\$'[1-i]\} - 1$
$= (51,294[0.1900+0.5800]+5,529$
$\quad /\{[1-0.3000][1-0.3000]\})$
$\quad /\{48,851[1-0.0100]\} - 1$
$= 48,851$ millions USD

Law-3714:

If both (**f**= F/S= Fixed Portion= 58.00%), (**R**= Retained Earnings= 5,529 millions USD), (**s**= [S/S']-1= Sales Growth= 5.00%), (**d**= D/A− Dividend Portion or Payout= 30.00%), (**t**= T/B= Tax Rate= 30.00%), (**$'**= Sales of Past Year= 48,851 millions USD), (**$**= Sales or Revenues= 51,294 millions USD), and (**v**= V/S= Variable Portion= 19.00%), are known, then it's (**i**= Interest Portion planned), is:

$i = 1-(\$[v+f]+R/\{[1-t][1-d]\})/\{\$'[1+s]\}$
$= 1-(51,294[0.1900+0.5800]+5,529$
$\quad /\{[1-0.3000][1-0.3000]\})$
$\quad /\{48,851[1+0.0500]\}$
$= 1.00\%$

Law-3715:
If both (**f**= F/S= Fixed Portion= 58.00%), (**R**= Retained Earnings= 5,529 millions USD), (**s**= [S/S']-1= Sales Growth= 5.00%), (**d**= D/A= Dividend Portion or Payout= 30.00%), (**t**= T/B= Tax Rate= 30.00%), (**$'**= Sales of Past Year= 48,851 millions USD), (**i**= I/S= Interest Portion= 1.00%), and (**v**= V/S= Variable Portion= 19.00%), are known, then it's (**$**= Sales or Revenues planned), is:

$$\$ = (\$'[1+s][1-i]-R/\{[1-t][1-d]\})/[v+f]$$
$$= (48,851[1+0.0500][1-0.0100]-5,529$$
$$/\{[1-0.3000][1-0.3000]\})$$
$$/[0.1900+0.5800]$$
$$= \underline{51,294} \text{ millions USD}$$

Law-3716:
If both (**f**= F/S= Fixed Portion= 58.00%), (**R**= Retained Earnings= 5,529 millions USD), (**s**= [S/S']-1= Sales Growth= 5.00%), (**d**= D/A= Dividend Portion or Payout= 30.00%), (**t**= T/B= Tax Rate= 30.00%), (**$'**= Sales of Past Year= 48,851 millions USD), (**i**= I/S= Interest Portion= 1.00%), and (**$**= Sales or Revenues= 51,294 millions USD), are known, then it's (**v**= Variable Portion planned), is:

$$v = (\$'[1+s][1-i]-R/\{[1-t][1-d]\})/\$-f$$
$$= (48,851[1+0.0500][1-0.0100]-5,529$$
$$/\{[1-0.3000][1-0.3000]\})$$
$$/51,294-0.5800$$
$$= \underline{19.00}\%$$

Law-3717:

If both (v= V/S= Variable Portion= 19.00%), (R= Retained Earnings= 5,529 millions USD), (s= [S/S']-1= Sales Growth= 5.00%), (d= D/A= Dividend Portion or Payout= 30.00%), (t= T/B= Tax Rate= 30.00%), ($\$'$= Sales of Past Year= 48,851 millions USD), (i= I/S= Interest Portion= 1.00%), and ($\$$= Sales or Revenues= 51,294 millions USD), are known, then it's (f= Fixed Portion planned), is:

$f = (\$'[1+s][1-i]-R/\{[1-t][1-d]\})/\$-v$
= (48,851[1+0.0500][1-0.0100]-5,529
/{[1-0.3000][1-0.3000]})
/51,294-0.1900
= 58.00%

Law-3718:

If both (v= V/S= Variable Portion= 19.00%), (R= Retained Earnings= 5,529 millions USD), (s= [S/S']-1= Sales Growth= 5.00%), (d= D/A= Dividend Portion or Payout= 30.00%), (f= F/S= Fixed Portion= 58.00%), ($\$'$= Sales of Past Year= 48,851 millions USD), (i= I/S= Interest Portion= 1.00%), and ($\$$= Sales or Revenues= 51,294 millions USD), are known, then it's (t= Tax Rate planned), is:

$t = 1-R/([1-d]\{\$'[1+s][1-i]-\$[v+f]\})$
= 1-5,529/([1-0.3000]{48,851
[1+0.0500][1-0.0100]-51,294
[0.1900+0.5800]})
= 30.00%

Law-3719:
If both (v= V/S= Variable Portion= 19.00%), (R= Retained Earnings= 5,529 millions USD), (s= [S/S']-1= Sales Growth= 5.00%), (t= T/B= Tax Rate= 30.00%), (f= F/S= Fixed Portion= 58.00%), ($\$'$= Sales of Past Year= 48,851 millions USD), (i= I/S= Interest Portion= 1.00%), and ($\$$= Sales or Revenues= 51,294 millions USD), are known, then it's (d= Dividend Portion or Payout planned), is:

d= 1-R/([1-t]{$\$'$[1+$s$][1-$i$]-$\$$[v+f]})
= 1-5,529/([1-0.3000]{48,851
[1+0.0500][1-0.0100]-51,294
[0.1900+0.5800]})
= 30.00%

Law-3720:
If both (v= V/S= Variable Portion= 19.00%), (s= [S/S']-1= Sales Growth= 5.00%), (t= T/B= Tax Rate= 30.00%), (d= D/A= Dividend Portion or Payout= 30.00%), (f= F/S= Fixed Portion= 58.00%), ($\$'$= Sales of Past Year= 48,851 millions USD), (I= Interest Expense= 513 millions USD), and ($\$$= Sales or Revenues= 51,294 millions USD), are known, then it's (R= Retained Earnings planned), is:

R= [1-t][1-d]{$\$'$[1+$s$]-$\$v$-$\$f$[1+$s$]-$I$}
= [1-t][1-d]{$\$'$[1+$s$][1-$f$]-$\$v$-$I$}
= [1-0.3000][1-0.3000]{48,851
[1+0.0500][1-0.5800]-51,294
*0.1900-513]}
= 5,529 millions USD

Law-3721:

If both (v= V/S= Variable Portion= 19.00%), (s= [S/S']-1= Sales Growth= 5.00%), (t= T/B= Tax Rate= 30.00%), (d= D/A= Dividend Portion or Payout= 30.00%), (f= F/S= Fixed Portion= 58.00%), (R= Retained Earnings= 5,529 millions USD), (I=Interest Expense= 513 millions USD), and (S= Sales or Revenues= 51,294 millions USD), are known, then it's (S'= Sales Past), must be:

S'= (Sv+I+R/{[1-t][1-d]})/{[1+s][1-f]}
= (51,294*0.1900+513+5,529
/{[1-0.3000][1-0.3000]})
/{[1+0.0500][1-0.5800]}
= 48,851 millions USD

Law-3722:

If both (v= V/S= Variable Portion= 19.00%), (S'= Sales of Past Year= 48,851 millions USD), (t= T/B= Tax Rate= 30.00%), (d= D/A= Dividend Portion or Payout= 30.00%), (f= F/S= Fixed Portion= 58.00%), (R= Retained Earnings= 5,529 millions USD), (I=Interest Expense= 513 millions USD), and (S= Sales or Revenues= 51,294 millions USD), are known, then it's (s= Sales Growth planned), is:

s= (Sv+I+R/{[1-t][1-d]})/{S'[1-f]}-1
= (51,294*0.1900+513+5,529
/{[1-0.3000][1-0.3000]})
/{48,851[1-0.5800]}-1
= 5.00%

Law-3723:
If both (v= V/S= Variable Portion= 19.00%), ($'= Sales of Past Year= 48,851 millions USD), (t= T/B= Tax Rate= 30.00%), (d= D/A= Dividend Portion or Payout= 30.00%), (s= [S/S']-1= Sales Growth= 5.00%), (R= Retained Earnings= 5,529 millions USD), (I=Interest Expense= 513 millions USD), and ($= Sales or Revenues= 51,294 millions USD), are known, then it's (f= Fixed Portion planned), is:

$$f = 1-(\$v+I+R/\{[1-t][1-d]\})/\{\$'[1+s]\}$$
$$= 1-(51,294*0.1900+513+5,529$$
$$/\{[1-0.3000][1-0.3000]\})$$
$$/\{48,851[1+0.0500]\}$$
$$= 58.00\%$$

Law-3724:
If both (v= V/S= Variable Portion= 19.00%), ($'= Sales of Past Year= 48,851 millions USD), (t= T/B= Tax Rate= 30.00%), (d= D/A= Dividend Portion or Payout= 30.00%), (s= [S/S']-1= Sales Growth= 5.00%), (R= Retained Earnings= 5,529 millions USD), (I=Interest Expense= 513 millions USD), and (f= F/S= Fixed Portion= 58.00%), are known, then it's ($= Sales or Revenues planned), is:

$$\$ = (\$'[1+s][1-f]-I-R/\{[1-t][1-d]\})/v$$
$$= (48,851[1+0.0500][1-0.5800]-513$$
$$-5,529/\{[1-0.3000][1-0.3000]\})$$
$$/0.1900$$
$$= 51,294 \text{ millions USD}$$

Law-3725:

If both ($= Sales or Revenues= 51,294 millions USD), ($'= Sales of Past Year= 48,851 millions USD), (t= T/B= Tax Rate= 30.00%), (d= D/A= Dividend Portion or Payout= 30.00%), (s= [S/S']-1= Sales Growth= 5.00%), (R= Retained Earnings= 5,529 millions USD), (I=Interest Expense= 513 millions USD), and (f= F/S= Fixed Portion= 58.00%), are known, then it's (v= Variable Portion planned), is:

$v = (\$'[1+s][1-f]-I-R/\{[1-t][1-d]\})/\$$
$= (48,851[1+0.0500][1-0.5800]-513$
$\quad -5,529/\{[1-0.3000][1-0.3000]\})$
$\quad /51,294$
$= \underline{19.00\%}$

Law-3726:

If both ($= Sales or Revenues= 51,294 millions USD), ($'= Sales of Past Year= 48,851 millions USD), (v= V/S= Variable Portion= 19.00%), (t= T/B= Tax Rate= 30.00%), (d= D/A= Dividend Portion or Payout= 30.00%), (s= [S/S']-1= Sales Growth= 5.00%), (R= Retained Earnings= 5,529 millions USD), and (f= F/S= Fixed Portion= 58.00%), are known, then it's (I= Interest Expense planned), is:

$I = \$'[1+s][1-f]-\$v-R/\{[1-t][1-d]\}$
$= 48,851[1+0.0500][1-0.5800]- 51,294$
$\quad *0.1900-5,529/\{[1-0.3000]$
$\quad [1-0.3000]\}$
$= \underline{513}$ millions USD

Law-3727:
If both ($= Sales or Revenues= 51,294 millions USD), ($'= Sales of Past Year= 48,851 millions USD), (v= V/S= Variable Portion= 19.00%), (I= Interest Expense= 513 millions USD), (d= D/A= Dividend Portion or Payout= 30.00%), (s= [S/S']-1= Sales Growth= 5.00%), (R= Retained Earnings= 5,529 millions USD), and (f= F/S= Fixed Portion= 58.00%), are known, then it's (t= Tax Rate planned), is:
$$t= 1-R/([1-d]\{\$'[1+s][1-f]-\$v-I\}$$
$$= 1-5,529/\{[1-0.3000]\{48,851$$
$$[1+0.0500][1-0.5800]$$
$$-51,294*0.1900-513]\}$$
$$= 30.00\%$$

Law-3728:
If both ($= Sales or Revenues= 51,294 millions USD), ($'= Sales of Past Year= 48,851 millions USD), (v= V/S= Variable Portion= 19.00%), (I= Interest Expense= 513 millions USD), (t= T/B= Tax Rate= 30.00%), (s= [S/S']-1= Sales Growth= 5.00%), (R= Retained Earnings= 5,529 millions USD), and (f= F/S= Fixed Portion= 58.00%), are known, then it's (d= Dividend Portion or Payout planned), is:
$$d= 1-R/([1-t]\{\$'[1+s][1-f]-\$v-I\}$$
$$= 1-5,529/\{[1-0.3000]\{48,851$$
$$[1+0.0500][1-0.5800]$$
$$-51,294*0.1900-513]\}$$
$$= 30.00\%$$

Law-3729:
If both ($= Sales or Revenues= 51,294 millions USD), ($'= Sales of Past Year= 48,851 millions USD), (i= I/S= Interest Portion= 1.00%), (v= V/S= Variable Portion= 19.00%), (d= D/A= Dividend Portion or Payout= 30.00%), (t= T/B= Tax Rate= 30.00%), (s= [S/S']-1= Sales Growth= 5.00%), and (f= F/S= Fixed Portion= 58.00%), are known, then it's (R= Retained Earnings planned), is:

$$R = [1-t][1-d]\{\$'[1+s]-\$v-\$'f\{1+s\}-\$i\}$$
$$= [1-t][1-d]\{\$'[1+s][1-f]-\$[v+i]\}$$
$$= [1-0.3000][1-0.3000]\{48,851$$
$$[1+0.0500][1-0.5800]$$
$$-51,294[0.1900+0.0100]\}$$
$$= 5,529 \text{ millions USD}$$

Law-3730:
If both ($= Sales or Revenues= 51,294 millions USD), (R= Retained Earnings= 5,529 millions USD), (i= I/S= Interest Portion= 1.00%), (v= V/S= Variable Portion= 19.00%), (d= D/A= Dividend Portion or Payout= 30.00%), (t= T/B= Tax Rate= 30.00%), (s= [S/S']-1= Sales Growth= 5.00%), and (f= F/S= Fixed Portion= 58.00%), are known, then it's ($'= Sales Past), must be:

$$\$' = (\$[v+i]+R/\{[1-t][1-d]\})/\{[1+s][1-f]\}$$
$$= (51,294[0.1900+0.0100]+5,529$$
$$/\{[1-0.3000][1-0.3000]\})$$
$$/\{[1+0.0500][1-0.5800]\}$$
$$= 48,851 \text{ millions USD}$$

Law-3731:

If both ($= Sales or Revenues= 51,294 millions USD), (**R**= Retained Earnings= 5,529 millions USD), (**i**= I/S= Interest Portion= 1.00%), (**v**= V/S= Variable Portion= 19.00%), (**d**= D/A= Dividend Portion or Payout= 30.00%), (**t**= T/B= Tax Rate= 30.00%), ($'= Sales of Past Year= 48,851 millions USD), and (**f**= F/S= Fixed Portion= 58.00%), are known, then it's (**s**= Sales Growth planned), is:

$$s = (\$[v+i]+R/\{[1-t][1-d]\})/\{\$'[1-f]\}-1$$
$$= (51,294[0.1900+0.0100\}+5,529$$
$$/\{[1-0.3000][1-0.3000]\})$$
$$/\{48,851[1-0.5800]\}-1$$
$$= \underline{5.00\%}$$

Law-3732:

If both ($= Sales or Revenues= 51,294 millions USD), (**R**= Retained Earnings= 5,529 millions USD), (**i**= I/S= Interest Portion= 1.00%), (**v**= V/S= Variable Portion= 19.00%), (**d**= D/A= Dividend Portion or Payout= 30.00%), (**t**= T/B= Tax Rate= 30.00%), ($'= Sales of Past Year= 48,851 millions USD), and (**s**= [S/S']-1= Sales Growth= 5.00%), are known, then it's (**f**= Fixed Portion planned), is:

$$f = 1-(\$[v+i]+R/\{[1-t][1-d]\})/\{\$'[1+s]\}$$
$$= 1-(51,294[0.1900+0.0100\}+5,529$$
$$/\{[1-0.3000][1-0.3000]\})$$
$$/\{48,851[1+0.0500]\}$$
$$= \underline{58.00\%}$$

Law-3733:
If both (**f**= F/S= Fixed Portion= 58.00%), (**R**= Retained Earnings= 5,529 millions USD), (**i**= I/S= Interest Portion= 1.00%), (**v**= V/S= Variable Portion= 19.00%), (**d**= D/A= Dividend Portion or Payout= 30.00%), (**t**= T/B= Tax Rate= 30.00%), (**$'**= Sales of Past Year= 48,851 millions USD), and (**s**= [S/S']-1= Sales Growth= 5.00%), are known, then it's (**$**= Sales or Revenues planned), is:

$= ($'[1+s][1-f]-R/{[1-t][1-d]})/[v+i]
  = 48,851[1+0.0500][1-0.5800]-5,529
          /{[1-0.3000][1-0.3000]})
          /[0.1900+0.0100]
  = 51,294 millions USD

Law-3734:

If both (**f**= F/S= Fixed Portion= 58.00%), (**R**= Retained Earnings= 5,529 millions USD), (**i**= I/S= Interest Portion= 1.00%), (**$**= Sales or Revenues= 51,294 millions USD), (**d**= D/A= Dividend Portion or Payout= 30.00%), (**t**= T/B= Tax Rate= 30.00%), (**$'**= Sales of Past Year= 48,851 millions USD), and (**s**= [S/S']-1= Sales Growth= 5.00%), are known, then it's (**v**= Variable Portion planned), is:

$$v = (\$'[1+s][1-f]-R/\{[1-t][1-d]\})/\$-i$$
$$= 48,851[\ 1+0.0500][1-0.5800]-5,529 /\{[1-0.3000][1-0.3000]\})$$
$$/51,294-0.0100$$
$$= \underline{19.00\%}$$

Finance Construction-4, *Tim Asikin, Steve Asikin, Indra Senihardja*

Law-3735:

If both (**f**= F/S= Fixed Portion= 58.00%), (**R**= Retained Earnings= 5,529 millions USD), (**v**= V/S= Variable Portion= 19.00%), (**$**= Sales or Revenues= 51,294 millions USD), (**d**= D/A= Dividend Portion or Payout= 30.00%), (**t**= T/B= Tax Rate= 30.00%), (**$'**= Sales of Past Year= 48,851 millions USD), and (**s**= [S/S']-1= Sales Growth= 5.00%), are known, then it's (**i**= Interest Portion planned), is:

i= ($'[1+s][1-f]-R/\{[1-t][1-d]\})/$-v
 = 48,851[1+0.0500][1-0.5800]-5,529
     /\{[1-0.3000][1-0.3000]\})
     /51,294-0.1900
 = 1.00%

Law-3736:
If both (**f**= F/S= Fixed Portion= 58.00%), (**R**= Retained Earnings= 5,529 millions USD), (**v**= V/S= Variable Portion= 19.00%), (**$**= Sales or Revenues= 51,294 millions USD), (**d**= D/A= Dividend Portion or Payout= 30.00%), (**i**= I/S= Interest Portion= 1.00%), (**$'**= Sales of Past Year= 48,851 millions USD), and (**s**= [S/S']-1= Sales Growth= 5.00%), are known, then it's (**t**= Tax Rate planned), is:
$$\mathbf{t}= 1-\mathbf{R}/([1-\mathbf{d}]\{\mathbf{\$'}[1+\mathbf{s}][1-\mathbf{f}]-\mathbf{\$}[\mathbf{v}+\mathbf{i}]\})$$
$$= 1-5,529/([1-0.3000]\{48,851[1+0.0500][1-0.5800]-51,294[0.1900+0.0100]\})$$
$$= 30.00\%$$

Law-3737:
If both (**f**= F/S= Fixed Portion= 58.00%), (**R**= Retained Earnings= 5,529 millions USD), (**v**= V/S= Variable Portion= 19.00%), (**$**= Sales or Revenues= 51,294 millions USD), (**t**= T/B= Tax Rate= 30.00%), (**i**= I/S= Interest Portion= 1.00%), (**$'**= Sales of Past Year= 48,851 millions USD), and (**s**= [S/S']-1= Sales Growth= 5.00%), are known, then it's (**d**= Dividend Portion or Payout planned), is:
$$\mathbf{d}= 1-\mathbf{R}/([1-\mathbf{t}]\{\mathbf{\$'}[1+\mathbf{s}][1-\mathbf{f}]-\mathbf{\$}[\mathbf{v}+\mathbf{i}]\})$$
$$= 1-5,529/([1-0.3000]\{48,851[1+0.0500][1-0.5800]-51,294[0.1900+0.0100]\})$$
$$= 30.00\%$$

Law-3738:
If both (f= F/S= Fixed Portion= 58.00%), (v= V/S= Variable Portion= 19.00%), (S= Sales or Revenues= 51,294 millions USD), (t= T/B= Tax Rate= 30.00%), (d= D/A= Dividend Portion or Payout= 30.00%), (i= I/S= Interest Portion= 1.00%), (S'= Sales of Past Year= 48,851 millions USD), and (s= [S/S']-1= Sales Growth= 5.00%), are known, then it's (R= Retained Earnings planned), is:

$R = [1-t][1-d]\{S'[1+s]-Sv-S'f[1+s]-S'i[1+s]\}$
$= [1-t][1-d]\{S'[1+s][1-f-i]-Sv\}$
$= 1\text{-}5{,}529/([1-0.3000]\{48{,}851$
$[1+0.0500][1-0.5800]$
$-51{,}294[0.1900+0.0100]\})$
$= \underline{30.00\%}$

Law-3739:
If both (f= F/S= Fixed Portion= 58.00%), (v= V/S= Variable Portion= 19.00%), (S= Sales or Revenues= 51,294 millions USD), (t= T/B= Tax Rate= 30.00%), (d= D/A= Dividend Portion or Payout= 30.00%), (i= I/S= Interest Portion= 1.00%), (R= Retained Earnings= 5,529 millions USD), and (s= [S/S']-1= Sales Growth= 5.00%), are known, then it's (S'= Sales Past), must be:

$S' = (Sv+R/\{[1-t][1-d]\})/\{[1+s][1-f-i]\}$
$= (51{,}294*0.1900+5{,}529/\{[1-0.3000]$
$[1-0.3000]\})/[1+0.0500]$
$[1-0.5800-0.0100]\}$
$= \underline{48{,}851} \text{ millions USD}$

Law-3740:

If both (**f**= F/S= Fixed Portion= <u>58.00</u>%), (**v**= V/S= Variable Portion= <u>19.00</u>%), (**$**= Sales or Revenues= <u>51,294</u> millions USD), (**t**= T/B= Tax Rate= <u>30.00</u>%), (**d**= D/A= Dividend Portion or Payout= 30.00%), (**i**= I/S= Interest Portion= <u>1.00</u>%), (**R**= Retained Earnings= <u>5,529</u> millions USD), and (**$'**= Sales of Past Year= <u>48,851</u> millions USD), are known, then it's (**s**= Sales Growth planned), is:

$$s = (\$v + R/\{[1-t][1-d]\})/\{\$'[1-f-i]\} - 1$$
$$= (51{,}294 * 0.1900 + 5{,}529/\{[1-0.3000]$$
$$[1-0.3000]\})/48{,}851[1-0.5800$$
$$-0.0100]\} - 1$$
$$= \underline{5.00}\%$$

Law-3741:

If both (**s**= [S/S']-1= Sales Growth= <u>5.00</u>%), (**v**= V/S= Variable Portion= <u>19.00</u>%), (**$**= Sales or Revenues= <u>51,294</u> millions USD), (**t**= T/B= Tax Rate= <u>30.00</u>%), (**d**= D/A= Dividend Portion or Payout= 30.00%), (**i**= I/S= Interest Portion= <u>1.00</u>%), (**R**= Retained Earnings= <u>5,529</u> millions USD), and (**$'**= Sales of Past Year= <u>48,851</u> millions USD), are known, then it's (**f**= Fixed Portion planned), is:

$$f = 1 - (\$v + R/\{[1-t][1-d]\})/\{\$'[1+s]\} - i$$
$$= 1 - (51{,}294 * 0.1900 + 5{,}529/\{[1-0.3000]$$
$$[1-0.3000]\})/48{,}851[1+0.0500]$$
$$-0.0100$$
$$= \underline{58.00}\%$$

Law-3742:

If both (s= [S/S']-1= Sales Growth= 5.00%), (v= V/S= Variable Portion= 19.00%), (S= Sales or Revenues= 51,294 millions USD), (t= T/B= Tax Rate= 30.00%), (d= D/A= Dividend Portion or Payout= 30.00%), (f= F/S= Fixed Portion= 58.00%), (R= Retained Earnings= 5,529 millions USD), and (S'= Sales of Past Year= 48,851 millions USD), are known, then it's (i= Interest Portion planned), is:

i= 1-(Sv+R/{[1-t][1-d]})/{S'[1+s]}-f
= 1-(51,294*0.1900+5,529/{[1-0.3000]
[1-0.3000]})/48,851[1+0.0500]
-0.5800
= 1.00%

Law-3743:

If both (s= [S/S']-1= Sales Growth= 5.00%), (v= V/S= Variable Portion= 19.00%), (i= I/S= Interest Portion= 1.00%), (t= T/B= Tax Rate= 30.00%), (d= D/A= Dividend Portion or Payout= 30.00%), (f= F/S= Fixed Portion= 58.00%), (R= Retained Earnings= 5,529 millions USD), and (S'= Sales of Past Year= 48,851 millions USD), are known, then it's (S= Sales or Revenues planned), is:

S= (S'[1+s][1-f-i]-R/{[1-t][1-d]})/v
= (48,851[1+0.0500][1-0.5800-0.0100]
-5,529/{[1-0.3000][1-0.3000]})
/0.1900
= 51,294 millions USD

Law-3744:
If both (s= [S/S']-1= Sales Growth= 5.00%), ($\$$= Sales or Revenues= 51,294 millions USD), (i= I/S= Interest Portion= 1.00%), (t= T/B= Tax Rate= 30.00%), (d= D/A= Dividend Portion or Payout= 30.00%), (f= F/S= Fixed Portion= 58.00%), (R= Retained Earnings= 5,529 millions USD), and ($\$'$= Sales of Past Year= 48,851 millions USD), are known, then it's (v= Variable Portion planned), is:

v= ($\$'$[1+$s$][1-$f$-$i$]-$R$/{[1-$t$][1-$d$]})/$\$$
= (48,851[1+0.0500][1-0.5800-0.0100]
   -5,529/{[1-0.3000][1-0.3000]})
   /51,294
= 19.00%

Law-3745:
If both (s= [S/S']-1= Sales Growth= 5.00%), ($\$$= Sales or Revenues= 51,294 millions USD), (i= I/S= Interest Portion= 1.00%), (v= V/S= Variable Portion= 19.00%), (d= D/A= Dividend Portion or Payout= 30.00%), (f= F/S= Fixed Portion= 58.00%), (R= Retained Earnings= 5,529 millions USD), and ($\$'$= Sales of Past Year= 48,851 millions USD), are known, then it's (t= Tax Rate planned), is:

t= 1-R/([1-d]{$\$'$[1+$s$][1-$f$-$i$]-$\$v$})
= 1-5,529/([1-0.3000]{48,851
   [1+0.0500][1-0.5800-0.0100]
   -51,294*0.1900})
= 30.00%

Law-3746:

If both (s= [S/S']-1= Sales Growth= 5.00%), ($\$$= Sales or Revenues= 51,294 millions USD), (i= I/S= Interest Portion= 1.00%), (v= V/S= Variable Portion= 19.00%), (t= T/B= Tax Rate= 30.00%), (f= F/S= Fixed Portion= 58.00%), (R= Retained Earnings= 5,529 millions USD), and ($\$'$= Sales of Past Year= 48,851 millions USD), are known, then it's (d= Dividend Portion or Payout planned), is:

d= 1-R/([1-t]{$\$'$[1+$s$][1-$f$-$i$]-$\$v$})
  = 1-5,529/([1-0.3000]{48,851
    [1+0.0500][1-0.5800-0.0100]
    -51,294*0.1900})
  = 30.00%

Law-3747:

If both (s= [S/S']-1= Sales Growth= 5.00%), (I= Interest Expense= 513 millions USD), (v= V/S= Variable Portion= 19.00%), (t= T/B= Tax Rate= 30.00%), (F= Fixed Cost= 29,750 millions USD), (d= D/A= Dividend Portion or Payout= 30.00%), and ($\$'$= Sales of Past Year= 48,851 millions USD), are known, then it's (R= Retained Earnings planned), is:

R= [1-t][1-d]{$\$'$[1+$s$]-$\$'v$[1+$s$]-$F$-$I$}
  = [1-t][1-d]{$\$'$[1+$s$][1-$v$]- F-I}
  = [1-0.3000][1-0.3000]{48,851
    [1+0.0500][1-0.1900]-29,750
    -513}
  = 5,529 millions USD

Law-3748:

If both (s= [S/S']-1= Sales Growth= 5.00%),
(I= Interest Expense= 513 millions USD), (v= V/S=
Variable Portion= 19.00%), (t= T/B= Tax Rate=
30.00%), (F= Fixed Cost= 29,750 millions USD),
(d= D/A= Dividend Portion or Payout= 30.00%), and
(R= Retained Earnings= 5,529 millions USD), are
known, then it's (S'= Sales Past), must be:

$$S' = (F+I+R/\{[1-t][1-d]\})/\{[1+s][1-v]\}$$
$$= (29,750+513+5,529/\{[1-0.3000]$$
$$[1-0.3000]\})/[1+0.0500]$$
$$[1-0.1900]\}$$
$$= \underline{48,851} \text{ millions USD}$$

Law-3749:

If both (S'= Sales of Past Year= 48,851 millions
USD), (I= Interest Expense= 513 millions USD), (v=
V/S= Variable Portion= 19.00%), (t= T/B= Tax
Rate= 30.00%), (F= Fixed Cost= 29,750 millions
USD), (d= D/A= Dividend Portion or Payout=
30.00%), and (R= Retained Earnings= 5,529 millions
USD), are known, then it's (s= Sales Growth
planned), is:

$$s = (F+I+R/\{[1-t][1-d]\})/\{S'[1-v]\}-1$$
$$= (29,750+513+5,529/\{[1-0.3000]$$
$$[1-0.3000]\})$$
$$/48,851[1-0.1900]\}-1$$
$$= \underline{5.00}\%$$

Law-3750:
If both ($'= Sales of Past Year= 48,851 millions USD), (I= Interest Expense= 513 millions USD), (s= [S/S']-1= Sales Growth= 5.00%), (t= T/B= Tax Rate= 30.00%), (F= Fixed Cost= 29,750 millions USD), (d= D/A= Dividend Portion or Payout= 30.00%), and (R= Retained Earnings= 5,529 millions USD), are known, then it's (v= Variable Portion planned), is:

$$v = 1-(F+I+R/\{[1-t][1-d]\})/\{\$'[1+s]\}$$
$$= 1-(29,750+513+5,529/\{[1-0.3000]$$
$$[1-0.3000]\})/48,851[1+0.0500]\}$$

$$= 19.00\%$$

Law-3751:
If both ($'= Sales of Past Year= 48,851 millions USD), (I= Interest Expense= 513 millions USD), (s= [S/S']-1= Sales Growth= 5.00%), (t= T/B= Tax Rate= 30.00%), (v= V/S= Variable Portion= 19.00%), (d= D/A= Dividend Portion or Payout= 30.00%), and (R= Retained Earnings= 5,529 millions USD), are known, then it's (F= Fixed Cost planned), is:

$$F = \$'[1+s][1-v]-I-R/\{[1-t][1-d]\}$$
$$= 48,851[1+0.0500]-513-5,529$$
$$/\{[1-0.3000][1-0.3000]\}$$
$$= 29,750 \text{ millions USD}$$

Law-3752:

If both ($\$'$= Sales of Past Year= 48,851 millions USD), (**F**= Fixed Cost= 29,750 millions USD), (**s**= [S/S']-1= Sales Growth= 5.00%), (**t**= T/B= Tax Rate= 30.00%), (**v**= V/S= Variable Portion= 19.00%), (**d**= D/A= Dividend Portion or Payout= 30.00%), and (**R**= Retained Earnings= 5,529 millions USD), are known, then it's (**I**= Interest Expense planned), is:

$$I = \$'[1+s][1-v] - F - R/\{[1-t][1-d]\}$$
$$= 48,851[1+0.0500] - 29,750 - 5,529$$
$$/\{[1-0.3000][1-0.3000]\}$$
$$= 513 \text{ millions USD}$$

Law-3753:

If both ($\$'$= Sales of Past Year= 48,851 millions USD), (**F**= Fixed Cost= 29,750 millions USD), (**s**= [S/S']-1= Sales Growth= 5.00%), (**I**= Interest Expense= 513 millions USD), (**v**= V/S= Variable Portion= 19.00%), (**d**= D/A= Dividend Portion or Payout= 30.00%), and (**R**= Retained Earnings= 5,529 millions USD), are known, then it's (**t**= Tax Rate planned), is:

$$t = 1 - R/([1-d]\{\$'[1+s][1-v] - F - I\})$$
$$= 1 - 5,529/([1-0.3000]\{48,851$$
$$[1+0.0500] - 29,750 - 513\})$$
$$= 30.00\%$$

Law-3754:

If both ($'$= Sales of Past Year= 48,851 millions USD), (**F**= Fixed Cost= 29,750 millions USD), (**s**= [S/S']-1= Sales Growth= 5.00%), (**I**= Interest Expense= 513 millions USD), (**v**= V/S= Variable Portion= 19.00%), (**t**= T/B= Tax Rate= 30.00%), and (**R**= Retained Earnings= 5,529 millions USD), are known, then it's (**d**= Dividend Portion or Payout planned), is:

$$d = 1-R/([1-t]\{\$'[1+s][1-v]-F-I\})$$
$$= 1-5,529/([1-0.3000]\{48,851[1+0.0500]-29,750-513\}$$
$$= 30.00\%$$

Law-3755:

If both ($'$= Sales of Past Year= 48,851 millions USD), (**F**= Fixed Cost= 29,750 millions USD), (**s**= [S/S']-1= Sales Growth= 5.00%), (**i**= I/S= Interest Portion= 1.00%), (**$**= Sales or Revenues= 51,294 millions USD), (**v**= V/S= Variable Portion= 19.00%), (**t**= T/B= Tax Rate= 30.00%), and (**d**= D/A= Dividend Portion or Payout= 30.00%), are known, then it's (**R**= Retained Earnings planned), is:

$$R = [1-t][1-d]\{\$'[1+s]-\$'v[1+s]-F-\$i\})$$
$$= [1-t][1-d]\{\$'[1+s][1-v]-\$i-F\}$$
$$= [1-0.3000][1-0.3000]\{48,851[1+0.0500][1-0.1900]$$
$$-51,294*0.0100 -29,750\}$$
$$= 5,529 \text{ millions USD}$$

Law-3756:

If both (**R**= Retained Earnings= 5,529 millions USD), (**F**= Fixed Cost= 29,750 millions USD), (**s**= [S/S']-1= Sales Growth= 5.00%), (**i**= I/S= Interest Portion= 1.00%), (**$**= Sales or Revenues= 51,294 millions USD), (**v**= V/S= Variable Portion= 19.00%), (**t**= T/B= Tax Rate= 30.00%), and (**d**= D/A= Dividend Portion or Payout= 30.00%), are known, then it's (**$'**= Sales Past), must be:

$$\$' = (F+\$i+R/\{[1-t][1-d]\})/\{[1+s][1-v]\}$$
$$= (29{,}750+51{,}294*0.0100]+5{,}529$$
$$/\{[1-0.3000][1-0.3000]\})$$
$$/\{[1+0.0500][1-0.1900]\}$$
$$= 48{,}851 \text{ millions USD}$$

Law-3757:

If both (**R**= Retained Earnings= 5,529 millions USD), (**F**= Fixed Cost= 29,750 millions USD), (**$'**= Sales of Past Year= 48,851 millions USD), (**i**= I/S= Interest Portion= 1.00%), (**$**= Sales or Revenues= 51,294 millions USD), (**v**= V/S= Variable Portion= 19.00%), (**t**= T/B= Tax Rate= 30.00%), and (**d**= D/A= Dividend Portion or Payout= 30.00%), are known, then it's (**s**= Sales Growth planned), is:

$$s = (F+\$i+R/\{[1-t][1-d]\})/\{\$'[1-v]\}-1$$
$$= (29{,}750+51{,}294*0.0100+5{,}529$$
$$/\{[1-0.3000][1-0.3000]\})$$
$$/\{48{,}851[1-0.1900]\}-1$$
$$= 5.00\%$$

Law-3758:
If both (**R**= Retained Earnings= 5,529 millions USD), (**F**= Fixed Cost= 29,750 millions USD), (**$'**= Sales of Past Year= 48,851 millions USD), (**i**= I/S= Interest Portion= 1.00%), (**$**= Sales or Revenues= 51,294 millions USD), (**s**= [S/S']-1= Sales Growth= 5.00%), (**t**= T/B= Tax Rate= 30.00%), and (**d**= D/A= Dividend Portion or Payout= 30.00%), are known, then it's (**v**= Variable Portion planned), is:

$$v = 1-(F+\$i+R/\{[1-t][1-d]\})/\{\$'[1+s]\}$$
$$= 1-(29,750+51,294*0.0100+5,529$$
$$/\{[1-0.3000][1-0.3000]\})$$
$$/\{48,851[1+0.0500]\}$$
$$= 19.00\%$$

Law-3759:
If both (**R**= Retained Earnings= 5,529 millions USD), (**$'**= Sales of Past Year= 48,851 millions USD), (**i**= I/S= Interest Portion= 1.00%), (**$**= Sales or Revenues= 51,294 millions USD), (**s**= [S/S']-1= Sales Growth= 5.00%), (**v**= V/S= Variable Portion= 19.00%), (**t**= T/B= Tax Rate= 30.00%), and (**d**= D/A= Dividend Portion or Payout= 30.00%), are known, then it's (**F**= Fixed Cost planned), is:

$$F = \$'[1+s][1-v]-\$i+R/\{[1-t][1-d]\}$$
$$= 48,851[1+0.0500][1-0.1900]-51,294$$
$$*0.0100-5,529$$
$$/\{[1-0.3000][1-0.3000]\}$$
$$= 29,750 \text{ millions USD}$$

Law-3760:
If both (**R**= Retained Earnings= 5,529 millions USD), (**$'**= Sales of Past Year= 48,851 millions USD), (**i**= I/S= Interest Portion= 1.00%), (**F**= Fixed Cost= 29,750 millions USD), (**s**= [S/S']-1= Sales Growth= 5.00%), (**v**= V/S= Variable Portion= 19.00%), (**t**= T/B= Tax Rate= 30.00%), and (**d**= D/A= Dividend Portion or Payout= 30.00%), are known, then it's (**$**= Sales or Revenues planned), is:
$$\$= (\$'[1+s][1-v]-F+R/\{[1-t][1-d]\})/i$$
$$= (48,851[1+0.0500][1-0.1900]-29,750 -5,529/\{[1-0.3000][1-0.3000]\})/0.0100$$
$$= 51,294 \text{ millions USD}$$

Law-3761:
If both (**R**= Retained Earnings= 5,529 millions USD), (**$'**= Sales of Past Year= 48,851 millions USD), (**$**= Sales or Revenues= 51,294 millions USD), (**F**= Fixed Cost= 29,750 millions USD), (**s**= [S/S']-1= Sales Growth= 5.00%), (**v**= V/S= Variable Portion= 19.00%), (**t**= T/B= Tax Rate= 30.00%), and (**d**= D/A= Dividend Portion or Payout= 30.00%), are known, then it's (**i**= Interest Portion planned), is:
$$i= (\$'[1+s][1-v]-F+R/\{[1-t][1-d]\})/\$$$
$$= (48,851[1+0.0500][1-0.1900]-29,750 -5,529/\{[1-0.3000][1-0.3000]\})/51,294$$
$$= 1.00\%$$

Law-3762:
If both (**R**= Retained Earnings= 5,529 millions USD), (**$'**= Sales of Past Year= 48,851 millions USD), (**$**= Sales or Revenues= 51,294 millions USD), (**F**= Fixed Cost= 29,750 millions USD), (**s**= [S/S']-1= Sales Growth= 5.00%), (**v**= V/S= Variable Portion= 19.00%), (**i**= I/S= Interest Portion= 1.00%), and (**d**= D/A= Dividend Portion or Payout= 30.00%), are known, then it's (**t**= Tax Rate planned), is:
$t = 1 - R/([1-d]\{\$'[1+s][1-v] - F - \$i\})$
$= 1 - 5{,}529/([1-0.3000]\{48{,}851$
$[1+0.0500][1-0.1900] - 29{,}750$
$-51{,}294 * 0.0100\})$
$= 30.00\%$

Law-3763:
If both (**R**= Retained Earnings= 5,529 millions USD), (**$'**= Sales of Past Year= 48,851 millions USD), (**$**= Sales or Revenues= 51,294 millions USD), (**F**= Fixed Cost= 29,750 millions USD), (**s**= [S/S']-1= Sales Growth= 5.00%), (**v**= V/S= Variable Portion= 19.00%), (**i**= I/S= Interest Portion= 1.00%), and (**t**= T/B= Tax Rate= 30.00%), are known, then it's (**d**= Dividend Portion or Payout planned), is:
$d = 1 - R/([1-t]\{\$'[1+s][1-v] - F - \$i\})$
$= 1 - 5{,}529/([1-0.3000]\{48{,}851$
$[1+0.0500][1-0.1900] - 29{,}750$
$-51{,}294 * 0.0100\})$
$= 30.00\%$

Law-3764:
If both (**d**= D/A= Dividend Portion or Payout= 30.00%), (**$'**= Sales of Past Year= 48,851 millions USD), (**$**= Sales or Revenues= 51,294 millions USD), (**F**= Fixed Cost= 29,750 millions USD), (**s**= [S/S']-1= Sales Growth= 5.00%), (**v**= V/S= Variable Portion= 19.00%), (**i**= I/S= Interest Portion= 1.00%), and (**t**= T/B= Tax Rate= 30.00%), are known, then it's (**R**= Retained Earnings planned), is:

$$R = [1-t][1-d]\{\$'[1+s]-\$'v[1+s]-F-\$'i[1+s]\}$$
$$= [1-t][1-d]\{\$'[1+s][1-v-i]-F\}$$
$$= [1-0.3000][1-0.3000]\{48,851$$
$$[1+0.0500][1-0.1900-0.0100]$$
$$-29,750\}$$
$$= 5,529 \text{ millions USD}$$

Law-3765:
If both (**d**= D/A= Dividend Portion or Payout= 30.00%), (**R**= Retained Earnings= 5,529 millions USD), (**$**= Sales or Revenues= 51,294 millions USD), (**F**= Fixed Cost= 29,750 millions USD), (**s**= [S/S']-1= Sales Growth= 5.00%), (**v**= V/S= Variable Portion= 19.00%), (**i**= I/S= Interest Portion= 1.00%), and (**t**= T/B= Tax Rate= 30.00%), are known, then it's (**$'**= Sales Past), must be:

$$\$' = (F+R/\{[1-t][1-d]\})/\{[1+s][1-v-i]\}$$
$$= (29,750+5,529/\{[1-0.3000]$$
$$[1-0.3000]\})/\{[1+0.0500]$$
$$[1-0.1900-0.0100]\}$$
$$= 48,851 \text{ millions USD}$$

Law-3766:
If both (**d**= D/A= Dividend Portion or Payout= 30.00%), (**R**= Retained Earnings= 5,529 millions USD), (**$**= Sales or Revenues= 51,294 millions USD), (**F**= Fixed Cost= 29,750 millions USD), (**$'**= Sales of Past Year= 48,851 millions USD), (**v**= V/S= Variable Portion= 19.00%), (**i**= I/S= Interest Portion= 1.00%), and (**t**= T/B= Tax Rate= 30.00%), are known, then it's (**s**= Sales Growth planned), is:
$$s = (F+R/\{[1-t][1-d]\})/\{\$'[1-v-i]\}-1$$
$$= (29{,}750+5{,}529/\{[1-0.3000][1-0.3000]\})/\{48{,}851[1-0.1900-0.0100]\}-1$$
$$= 5.00\%$$

Law-3767:
If both (**d**= D/A= Dividend Portion or Payout= 30.00%), (**R**= Retained Earnings= 5,529 milllons USD), (**$**= Sales or Revenues= 51,294 millions USD), (**F**= Fixed Cost= 29,750 millions USD), (**$'**= Sales of Past Year= 48,851 millions USD), (**s**= [S/S']-1= Sales Growth= 5.00%), (**i**= I/S= Interest Portion= 1.00%), and (**t**= T/B= Tax Rate= 30.00%), are known, then it's (**v**= Variable Portion planned), is:
$$v = 1-(F+R/\{[1-t][1-d]\})/\{\$'[1+s]\}-i$$
$$= 1-(29{,}750+5{,}529/\{[1-0.3000][1-0.3000]\})/\{48{,}851[1+0.0500]\}-0.0100$$
$$= 19.00\%$$

Law-3768:

If both (**d**= D/A=Dividend Portion or Payout= 30.00%), (**R**= Retained Earnings= 5,529 millions USD), (**$**= Sales or Revenues= 51,294 millions USD), (**F**= Fixed Cost= 29,750 millions USD), (**$'**= Sales of Past Year= 48,851 millions USD), (**s**= [S/S']-1= Sales Growth= 5.00%), (**v**= V/S= Variable Portion= 19.00%), and (**t**= T/B= Tax Rate= 30.00%), are known, then it's (**i**= Interest Portion planned), is:

$$i = 1-(F+R/\{[1-t][1-d]\})/\{\$'[1+s]\}-v$$
$$= 1-(29,750+5,529/\{[1-0.3000][1-0.3000]\})/\{48,851[1+0.0500]\}-0.1900$$
$$= \underline{1.00}\%$$

Law-3769:

If both (**d**= D/A= Dividend Portion or Payout= 30.00%), (**R**= Retained Earnings= 5,529 millions USD), (**$**= Sales or Revenues= 51,294 millions USD), (**i**= I/S= Interest Portion= 1.00%), (**$'**= Sales of Past Year= 48,851 millions USD), (**s**= [S/S']-1= Sales Growth= 5.00%), (**v**= V/S= Variable Portion= 19.00%), and (**t**= T/B= Tax Rate= 30.00%), are known, then it's (**F**= Fixed Cost planned), is:

$$F = \$'[1+s][1-v-i]-R/\{[1-t][1-d]\}$$
$$= 48,851[1+0.0500][1-0.1900-0.0100]$$
$$-5,529/\{[1-0.3000][1-0.3000]\}$$
$$= \underline{29,750} \text{ millions USD}$$

Law-3770:
If both (**d**= D/A= Dividend Portion or Payout= 30.00%), (**R**= Retained Earnings= 5,529 millions USD), (**$**= Sales or Revenues= 51,294 millions USD), (**i**= I/S= Interest Portion= 1.00%), (**$'**= Sales of Past Year= 48,851 millions USD), (**s**= [S/S']-1= Sales Growth= 5.00%), (**v**= V/S= Variable Portion= 19.00%), and (**F**= Fixed Cost= 29,750 millions USD), are known, then it's (**t**= Tax Rate planned), is:

t= 1-**R**/([1-**d**]{**$'**[1+**s**][1-**v**-**i**]-**F**})
  = 1-5,529/([1-0.3000]{48,851
    [1+0.0500][1-0.1900-0.0100]
    -29,750})
  = 30.00%

Law-3771:
If both (**t**= T/B= Tax Rate= 30.00%), (**R**= Retained Earnings= 5,529 millions USD), (**$**= Sales or Revenues= 51,294 millions USD), (**i**= I/S= Interest Portion= 1.00%), (**$'**= Sales of Past Year= 48,851 millions USD), (**s**= [S/S']-1= Sales Growth= 5.00%), (**v**= V/S= Variable Portion= 19.00%), and (**F**= Fixed Cost= 29,750 millions USD), are known, then it's (**d**= Dividend Portion or Payout planned), is:

d= 1-**R**/([1-**t**]{**$'**[1+**s**][1-**v**-**i**]-**F**})
  = 1-5,529/([1-0.3000]{48,851
    [1+0.0500][1-0.1900-0.0100]
    -29,750})
  = 30.00%

Law-3772:
If both (**t**= T/B= Tax Rate= 30.00%), (**d**= D/A= Dividend Portion or Payout= 30.00%), (**$**= Sales or Revenues= 51,294 millions USD), (**I**= Interest Expense= 513 millions USD), (**$'**= Sales of Past Year= 48,851 millions USD), (**s**= [S/S']-1= Sales Growth= 5.00%), (**v**= V/S= Variable Portion= 19.00%), and (**f**=F/S= Fixed Portion= 58.00%), are known, then it's (**R**= Retained Earnings planned), is:
$$R = [1-t][1-d]\{\$'[1+s]-\$'v[1+s]-\$f-I\}$$
$$= [1-t][1-d]\{\$'[1+s][1-v]-\$f-I\}$$
$$= [1-0.3000][1-0.3000]\{48,851[1+0.0500][1-0.1900]$$
$$-51,294*0.5800-513]\}$$
$$= 5,529 \text{ millions USD}$$

Law-3773:
If both (**t**= T/B= Tax Rate= 30.00%), (**d**= D/A= Dividend Portion or Payout= 30.00%), (**$**= Sales or Revenues= 51,294 millions USD), (**I**= Interest Expense= 513 millions USD), (**R**= Retained Earnings= 5,529 millions USD), (**s**= [S/S']-1= Sales Growth= 5.00%), (**v**= V/S= Variable Portion= 19.00%), and (**f**=F/S= Fixed Portion= 58.00%), are known, then it's (**$'**= Sales Past), must be:
$$\$' = (\$f+I+R/\{[1-t][1-d]\})/\{[1+s][1-v]\}$$
$$= (51,294*0.5800+513+5,529$$
$$/\{[1-0.3000][1-0.3000]\})$$
$$/\{[1+0.0500][1-0.1900]\}$$
$$= 48,851 \text{ millions USD}$$

Law-3774:

If both (**t**= T/B= Tax Rate= 30.00%), (**d**= D/A= Dividend Portion or Payout= 30.00%), (**$**= Sales or Revenues= 51,294 millions USD), (**I**= Interest Expense= 513 millions USD), (**$'**= Sales of Past Year= 48,851 millions USD),(**R**= Retained Earnings= 5,529 millions USD),   (**v**= V/S= Variable Portion= 19.00%), and (**f**= F/S= Fixed Portion= 58.00%), are known, then it's (**s**= Sales Growth planned), is:

$s = (\$f + I + R/\{[1-t][1-d]\})/\{\$'[1-v]\} - 1$
$= (51,294*0.5800+513+5,529$
$/\{[1-0.3000][1-0.3000]\})$
$/\{48,851[1-0.1900]\} - 1$
$= 5.00\%$

Law-3775:

If both (**t**= T/B= Tax Rate= 30.00%), (**d**= D/A= Dividend Portion or Payout= 30.00%), (**$**= Sales or Revenues= 51,294 millions USD), (**I**= Interest Expense= 513 millions USD), (**$'**= Sales of Past Year= 48,851 millions USD),(**R**= Retained Earnings= 5,529 millions USD), (**s**= [S/S']-1= Sales Growth= 5.00%), and (**f**= F/S= Fixed Portion= 58.00%), are known, then it's (**v**= Variable Portion planned), is:

$v = 1 - (\$f + I + R/\{[1-t][1-d]\})/\{\$'[1+s]\}$
$= 1 - (51,294*0.5800+513+5,529$
$/\{[1-0.3000][1-0.3000]\})$
$/\{48,851[1+0.0500]\}$
$= 19.00\%$

Law-3776:
If both (**t**= T/B= Tax Rate= 30.00%), (**d**= D/A= Dividend Portion or Payout= 30.00%), (**v**= V/S= Variable Portion= 19.00%), (**I**= Interest Expense= 513 millions USD), (**S'**= Sales of Past Year= 48,851 millions USD),(**R**= Retained Earnings= 5,529 millions USD), (**s**= [S/S']-1= Sales Growth= 5.00%), and(**f**= F/S= Fixed Portion= 58.00%), are known, then it's (**S**= Sales or Revenues planned), is:

$$S = (S'[1+s][1-v]-I-R/\{[1-t][1-d]\})/f$$
$$= (48,851[1+0.0500][1-0.1900]-513$$
$$-5,529/\{[1-0.3000][1-0.3000]\})$$
$$/0.5800$$
$$= 51,294 \text{ millions USD}$$

Law-3777:
If both (**t**= T/B= Tax Rate= 30.00%), (**d**= D/A= Dividend Portion or Payout= 30.00%), (**v**= V/S= Variable Portion= 19.00%), (**I**= Interest Expense= 513 millions USD), (**S'**= Sales of Past Year= 48,851 millions USD),(**R**= Retained Earnings= 5,529 millions USD), (**s**= [S/S']-1= Sales Growth= 5.00%), and (**S**= Sales or Revenues= 51,294 millions USD), are known, then it's (**f**= Fixed Portion planned), is:

$$f = (S'[1+s][1-v]-I-R/\{[1-t][1-d]\})/S$$
$$= (48,851[1+0.0500][1-0.1900]-513$$
$$-5,529/\{[1-0.3000][1-0.3000]\})$$
$$/51,294$$
$$= 58.00\%$$

Law-3778:

If both (**t**= T/B= Tax Rate= 30.00%), (**d**= D/A= Dividend Portion or Payout= 30.00%), (**v**= V/S= Variable Portion= 19.00%), (**f**= F/S= Fixed Portion= 58.00%), (**$'**= Sales of Past Year= 48,851 millions USD), (**R**= Retained Earnings= 5,529 millions USD), (**s**= [S/S']-1= Sales Growth= 5.00%), and (**$**= Sales or Revenues= 51,294 millions USD), are known, then it's (**I**= Interest Expense planned), is:

I= **$'**[1+**s**][1-**v**]-**$f**-**R**/{[1-**t**][1-**d**]}
= 48,851[1+0.0500][1-0.1900]-51,294
*0.5800-5,529
/{[1-0.3000][1-0.3000]}
= 513 millions USD

Law-3779:

If both (**I**= Interest Expense= 513 millions USD), (**d**= D/A= Dividend Portion or Payout= 30.00%), (**v**= V/S= Variable Portion= 19.00%), (**f**= F/S= Fixed Portion= 58.00%), (**$'**= Sales of Past Year= 48,851 millions USD), (**R**= Retained Earnings= 5,529 millions USD), (**s**= [S/S']-1= Sales Growth= 5.00%), and (**$**= Sales or Revenues= 51,294 millions USD), are known, then it's (**t**= Tax Rate planned), is:

t= 1-**R**/([1-**d**]{**$'**[1+**s**][1-**v**]-**$f**-**I**})
= 1-5,529/([1-0.3000]{48,851
[1+0.0500][1-0.1900]
-51,294*0.5800-513})
= 30.00%

Law-3780:
If both (**I**= Interest Expense= 513 millions USD), (**t**= T/B= Tax Rate= 30.00%), (**v**= V/S= Variable Portion= 19.00%), (**f**= F/S= Fixed Portion= 58.00%), (**$'**= Sales of Past Year= 48,851 millions USD), (**R**= Retained Earnings= 5,529 millions USD), (**s**= [S/S']-1= Sales Growth= 5.00%), and (**$**= Sales or Revenues= 51,294 millions USD), are known, then it's (**d**= Dividend Portion or Payout planned), is:

$$d = 1 - R/([1-t]\{\$'[1+s][1-v]-\$f-I\})$$
$$= 1 - 5,529/([1-0.3000]\{48,851$$
$$[1+0.0500][1-0.1900]$$
$$-51,294*0.5800-513\})$$
$$= 30.00\%$$

Law-3781:
If both (**i**= I/S= Interest Portion= 1.00%), (**t**= T/B= Tax Rate= 30.00%), (**v**= V/S= Variable Portion= 19.00%), (**f**= F/S= Fixed Portion= 58.00%), (**$'**= Sales of Past Year= 48,851 millions USD), (**d**= D/A=Dividend Portion or Payout= 30.00%), (**s**= [S/S']-1= Sales Growth= 5.00%), and (**$**= Sales or Revenues= 51,294 millions USD), are known, then it's (**R**= Retained Earnings planned), is:

$$R = [1-t][1-d]\{\$'[1+s]-\$'v[1+s]-\$f-\$i\}$$
$$= [1-t][1-d]\{\$'[1+s][1-v]-\$[f+i]\}$$
$$= [1-0.3000][1-0.3000]\{48,851$$
$$[1+0.0500][1-0.1900]$$
$$-51,294[0.5800+0.0100]\}$$
$$= 5,529 \text{ millions USD}$$

Law-3782:

If both (i= I/S= Interest Portion= 1.00%), (t= T/B= Tax Rate= 30.00%), (v= V/S= Variable Portion= 19.00%), (f= F/S= Fixed Portion= 58.00%), (R= Retained Earnings= 5,529 millions USD), (d= D/A=Dividend Portion or Payout= 30.00%), (s= [S/S']-1= Sales Growth= 5.00%), and (S= Sales or Revenues= 51,294 millions USD), are known, then it's (S'= Sales Past), must be:

$$S' = (S[f+i]+R[1-t][1-d])/\{[1+s][1-v]\}$$
$$= (51,294[0.5800+0.0100]+5,529$$
$$/\{[1-0.3000][1-0.3000]\})$$
$$/\{[1+0.0500][1-0.1900]\}$$
$$= 48,851 \text{ millions USD}$$

Law-3783:

If both (i= I/S= Interest Portion= 1.00%), (t= T/B= Tax Rate= 30.00%), (v= V/S= Variable Portion= 19.00%), (f= F/S= Fixed Portion= 58.00%), (R= Retained Earnings= 5,529 millions USD), (d= D/A= Dividend Portion or Payout= 30.00%), (S'= Sales of Past Year= 48,851 millions USD), and (S= Sales or Revenues= 51,294 millions USD), are known, then it's (s= Sales Growth planned), is:

$$s = (S[f+i]+R[1-t][1-d])/\{S'[1-v]\}-1$$
$$= (51,294[0.5800+0.0100]+5,529$$
$$/\{[1-0.3000][1-0.3000]\})$$
$$/\{48,851[1-0.1900]\}-1$$
$$= 5.00\%$$

Law-3784:

If both (i= I/S= Interest Portion= 1.00%), (t= T/B= Tax Rate= 30.00%), (s= [S/S']-1= Sales Growth= 5.00%), (f= F/S= Fixed Portion= 58.00%), (R= Retained Earnings= 5,529 millions USD), (d= D/A= Dividend Portion or Payout= 30.00%), (S'= Sales of Past Year= 48,851 millions USD), and (S= Sales or Revenues= 51,294 millions USD), are known, then it's (v= Variable Portion planned), is:

v= 1-($S[f+i]+R[1-t][1-d]$)/$\{S'[1+s]\}$
   = 1-(51,294[0.5800+0.0100]+5,529
       /{[1-0.3000][1-0.3000]})
       /{48,851[1+0.0500]}
   = 19.00%

Law-3785:

If both (i= I/S= Interest Portion= 1.00%), (t= T/B= Tax Rate= 30.00%), (s= [S/S']-1= Sales Growth= 5.00%), (f= F/S= Fixed Portion= 58.00%), (R= Retained Earnings= 5,529 millions USD), (v= V/S= Variable Portion= 19.00%), (d= D/A= Dividend Portion or Payout= 30.00%), and (S'= Sales of Past Year= 48,851 millions USD), are known, then it's (S= Sales or Revenues planned), is:

S= ($S'[1+s][1-v]-R[1-t][1-d]$)/$[f+i]$
   = (48,851[1+0.0500] [1-0.1900]-5,529
       /{[1-0.3000][1-0.3000]})
       /[0.5800+0.0100]
   = 51,294 millions USD

Law-3786:

If both (**i**= I/S= Interest Portion= <u>1.00</u>%), (**t**= T/B= Tax Rate= <u>30.00</u>%), (**s**= [S/S']-1= Sales Growth= <u>5.00</u>%), (**$**= Sales or Revenues= <u>51,294</u> millions USD), (**R**= Retained Earnings= <u>5,529</u> millions USD), (**v**= V/S= Variable Portion= <u>19.00</u>%), (**d**= D/A= Dividend Portion or Payout= 30.00%), and (**$'**= Sales of Past Year= <u>48,851</u> millions USD), are known, then it's (**f**= Fixed Portion planned), is:

$$f = (\$'[1+s][1-v]-R[1-t][1-d])/\$-i$$
$$= (48,851[1+0.0500][1-0.1900]-5,529$$
$$/\{[1-0.3000][1-0.3000]\})$$
$$/51,294-0.0100$$
$$= \underline{58.00}\%$$

Law-3787:

If both (**f**= F/S= Fixed Portion= <u>58.00</u>%), (**t**= T/B= Tax Rate= <u>30.00</u>%), (**s**= [S/S']-1= Sales Growth= <u>5.00</u>%), (**$**= Sales or Revenues= <u>51,294</u> millions USD), (**R**= Retained Earnings= <u>5,529</u> millions USD), (**v**= V/S= Variable Portion= <u>19.00</u>%), (**d**= D/A= Dividend Portion or Payout= 30.00%), and (**$'**= Sales of Past Year= <u>48,851</u> millions USD), are known, then it's (**i**= Interest Portion planned), is:

$$i = (\$'[1+s][1-v]-R[1-t][1-d])/\$-f$$
$$= (48,851[1+0.0500][1-0.1900]-5,529$$
$$/\{[1-0.3000][1-0.3000]\})$$
$$/51,294-0.5800$$
$$= \underline{1.00}\%$$

Law-3788:

If both (**f**= F/S= Fixed Portion= 58.00%), (**i**= I/S= Interest Portion= 1.00%), (**s**= [S/S']-1= Sales Growth= 5.00%), (**$**= Sales or Revenues= 51,294 millions USD), (**R**= Retained Earnings= 5,529 millions USD), (**v**= V/S= Variable Portion= 19.00%), (**d**= D/A= Dividend Portion or Payout= 30.00%), and (**$'**= Sales of Past Year= 48,851 millions USD), are known, then it's (**t**= Tax Rate planned), is:

$$t = 1 - R/([1-d]\{\$'[1+s][1-v] - \$[f+i]\})$$
$$= 1 - 5,529/([1-0.3000]\{48,851[1+0.0500][1-0.1900] - 51,294[0.5800+0.0100]\})$$
$$= 30.00\%$$

Law-3789:

If both (**f**= F/S= Fixed Portion= 58.00%), (**i**= I/S= Interest Portion= 1.00%), (**s**= [S/S']-1= Sales Growth= 5.00%), (**$**= Sales or Revenues= 51,294 millions USD), (**R**= Retained Earnings= 5,529 millions USD), (**v**= V/S= Variable Portion= 19.00%), (**t**= T/B= Tax Rate= 30.00%), and (**$'**= Sales of Past Year= 48,851 millions USD), are known, then it's (**d**= Dividend Portion or Payout planned), is:

$$d = 1 - R/([1-t]\{\$'[1+s][1-v] - \$[f+i]\})$$
$$= 1 - 5,529/([1-0.3000]\{48,851[1+0.0500][1-0.1900] - 51,294[0.5800+0.0100]\})$$
$$= 30.00\%$$

Law-3790:
If both (**f**= F/S= Fixed Portion= 58.00%), (**i**= I/S= Interest Portion= 1.00%), (**s**= [S/S']-1= Sales Growth= 5.00%), (**$**= Sales or Revenues= 51,294 millions USD), (**d**= D/A= Dividend Portion or Payout= 30.00%), (**v**= V/S= Variable Portion= 19.00%), (**t**= T/B= Tax Rate= 30.00%), and (**$'**= Sales of Past Year= 48,851 millions USD),are known, then it's (**R**= Retained Earnings planned), is:
  R= [1-**t**][1-**d**]{**$'**[1+**s**]-**$'v**[1+**s**]-**$f**-**$'i**[1+**s**]}
     = [1-**t**][1-**d**]{**$'**[1+**s**][1-**v**-**i**]-**$f**}
     = [1-0.3000][1-0.3000]{48,851
          [1+0.0500][1-0.1900-0.0100]
          -51,294*0.5800}
     = 5,529 millions USD

Law-3791:
If both (**f**= F/S= Fixed Portion= 58.00%), (**i**= I/S= Interest Portion= 1.00%), (**s**= [S/S']-1= Sales Growth= 5.00%), (**$**= Sales or Revenues= 51,294 millions USD), (**d**= D/A= Dividend Portion or Payout= 30.00%), (**v**= V/S= Variable Portion= 19.00%), (**t**= T/B= Tax Rate= 30.00%), and (**R**= Retained Earnings= 5,529 millions USD), are known, then it's (**$'**= Sales Past), must be:
  $'**= (**$f+**R**/{[1-**t**][1-**d**]})/{[1+**s**][1-**v**-**i**]}
     = (51,294*0.5800+5,529/{[1-0.3000]
          [1-0.3000]})/{[1+0.0500]
          [1-0.1900-0.0100]}
     = 48,851 millions USD

Law-3792:

If both (f= F/S=Fixed Portion= 58.00%), (i= I/S= Interest Portion= 1.00%), ($\$'$= Sales of Past Year= 48,851 millions USD), ($\$$= Sales or Revenues= 51,294 millions USD), (d= D/A= Dividend Portion or Payout= 30.00%), (v= V/S= Variable Portion= 19.00%), (t= T/B= Tax Rate= 30.00%), and (R= Retained Earnings= 5,529 millions USD), are known, then it's (s= Sales Growth planned), is:

$s = (\$f + R/\{[1-t][1-d]\})/\{\$'[1-v-i]\} - 1$
$= (51,294*0.5800+5,529/\{[1-0.3000][1-0.3000]\})/\{48,851[1-0.1900-0.0100]\}-1$
$= 5.00\%$

Law-3793:

If both (f= F/S= Fixed Portion= 58.00%), (i= I/S= Interest Portion= 1.00%), ($\$'$= Sales of Past Year= 48,851 millions USD), ($\$$= Sales or Revenues= 51,294 millions USD), (d= D/A= Dividend Portion or Payout= 30.00%), (s= [S/S']-1= Sales Growth= 5.00%), (t= T/B= Tax Rate= 30.00%), and (R= Retained Earnings= 5,529 millions USD), are known, then it's (v= Variable Portion planned), is:

$v = 1 - (\$f + R/\{[1-t][1-d]\})/\{\$'[1+s]\} - i$
$= 1-(51,294*0.5800+5,529/\{[1-0.3000][1-0.3000]\})/\{48,851[1+0.0500]\}-0.0100$
$= 19.00\%$

Law-3794:

If both (**f**= F/S= Fixed Portion= 58.00%), (**v**= V/S= Variable Portion= 19.00%), (**$'**= Sales of Past Year= 48,851 millions USD), (**$**= Sales or Revenues= 51,294 millions USD), (**d**= D/A= Dividend Portion or Payout= 30.00%), (**s**= [S/S']-1= Sales Growth= 5.00%), (**t**= T/B= Tax Rate= 30.00%), and (**R**= Retained Earnings= 5,529 millions USD),are known, then it's (**i**= Interest Portion planned), is:

$\quad$ **i**= 1-(**$f**+**R**/{[1-**t**][1-**d**]})/{**$'**[1+**s**]}-**v**
$\quad\quad$ = 1-(51,294*0.5800+5,529/{[1-0.3000]
$\quad\quad\quad$ [1-0.3000]})/{48,851
$\quad\quad\quad$ [1+0.0500]}-0.1900
$\quad\quad$ = 1.00%

Law-3795:

If both (**f**= F/S= Fixed Portion= 58.00%), (**v**= V/S= Variable Portion= 19.00%), (**$'**= Sales of Past Year= 48,851 millions USD), (**I**= I/S= Interest Portion= 1.00%), (**d**= D/A= Dividend Portion or Payout= 30.00%), (**s**= [S/S']-1= Sales Growth= 5.00%), (**t**= T/B= Tax Rate= 30.00%), and (**R**= Retained Earnings= 5,529 millions USD),are known, then it's (**$**= Sales or Revenues planned), is:

$\quad$ **$**= (**$'**[1+**s**][1-**v**-**i**]-**R**/{[1-**t**][1-**d**]})/**f**
$\quad\quad$ = (48,851[1+0.0500][1-0.1900-0.0100]
$\quad\quad\quad$ -5,529/{[1-0.3000][1-0.3000]})
$\quad\quad\quad$ /0.5800
$\quad\quad$ = 51,294 millions USD

Law-3796:

If both ($= Sales or Revenues= 51,294 millions USD), (v= V/S= Variable Portion= 19.00%), ($'= Sales of Past Year= 48,851 millions USD), (i= I/S= Interest Portion= 1.00%), (d= D/A= Dividend Portion or Payout= 30.00%), (s= [S/S']-1= Sales Growth= 5.00%), (t= T/B= Tax Rate= 30.00%), and (R= Retained Earnings= 5,529 millions USD),are known, then it's (f= Fixed Portion planned), is:

$$f = (\$'[1+s][1-v-i]-R/\{[1-t][1-d]\})/\$$$
$$= (48,851[1+0.0500][1-0.1900-0.0100]$$
$$-5,529/\{[1-0.3000][1-0.3000]\})$$
$$/51,294$$
$$= 58.00\%$$

Law-3797:

If both ($= Sales or Revenues= 51,294 millions USD), (v= V/S= Variable Portion= 19.00%), ($'= Sales of Past Year= 48,851 millions USD), (i= I/S= Interest Portion= 1.00%), (d= D/A= Dividend Portion or Payout= 30.00%), (s= [S/S']-1= Sales Growth= 5.00%), (f= F/S= Fixed Portion= 58.00%), and (R= Retained Earnings= 5,529 millions USD),are known, then it's (t= Tax Rate planned), is:

$$t = 1-R/([1-d]\{\$'[1+s][1-v-i]-\$f\})$$
$$= 1-5,529/([1-0.3000]\{48,851$$
$$[1+0.0500][1-0.1900-0.0100]$$
$$-51,294*0.5800]\})$$
$$= 30.00\%$$

Law-3798:
If both ($= Sales or Revenues= 51,294 millions USD), (v= V/S= Variable Portion= 19.00%), ($'= Sales of Past Year= 48,851 millions USD), (i= I/S= Interest Portion= 1.00%), (t= T/B= Tax Rate= 30.00%), (s= [S/S']-1= Sales Growth= 5.00%), (f= F/S= Fixed Portion= 58.00%), and (R= Retained Earnings= 5,529 millions USD), are known, then it's (d= Dividend Portion or Payout planned), is:

d= 1-R/([1-t]{$'[1+s][1-v-i]-$f})
　= 1-5,529/([1-0.3000]{48,851
　　　　[1+0.0500][1-0.1900-0.0100]
　　　　-51,294*0.5800]})
　= 30.00%

Law-3799:
If both (v= V/S= Variable Portion= 19.00%), ($'= Sales of Past Year= 48,851 millions USD), (I= Interest Expense= 513 millions USD), (t= T/B= Tax Rate= 30.00%), (s= [S/S']-1= Sales Growth= 5.00%), (f= F/S= Fixed Portion= 58.00%), and (d= D/A= Dividend Portion or Payout= 30.00%), are known, then it's (R= Retained Earnings planned), is:

R= [1-t][1-d]{$'[1+s]-$'v[1+s]-$'f[1+s]-I}
　= [1-t][1-d]{$'[1+s][1-v-f]-I}
　= [1-0.3000][1-0.3000]{48,851
　　　　[1+0.0500][1-0.1900-0.5800]
　　　　-513}
　= 5,529 millions USD

Law-3800:

If both (**v**= V/S= Variable Portion= 19.00%), (**R**= Retained Earnings= 5,529 millions USD), (**I**= Interest Expense= 513 millions USD), (**t**= T/B= Tax Rate= 30.00%), (**s**= [S/S']-1= Sales Growth= 5.00%), (**f**= F/S= Fixed Portion= 58.00%), and (**d**= D/A= Dividend Portion or Payout= 30.00%), are known, then it's (**$'**= Sales Past), must be:

$' = (**I**+**R**/{[1-**t**][1-**d**]})/{[1+**s**][1-**v**-**f**]}
  = (513+5,529/{[1-0.3000][1-0.3000]})
     /{[1+0.0500][1-0.1900-0.5800]}
  = 48,851 millions USD

Law-3801:

If both (**v**= V/S= Variable Portion= 19.00%), (**R**= Retained Earnings= 5,529 millions USD), (**I**= Interest Expense= 513 millions USD), (**t**= T/B= Tax Rate= 30.00%), (**$'**= Sales of Past Year= 48,851 millions USD), (**f**= F/S= Fixed Portion= 58.00%), and (**d**= D/A= Dividend Portion or Payout= 30.00%), are known, then it's (**s**= Sales Growth planned), is:

s= (**I**+**R**/{[1-**t**][1-**d**]})/{**$'**[1-**v**-**f**]}-1
  = (513+5,529/{[1-0.3000][1-0.3000]})
     /{48,851[1-0.1900-0.5800]}-1
  = 5.00%

Law-3802:
If both (s= [S/S']-1= Sales Growth= 5.00%), (R= Retained Earnings= 5,529 millions USD), (I= Interest Expense= 513 millions USD), (t= T/B= Tax Rate= 30.00%), (S'= Sales of Past Year= 48,851 millions USD), (f= F/S= Fixed Portion= 58.00%), and (d= D/A= Dividend Portion or Payout= 30.00%), are known, then it's (v= Variable Portion planned), is:

v= 1-f-(I+R/{[1-t][1-d]})/{S'[1+s]}
= 1-0.5800-(513+5,529/{[1-0.3000]
[1-0.3000]})/{48,851
[1+0.0500]}
= 19.00%

Law-3803:
If both (s= [S/S']-1= Sales Growth= 5.00%), (R= Retained Earnings= 5,529 millions USD), (I= Interest Expense= 513 millions USD), (t= T/B= Tax Rate= 30.00%), (S'= Sales of Past Year= 48,851 millions USD), (v= V/S= Variable Portion= 19.00%), and (d= D/A= Dividend Portion or Payout= 30.00%), are known, then it's (f= Fixed Portion planned), is:

f= 1-v-(I+R/{[1-t][1-d]})/{S'[1+s]}
= 1-0.1900-(513+5,529/{[1-0.3000]
[1-0.3000]})
/{48,851[1+0.0500]}
= 58.00%

Law-3804:
If both (**s**= [S/S']-1= Sales Growth= 5.00%), (**R**= Retained Earnings= 5,529 millions USD), (**f**= F/S= Fixed Portion= 58.00%), (**t**= T/B= Tax Rate= 30.00%), (**S'**= Sales of Past Year= 48,851 millions USD), (**v**= V/S= Variable Portion= 19.00%), and (**d**= D/A= Dividend Portion or Payout= 30.00%), are known, then it's (**I**= Interest Expense planned), is:
$$I = S'[1+s][1-v-f] - R/\{[1-t][1-d]\}$$
$$= 48,851[1+0.0500][1-0.1900-0.5800]$$
$$\quad -5,529/\{[1-0.3000][1-0.3000]\}$$
$$= \underline{513} \text{ millions USD}$$

Law-3805:
If both (**s**= [S/S']-1= Sales Growth= 5.00%), (**R**= Retained Earnings= 5,529 millions USD), (**f**= F/S= Fixed Portion= 58.00%), (**I**= Interest Expense= 513 millions USD), (**S'**= Sales of Past Year= 48,851 millions USD), (**v**= V/S= Variable Portion= 19.00%), and (**d**= D/A= Dividend Portion or Payout= 30.00%), are known, then it's (**t**= Tax Rate planned), is:
$$t = 1 - R/([1-d]\{S'[1+s][1-v-f] - I\}$$
$$= 1 - 5,529/([1-0.3000]\{48,851$$
$$\quad [1+0.0500][1-0.1900-0.5800]$$
$$\quad -513\})$$
$$= \underline{30.00\%}$$

Law-3806:
If both ($= [S/S']-1=$ Sales Growth= 5.00%), (R= Retained Earnings= 5,529 millions USD), (f= F/S= Fixed Portion= 58.00%), (I= Interest Expense= 513 millions USD), ($'= Sales of Past Year= 48,851 millions USD), (v= V/S= Variable Portion= 19.00%), and (t= T/B= Tax Rate= 30.00%), are known, then it's (d= Dividend Portion or Payout planned), is:
d= 1-R/([1-t]{$'[1+s][1-v-f]-I}
 = 1-5,529/([1-0.3000]{48,851
  [1+0.0500][1-0.1900-0.5800]
  -513})
 = 30.00%

Law-3807:
If both ($= [S/S']-1= Sales Growth= 5.00%), (d= D/A= Dividend Portion or Payout= 30.00%), (f= F/S= Fixed Portion= 58.00%), (i= I/S= Interest Portion= 1.00%), ($= Sales or Revenues= 51,294 millions USD), ($'= Sales of Past Year= 48,851 millions USD), (v= V/S= Variable Portion= 19.00%), and (t= T/B= Tax Rate= 30.00%), are known, then it's (R= Retained Earnings planned), is:
R= [1-t][1-d]{$'[1+s]-$'v[1+s]-$'f[1+s]-$i}
 = [1-t][1-d]{$'[1+s][1-v-f]-$i}
 = [1-0.3000][1-0.3000]{48,851
  [1+0.0500][1-0.1900-0.5800]
  -51,294*0.0100}
 = 5,529 millions USD

Law-3808:

If both (**s**= [S/S']-1= Sales Growth= 5.00%), (**d**= D/A= Dividend Portion or Payout= 30.00%), (**f**= F/S=Fixed Portion= 58.00%), (**i**= I/S= Interest Portion= 1.00%), (**$**= Sales or Revenues= 51,294 millions USD), (**R**= Retained Earnings= 5,529 millions USD), (**v**= V/S= Variable Portion= 19.00%), and (**t**= T/B= Tax Rate= 30.00%), are known, then it's (**$'**= Sales Past), must be:

$$\$' = (\$i+R/\{[1-t][1-d]\})/\{[1+s][1-v-f]\}$$
$$= (51{,}294*0.0100+5{,}529/\{[1-0.3000][1-0.3000]\})/\{[1+0.0500][1-0.1900-0.5800]\}$$
$$= 48{,}851 \text{ millions USD}$$

Law-3809:

If both (**$'**= Sales of Past Year= 48,851 millions USD), (**d**= D/A= Dividend Portion or Payout= 30.00%), (**f**= F/S= Fixed Portion= 58.00%), (**i**= I/S= Interest Portion= 1.00%), (**$**= Sales or Revenues= 51,294 millions USD), (**R**= Retained Earnings= 5,529 millions USD), (**v**= V/S= Variable Portion= 19.00%), and (**t**= T/B= Tax Rate= 30.00%), are known, then it's (**s**= Sales Growth planned), is:

$$s = (\$i+R/\{[1-t][1-d]\})/\{\$'[1-v-f]\}-1$$
$$= (51{,}294*0.0100+5{,}529/\{[1-0.3000][1-0.3000]\})/\{48{,}851[1-0.1900-0.5800]\}-1$$
$$= 5.00\%$$

Law-3810:
If both ($'= Sales of Past Year= 48,851 millions USD), (d= D/A= Dividend Portion or Payout= 30.00%), (f= F/S= Fixed Portion= 58.00%), (i= I/S= Interest Portion= 1.00%), ($= Sales or Revenues= 51,294 millions USD), (R= Retained Earnings= 5,529 millions USD), (s= [S/S']-1= Sales Growth= 5.00%), and (t= T/B= Tax Rate= 30.00%), are known, then it's (v= Variable Portion planned), is:
v= 1-($i+R/{[1-t][1-d]})/{$'[1+s]}-f
  = 1-(51,294*0.0100+5,529/{[1-0.3000]
       [1-0.3000]})/{48,851
       [1+0.0500]}-0.5800
  = 19.00%

Law-3811:
If both ($'= Sales of Past Year= 48,851 millions USD), (d= D/A= Dividend Portion or Payout= 30.00%), (v= V/S= Variable Portion= 19.00%), (i= I/S= Interest Portion= 1.00%), ($= Sales or Revenues= 51,294 millions USD), (R= Retained Earnings= 5,529 millions USD), (s= [S/S']-1= Sales Growth= 5.00%), and (t= T/B= Tax Rate= 30.00%), are known, then it's (f= Fixed Portion planned), is:
f= 1-($i+R/{[1-t][1-d]})/{$'[1+s]}-v
  = 1-(51,294*0.0100+5,529/{[1-0.3000]
       [1-0.3000]})/{48,851
       [1+0.0500]}-0.1900
  = 58.00%

Law-3812:
If both ($'= Sales of Past Year= 48,851 millions USD), (d= D/A= Dividend Portion or Payout= 30.00%), (v= V/S= Variable Portion= 19.00%), (i= I/S= Interest Portion= 1.00%), (f= F/S= Fixed Portion= 58.00%), (R= Retained Earnings= 5,529 millions USD), (s= [S/S']-1= Sales Growth= 5.00%), and (t= T/B= Tax Rate= 30.00%), are known, then it's ($= Sales or Revenues planned), is:

$= $'[1+s][1-v-f]-R/{[1-t][1-d]}/i
= 48,851[1+0.0500][1-0.1900-0.5800]
   -5,529/{[1-0.3000][1-0.3000]})
   /0.0100
= 51,294 millions USD

Law-3813:
If both ($'= Sales of Past Year= 48,851 millions USD), (d= D/A= Dividend Portion or Payout= 30.00%), (v= V/S= Variable Portion= 19.00%), ($= Sales or Revenues= 51,294 millions USD), (f= F/S= Fixed Portion= 58.00%), (R= Retained Earnings= 5,529 millions USD), (s= [S/S']-1= Sales Growth= 5.00%), and (t= T/B= Tax Rate= 30.00%), are known, then it's (i= Interest Portion planned), is:

i= $'[1+s][1-v-f]-R/{[1-t][1-d]}/$
= 48,851[1+0.0500][1-0.1900-0.5800]
   -5,529/{[1-0.3000][1-0.3000]})
   /51,294
= 1.00%

Law-3814:

If both ($'= Sales of Past Year= 48,851 millions USD), (d= D/A= Dividend Portion or Payout= 30.00%), (v= V/S= Variable Portion= 19.00%), ($= Sales or Revenues= 51,294 millions USD), (f= F/S= Fixed Portion= 58.00%), (R= Retained Earnings= 5,529 millions USD), (s= [S/S']-1= Sales Growth= 5.00%), and (i= I/S= Interest Portion= 1.00%), are known, then it's (t= Tax Rate planned), is:

$$t = 1 - R/([1-d]\{\$'[1+s][1-v-f]-\$i\})$$
$$= 1 - 5{,}529/([1-0.3000]\{48{,}851$$
$$[1+0.0500][1-0.1900-0.5800]$$
$$-51{,}294*0.0100\})$$
$$= 30.00\%$$

Law-3815:

If both ($'= Sales of Past Year= 48,851 millions USD), (t= T/B= Tax Rate= 30.00%), (v= V/S= Variable Portion= 19.00%), ($= Sales or Revenues= 51,294 millions USD), (f= F/S=Fixed Portion= 58.00%), (R= Retained Earnings= 5,529 millions USD), (s= [S/S']-1= Sales Growth= 5.00%), and (i= I/S= Interest Portion= 1.00%), are known, then it's (d= Dividend Portion or Payout planned), is:

$$d = 1 - R/([1-t]\{\$'[1+s][1-v-f]-\$i\})$$
$$= 1 - 5{,}529/([1-0.3000]\{48{,}851$$
$$[1+0.0500][1-0.1900-0.5800]$$
$$-51{,}294*0.0100\})$$
$$= 30.00\%$$

Law-3816:

If both ($= Sales or Revenues= 51,294 millions USD), (v= V/S= Variable Portion= 19.00%), ($'= Sales of Past Year= 48,851 millions USD), (i= I/S= Interest Portion= 1.00%), (t= T/B= Tax Rate= 30.00%), (s= [S/S']-1= Sales Growth= 5.00%), (f= F/S=Fixed Portion= 58.00%), and (d= D/A= Dividend Portion or Payout= 30.00%), are known, then it's (R= Retained Earnings planned), is:

R = [1-t][1-d]{$'[1+s]-$'v[1+s]
     -$'f[1+s]-$'i[1+s]}
   = [1-t][1-d]{$'[1+s][1-v-f-i]}
   = [1-0.3000][1-0.3000]{48,851
     [1+0.0500][1-0.1900-0.5800
     -0.0100]}
   = 5,529 millions USD

Law-3817:
If both ($= Sales or Revenues= 51,294 millions USD), (v= V/S= Variable Portion= 19.00%), (R= Retained Earnings= 5,529 millions USD), (i= I/S= Interest Portion= 1.00%), (t= T/B= Tax Rate= 30.00%), (s= [S/S']-1= Sales Growth= 5.00%), (f= F/S= Fixed Portion= 58.00%), and (d= D/A= Dividend Portion or Payout= 30.00%), are known, then it's ($'= Sales Past), must be:

$$\$' = (R/\{[1-t][1-d]\})/\{[1+s][1-v-f-i]\}$$
$$= (5{,}529/\{[1-0.3000][1-0.3000]\})$$
$$/\{[1+0.0500][1-0.1900-0.5800$$
$$-0.0100]\}$$
$$= 48{,}851 \text{ millions USD}$$

Law-3818:
If both ($= Sales or Revenues= 51,294 millions USD), (v= V/S= Variable Portion= 19.00%), (R= Retained Earnings= 5,529 millions USD), (i= I/S= Interest Portion= 1.00%), (t= T/B= Tax Rate= 30.00%), ($'= Sales of Past Year= 48,851 millions USD), (f= F/S= Fixed Portion= 58.00%), and (d= D/A= Dividend Portion or Payout= 30.00%), are known, then it's (s= Sales Growth planned), is:

$$s = (R/\{[1-t][1-d]\})/\{\$'[1-v-f-i]\} - 1$$
$$= (5{,}529/\{[1-0.3000][1-0.3000]\})$$
$$/\{48{,}851[1-0.1900-0.5800$$
$$-0.0100]\} - 1$$
$$= 5.00\%$$

Law-3819:
If both ($= Sales or Revenues= 51,294 millions USD), (s= [S/S']-1= Sales Growth= 5.00%), (R= Retained Earnings= 5,529 millions USD), (i= I/S= Interest Portion= 1.00%), (t= T/B= Tax Rate= 30.00%), ($'= Sales of Past Year= 48,851 millions USD), (f= F/S= Fixed Portion= 58.00%), and (d= D/A= Dividend Portion or Payout= 30.00%), are known, then it's (v= Variable Portion planned), is:

$v = 1 - f - i - (R/\{[1-t][1-d]\})/\{\$'[1+s]\}$
$= 1 - 0.5800 - 0.0100 - (5,529/\{[1-0.3000][1-0.3000]\})/\{48,851[1+0.0500]\}$
$= 19.00\%$

Law-3820:
If both ($= Sales or Revenues= 51,294 millions USD), (s= [S/S']-1= Sales Growth= 5.00%), (R= Retained Earnings= 5,529 millions USD), (i= I/S= Interest Portion= 1.00%), (t= T/B= Tax Rate= 30.00%), ($'= Sales of Past Year= 48,851 millions USD), (v= V/S= Variable Portion= 19.00%), and (d= D/A= Dividend Portion or Payout= 30.00%), are known, then it's (f= Fixed Portion planned), is:

$f = 1 - v - i - (R/\{[1-t][1-d]\})/\{\$'[1+s]\}$
$= 1 - 0.1900 - 0.0100 - (5,529/\{[1-0.3000][1-0.3000]\})/\{48,851[1+0.0500]\}$
$= 58.00\%$

Law-3821:

If both ($= Sales or Revenues= 51,294 millions USD), (s= [S/S']-1= Sales Growth= 5.00%), (R= Retained Earnings= 5,529 millions USD), (f= F/S= Fixed Portion= 58.00%), (t= T/B= Tax Rate= 30.00%), ($'= Sales of Past Year= 48,851 millions USD), (v= V/S= Variable Portion= 19.00%), and (d= D/A= Dividend Portion or Payout= 30.00%), are known, then it's (i= Interest Portion planned), is:

$i = 1-v-f-(R/\{[1-t][1-d]\})/\{\$'[1+s]\}$
$= 1-0.1900-0.5800-(5,529/\{[1-0.3000]$
$[1-0.3000]\})$
$/\{48,851[1+0.0500]\}$
$= \underline{1.00\%}$

Law-3822:

If both ($= Sales or Revenues= 51,294 millions USD), (s= [S/S']-1= Sales Growth= 5.00%), (R= Retained Earnings= 5,529 millions USD), (f= F/S= Fixed Portion= 58.00%), (i= I/S= Interest Portion= 1.00%), ($'= Sales of Past Year= 48,851 millions USD), (v= V/S= Variable Portion= 19.00%), and (d= D/A= Dividend Portion or Payout= 30.00%), are known, then it's (t= Tax Rate planned), is:

$t = 1-R/([1-d]\{\$'[1+s][1-v-f-i]\})$
$= 1-5,529/([1-0.3000]\{48,851$
$[1+0.0500][1-0.1900-0.5800$
$-0.0100\})$
$= \underline{30.00\%}$

Law-3823:

If both (**$** = Sales or Revenues= 51,294 millions USD), (**s**= [S/S']-1= Sales Growth= 5.00%), (**R**= Retained Earnings= 5,529 millions USD), (**f**= F/S= Fixed Portion= 58.00%), (**i**= I/S= Interest Portion= 1.00%), (**$'**= Sales of Past Year= 48,851 millions USD), (**v**= V/S= Variable Portion= 19.00%), and (**t**= T/B= Tax Rate= 30.00%), are known, then it's (**d**= Dividend Portion or Payout planned), is:

$$d = 1 - R/([1-t]\{\$'[1+s][1-v-f-i]\})$$
$$= 1 - 5,529/([1-0.3000]\{48,851$$
$$[1+0.0500][1-0.1900-0.5800$$
$$-0.0100]\})$$
$$= \underline{30.00\%}$$

CHAPTER-14:
E= EQUITY or CAPITAL Optimization

Based on 12/31/15 Pfizer Annual Report
(http://www.nasdaq.com/symbol/pfe/financials?query=income-statement)

Rule-3824:
If both (**U**= Utilized or Starting Capital= 64,720 millions USD), and (**R**= Retained Earnings= 5,529 millions USD), are known, then it's (**E**= Equity or Capital planned), is:
  E= U+R
    = 64,720+5,529
    = 70,249 millions USD

Rule-3825:
If both (**U**= Utilized or Starting Capital= 64,720 millions USD), and (**E**= Equity or Capital= 70,249 millions USD), are known, then it's (**R**= Retained Earnings planned), is:
  R= E-U
    = 70,249-64,720
    = 5,529 millions USD

Rule-3826:
If both (**R**= Retained Earnings= 5,529 millions USD), and (**E**= Equity or Capital= 70,249 millions USD), are known, then it's (**U**= Utilized or Starting Capital planned), is:
U= E-R
$$= 70,249-5,529$$
$$= \underline{64,720} \text{ millions USD}$$

Rule-3827:
If both (**D**= Dividend Paid= 2,370 millions USD), (**U**= Utilized or Starting Capital= 64,720 millions USD), and (**A**= After Tax Income= 7,899 millions USD), are known, then it's (**E**= Equity or Capital planned), is:
E= U+A-D
$$= 64,720+7,899-2,370$$
$$= \underline{70,249} \text{ millions USD}$$

Rule-3828:
If both (**D**= Dividend Paid= 2,370 millions USD), (**U**= Utilized or Starting Capital= 64,720 millions USD), and (**E**= Equity or Capital= 70,249 millions USD), are known, then it's (**A**= After Tax Income planned), is:
A= D+E-U
$$= 2,370+70,249-64,720$$
$$= \underline{7,899} \text{ millions USD}$$

Rule-3829:
If both (**A**= After Tax Income= 7,899 millions USD), (**U**= Utilized or Starting Capital= 64,720 millions USD), and (**E**= Equity or Capital= 70,249 millions USD), are known, then it's (**D**= Dividend Paid planned), is:
D= U+A-E
= 64,720+7,899-70,249
= 2,370 millions USD

Rule-3830:
If both (**A**= After Tax Income= 7,899 millions USD), (**D**= Dividend Paid= 2,370 millions USD), and (**E**= Equity or Capital= 70,249 millions USD), are known, then it's (**U**= Utilized or Starting Capital planned), is:
U= D+E-A
= 2,370+70,249-7,899
= 64,720 millions USD

Rule-3831:
If both (**U**= Utilized or Starting Capital= 64,720 millions USD), (**A**= After Tax Income= 7,899 millions USD), and (**d**= D/A= Dividend Portion or Payout= 30.00%), are known, then it's (**E**= Equity or Capital planned), is:
E= U+A-Ad= U+A[1-d]
= 64,720+7,899[1-0.3000]
= 70,249 millions USD

Rule-3832:
If both (**U**= Utilized or Starting Capital= 64,720 millions USD), (**E**= Equity or Capital= 70,249 millions USD),and (**d**= D/A= Dividend Portion or Payout= 30.00%), are known, then it's (**A**= After Tax Income planned), is:
A= [**E**-**U**]/[1-**d**]
 = [70,249-64,720]/[1-0.3000]
 = 7,899 millions USD

Rule-3833:
If both (**U**= Utilized or Starting Capital= 64,720 millions USD), (**E**= Equity or Capital= 70,249 millions USD),and (**A**= After Tax Income= 7,899 millions USD),are known, then it's (**d**= Dividend Portion or Payout planned), is:
d= 1-[**E**-**U**]/**A**
 = 1-[70,249-64,720]/7,899
 = 30.00%

Rule-3834:
If both (**d**= D/A= Dividend Portion or Payout= 30.00%), (**E**= Equity or Capital= 70,249 millions USD),and (**A**= After Tax Income= 7,899 millions USD),are known, then it's (**U**= Utilized or Starting Capital planned), is:
U= **E**-**A**[1-**d**]
 = 70,249-7,899[1-0.3000]
 = 64,720 millions USD

Rule-3835:
If both (**d**= D/A= Dividend Portion or Payout= 30.00%), (**T**= Tax Paid= 3,385 millions USD), (**U**= Utilized or Starting Capital= 64,720 millions USD), and (**B**= Before Tax Income= 11,285 millions USD), are known, then it's (**E**= Equity or Capital planned), is:

$$E = U+[1-d][B-T]$$
$$= 64,720+[1-0.3000][11,285-3,385]$$
$$= 70,249 \text{ millions USD}$$

Rule-3836:
If both (**d**= D/A= Dividend Portion or Payout= 30.00%), (**T**= Tax Paid= 3,385 millions USD), (**U**= Utilized or Starting Capital= 64,720 millions USD), and (**E**= Equity or Capital= 70,249 millions USD), are known, then it's (**B**= Before Tax Income planned), is:

$$B = T+[E-U]/[1-d]$$
$$= 3,385+[70,249-64,720]/[1-0.3000]$$
$$= 11,285 \text{ millions USD}$$

Rule-3837:
If both (**d**= D/A= Dividend Portion or Payout= 30.00%), (**B**= Before Tax Income= 11,285 millions USD), (**U**= Utilized or Starting Capital= 64,720 millions USD), and (**E**= Equity or Capital= 70,249 millions USD), are known, then it's (**T**= Tax Paid planned), is:

$$T = B-[E-U]/[1-d]$$
$$= 11,285-[70,249-64,720]/[1-0.3000]$$
$$= 3,385 \text{ millions USD}$$

Rule-3838:
If both (**T**= Tax Paid= 3,385 millions USD), (**B**= Before Tax Income= 11,285 millions USD), (**U**= Utilized or Starting Capital= 64,720 millions USD), and (**E**= Equity or Capital= 70,249 millions USD), are known, then it's (**d**= Dividend Portion or Payout planned), is:

$$d = 1-[E-U]/[B-T]$$
$$= 1-[70,249-64,720]/[11,285-3,385]$$
$$= 30.00\%$$

Rule-3839:
If both (**T**= Tax Paid= 3,385 millions USD), (**B**= Before Tax Income= 11,285 millions USD), (**d**= D/A= Dividend Portion or Payout= 30.00%), and (**E**= Equity or Capital= 70,249 millions USD), are known, then it's (**U**= Utilized or Starting Capital planned), is:
$$U = E-[1-d][B-T]$$
$$= 70,249-[1-0.3000][11,285-3,385]$$
$$= 64,720 \text{ millions USD}$$

Rule-3840:
If both (**t**= T/B= Tax Rate= 30.00%), (**B**= Before Tax Income= 11,285 millions USD), (**d**= D/A= Dividend Portion or Payout= 30.00%), and (**U**= Utilized or Starting Capital= 64,720 millions USD), are known, then it's (**E**= Equity or Capital planned), is:
$$E = U+[B-Bt][1-d] = U+B[1-t][1-d]$$
$$= 64,720+11,285[1-0.3000][1-0.3000]$$
$$= 70,249 \text{ millions USD}$$

Rule-3841:
If both (**t**= T/B= Tax Rate= 30.00%), (**E**= Equity or Capital= 70,249 millions USD), (**d**= D/A=Dividend Portion or Payout= 30.00%), and (**U**= Utilized or Starting Capital= 64,720 millions USD), are known, then it's (**B**= Before Tax Income planned), is:
B= [**E**-**U**]/{[1-**t**][1-**d**]}
    = [70,249-64,720]/{[1-0.3000][1-0.3000]}
    = 11,285 millions USD

Rule-3842:
If both (**B**= Before Tax Income= 11,285 millions USD), (**E**= Equity or Capital= 70,249 millions USD), (**d**= D/A= Dividend Portion or Payout= 30.00%), and (**U**= Utilized or Starting Capital= 64,720 millions USD), are known, then it's (**t**= Tax Rate planned), is:
t= 1-[**E**-**U**]/{**B**[1-**d**]}
    = 1-[70,249-64,720]/{11,285[1-0.3000]}

    = 30.00%

Rule-3843:
If both (**t**= T/B= Tax Rate= 30.00%), (**E**= Equity or Capital= 70,249 millions USD), (**B**= Before Tax Income= 11,285 millions USD), and (**U**= Utilized or Starting Capital= 64,720 millions USD), are known, then it's (**d**= Dividend Portion or Payout planned), is:
$$d = 1-[\mathbf{E}-\mathbf{U}]/\{\mathbf{B}[1-\mathbf{t}]\}$$
$$= 1-[70,249-64,720]/\{11,285[1-0.3000]\}$$
$$= 30.00\%$$

Rule-3844:
If both (**t**= T/B= Tax Rate= 30.00%), (**E**= Equity or Capital= 70,249 millions USD), (**B**= Before Tax Income= 11,285 millions USD), and (**d**= D/A= Dividend Portion or Payout= 30.00%), are known, then it's (**U**= Utilized or Starting Capital planned), is:
$$U = \mathbf{E}-\mathbf{B}[1-\mathbf{d}][1-\mathbf{t}]\}$$
$$= 70,249-11,285[1-0.3000][1-0.3000]$$
$$= 64,720 \text{ millions USD}$$

Rule-3845:
If both (**t**= T/B= Tax Rate= 30.00%), (**U**= Utilized or Starting Capital= 64,720 millions USD), (**O**= Operational Surplus= 11,798 millions USD), (**I**= Interest Expense= 513 millions USD), and (**d**= D/A= Dividend Portion or Payout= 30.00%), are known, then it's (**E**= Equity or Capital planned), is:

$$E = U+[1-d][1-t][O-I]$$
$$= 64,720+[1-0.3000][1-0.3000]$$
$$[11,798-513]$$
$$= 70,249 \text{ millions USD}$$

Rule-3846:
If both (**E**= Equity or Capital= 70,249 millions USD), (**U**= Utilized or Starting Capital= 64,720 millions USD), (**O**= Operational Surplus= 11,798 millions USD), (**I**= Interest Expense= 513 millions USD), and (**d**= D/A= Dividend Portion or Payout= 30.00%), are known, then it's (**t**= Tax Rate planned), is:

$$t = 1-[E-U]/\{[1-d][O-I]\}$$
$$= 1-[70,249-64,720]/\{[1-0.3000]$$
$$[11,798-513]\}$$
$$= 30.00\%$$

Rule-3847:
If both (**E**= Equity or Capital= 70,249 millions USD), (**U**= Utilized or Starting Capital= 64,720 millions USD), (**O**= Operational Surplus= 11,798 millions USD), (**I**= Interest Expense= 513 millions USD), and (**t**= T/B= Tax Rate= 30.00%), are known, then it's (**d**= Dividend Portion or Payout planned), is:

$$d = 1-[\mathbf{E}-\mathbf{U}]/\{[1-\mathbf{t}][\mathbf{O}-\mathbf{I}]\}$$
$$= 1-[70,249-64,720]/\{[1-0.3000][11,798-513]\}$$
$$= 30.00\%$$

Rule-3848:
If both (**E**= Equity or Capital= 70,249 millions USD), (**U**= Utilized or Starting Capital= 64,720 millions USD), (**d**= D/A= Dividend Portion or Payout= 30.00%), (**I**= Interest Expense= 513 millions USD), and (**t**= T/B= Tax Rate= 30.00%), are known, then it's (**O**= Operational Surplus planned), is:

$$\mathbf{O} = \mathbf{I}+[\mathbf{E}-\mathbf{U}]/\{[1-\mathbf{t}][1-\mathbf{d}]\}$$
$$= 513+[70,249-64,720]/\{[1-0.3000][1-0.3000]\}$$
$$= 11,798 \text{ millions USD}$$

Rule-3849:
If both (**E**= Equity or Capital= 70,249 millions USD), (**U**= Utilized or Starting Capital= 64,720 millions USD), (**d**= D/A= Dividend Portion or Payout= 30.00%), (**O**= Operational Surplus= 11,798 millions USD), and (**t**= T/B= Tax Rate= 30.00%), are known, then it's (**O**= Operational Surplus planned), is:
$$I = O-[E-U]/\{[1-t][1-d]\}$$
$$= 11,798-[70,249-64,720]/\{[1-0.3000][1-0.3000]\}$$
$$= 513 \text{ millions USD}$$

Rule-3850:
If both (**E**= Equity or Capital= 70,249 millions USD), (**I**= Interest Expense= 513 millions USD), (**d**= D/A= Dividend Portion or Payout= 30.00%), (**O**= Operational Surplus= 11,798 millions USD), and (**t**= T/B= Tax Rate= 30.00%), are known, then it's (**U**= Utilized or Starting Capital planned), is:
$$U = E-[O-I][1-t][1-d]$$
$$= 70,249-[11,798-513][1-0.3000][1-0.3000]$$
$$= 64,720 \text{ millions USD}$$

Rule-3851:
If both (**U**= Utilized or Starting Capital= 64,720 millions USD), (**i**= I/S= Interest Portion= 1.00%), (**$**= Sales or Revenues= 51,294 millions USD), (**d**= D/A= Dividend Portion or Payout= 30.00%), (**O**= Operational Surplus= 11,798 millions USD), and (**t**= T/B= Tax Rate= 30.00%), are known, then it's (**E**= Equity or Capital planned), is:

$$E = U + [1-t][1-d][O-\$i]$$
$$= 64,720 + [1-0.3000][1-0.3000]$$
$$[11,798-51,294*0.0100]$$
$$= 70,249 \text{ millions USD}$$

Rule-3852:
If both (**U**= Utilized or Starting Capital= 64,720 millions USD), (**i**= I/S= Interest Portion= 1.00%), (**$**= Sales or Revenues= 51,294 millions USD), (**d**= D/A= Dividend Portion or Payout= 30.00%), (**O**= Operational Surplus= 11,798 millions USD), and (**E**= Equity or Capital= 70,249 millions USD), are known, then it's (**t**= Tax Rate planned), is:

$$t = 1 - [E-U] / \{[1-d][O-\$i]\}$$
$$= 1 - [70,249-64,720] / \{[1-0.3000]$$
$$[11,798-51,294*0.0100]\}$$
$$= 30.00\%$$

Rule-3853:
If both (**U**= Utilized or Starting Capital= 64,720 millions USD), (**i**= I/S= Interest Portion= 1.00%), (**$**= Sales or Revenues= 51,294 millions USD), (**t**= T/B= Tax Rate= 30.00%), (**O**= Operational Surplus= 11,798 millions USD), and (**E**= Equity or Capital= 70,249 millions USD), are known, then it's (**d**= Dividend Portion or Payout planned), is:

$$\mathbf{d} = 1-[\mathbf{E}\text{-}\mathbf{U}]/\{[1\text{-}\mathbf{t}][\mathbf{O}\text{-}\mathbf{\$i}]\}$$
$$= 1-[70{,}249\text{-}64{,}720]/\{[1\text{-}0.3000]$$
$$[11{,}798\text{-}51{,}294*0.0100]\}$$
$$= 30.00\%$$

Rule-3854:
If both (**U**= Utilized or Starting Capital= 64,720 millions USD), (**i**= I/S= Interest Portion= 1.00%), (**$**= Sales or Revenues= 51,294 millions USD), (**t**= T/B= Tax Rate= 30.00%), (**d**= D/A= Dividend Portion or Payout= 30.00%), and (**E**= Equity or Capital= 70,249 millions USD), are known, then it's (**O**= Operational Surplus planned), is:

$$\mathbf{O} = \mathbf{\$i}+[\mathbf{E}\text{-}\mathbf{U}]/\{[1\text{-}\mathbf{t}][1\text{-}\mathbf{d}]\}$$
$$= 51{,}294*0.0100+[70{,}249\text{-}64{,}720]$$
$$/\{[1\text{-}0.3000][1\text{-}0.3000]\}$$
$$= 11{,}798 \text{ millions USD}$$

Rule-3855:
If both (**U**= Utilized or Starting Capital= 64,720 millions USD), (**i**= I/S= Interest Portion= 1.00%), (**O**= Operational Surplus= 11,798 millions USD), (**t**= T/B= Tax Rate= 30.00%), (**d**= D/A= Dividend Portion or Payout= 30.00%), and (**E**= Equity or Capital= 70,249 millions USD), are known, then it's (**$**= Sales or Revenues planned), is:

$$\$ = (\mathbf{O}\text{-}[\mathbf{E}\text{-}\mathbf{U}]/\{[1\text{-}\mathbf{t}][1\text{-}\mathbf{d}]\})/\mathbf{i}$$
$$= (11{,}798\text{-}[70{,}249\text{-}64{,}720]/\{[1\text{-}0.3000][1\text{-}0.3000]\})/\ 0.0100$$
$$= 51{,}294 \text{ millions USD}$$

Rule-3856:
If both (**U**= Utilized or Starting Capital= 64,720 millions USD), (**$**= Sales or Revenues= 51,294 millions USD), (**O**= Operational Surplus= 11,798 millions USD), (**t**= T/B= Tax Rate= 30.00%), (**d**= D/A= Dividend Portion or Payout= 30.00%), and (**E**= Equity or Capital= 70,249 millions USD), are known, then it's (**i**= Interest Portion planned), is:

$$\mathbf{i} = (\mathbf{O}\text{-}[\mathbf{E}\text{-}\mathbf{U}]/\{[1\text{-}\mathbf{t}][1\text{-}\mathbf{d}]\})/\$$$
$$= (11{,}798\text{-}[70{,}249\text{-}64{,}720]/\{[1\text{-}0.3000][1\text{-}0.3000]\})/\ 51{,}294$$
$$= 1.00\%$$

Rule-3857:
If both (**i**= I/S= Interest Portion= 1.00%), (**$**= Sales or Revenues= 51,294 millions USD), (**O**= Operational Surplus= 11,798 millions USD), (**t**= T/B= Tax Rate= 30.00%), (**d**= D/A= Dividend Portion or Payout= 30.00%), and (**E**= Equity or Capital= 70,249 millions USD), are known, then it's (**U**= Utilized or Starting Capital planned), is:

$$U = E - [O-\$i][1-t][1-d]$$
$$= 70,249 - [11,798 - 51,294*0.0100]$$
$$[1-0.3000][1-0.3000]\}$$
$$= 64,720 \text{ millions USD}$$

Rule-3858:
If both (**i**= I/S= Interest Portion= 1.00%), (**$'**= Sales of Past Year= 48,851 millions USD), (**s**= [S/S']-1= Sales Growth= 5.00%), (**O**= Operational Surplus= 11,798 millions USD), (**t**= T/B= Tax Rate= 30.00%), (**d**= D/A= Dividend Portion or Payout= 30.00%), and (**U**= Utilized or Starting Capital= 64,720 millions USD),are known, then it's (**E**= Equity or Capital planned), is:

$$E = U + [1-t][1-d]\{O-\$'i[1+s]\}$$
$$= 64,720 + [1-0.3000][1-0.3000]\{11,798$$
$$-48,851*0.0100[1+0.0500]\}$$
$$= 70,249 \text{ millions USD}$$

Rule-3859:
If both (**i**= I/S= Interest Portion= 1.00%), (**$'**= Sales of Past Year= 48,851 millions USD), (**$**= [S/S']-1= Sales Growth= 5.00%), (**E**= Equity or Capital= 70,249 millions USD), (**t**= T/B= Tax Rate= 30.00%), (**d**= D/A= Dividend Portion or Payout= 30.00%), and (**U**= Utilized or Starting Capital= 64,720 millions USD),are known, then it's (**O**= Operational Surplus planned), is:

$$O = \$'i[1+\$]+[E-U]/\{[1-t][1-d]\}$$
$$= 48,851*0.0100[1+0.0500]+[70,249 -64,720]/\{[1-0.3000][1-0.3000]\}$$
$$= 11,798 \text{ millions USD}$$

Rule-3860:
If both (**i**= I/S= Interest Portion= 1.00%), (**O**= Operational Surplus= 11,798 millions USD), (**$**= [S/S']-1= Sales Growth= 5.00%), (**E**= Equity or Capital= 70,249 millions USD), (**t**= T/B= Tax Rate= 30.00%), (**d**= D/A= Dividend Portion or Payout= 30.00%), and (**U**= Utilized or Starting Capital= 64,720 millions USD),are known, then it's (**$'**= Sales Past), must be:

$$\$' = (O-[E-U]/\{[1-t][1-d]\})/\{i[1+\$]\}$$
$$= (11,798- [70,249-64,720]/\{[1-0.3000][1-0.3000]\})/ \{0.0100[1+0.0500]\}$$
$$= 48,851 \text{ millions USD}$$

Rule-3861:
If both ($'= Sales of Past Year= 48,851 millions USD), (O= Operational Surplus= 11,798 millions USD), (s= [S/S']-1= Sales Growth= 5.00%), (E= Equity or Capital= 70,249 millions USD), (t= T/B= Tax Rate= 30.00%), (d= D/A= Dividend Portion or Payout= 30.00%), and (U= Utilized or Starting Capital= 64,720 millions USD),are known, then it's (i= Interest Portion planned), is:

$$i= (O-[E-U]/\{[1-t][1-d]\})/\{\$'[1+s]\}$$
$$= (11,798- [70,249-64,720]/\{[1-0.3000][1-0.3000]\})$$
$$/\{48,851[1+0.0500]\}$$
$$= 1.00\%$$

Rule-3862:
If both ($'= Sales of Past Year= 48,851 millions USD), (O= Operational Surplus= 11,798 millions USD), (i= I/S= Interest Portion= 1.00%), (E= Equity or Capital= 70,249 millions USD), (t= T/B= Tax Rate= 30.00%), (d= D/A= Dividend Portion or Payout= 30.00%), and (U= Utilized or Starting Capital= 64,720 millions USD),are known, then it's (s= Sales Growth planned), is:

$$s= (O-[E-U]/\{[1-t][1-d]\})/[\$'i]-1$$
$$= (11,798-[70,249-64,720]$$
$$/\{[1-0.3000][1-0.3000]\})$$
$$/[48,851*0.0100]-1$$
$$= 5.00\%$$

Rule-3863:
If both ($= Sales of Past Year= 48,851 millions USD), (O= Operational Surplus= 11,798 millions USD), (i= I/S= Interest Portion= 1.00%), (E= Equity or Capital= 70,249 millions USD), (s= [S/S']-1= Sales Growth= 5.00%), (d= D/A= Dividend Portion or Payout= 30.00%), and (U= Utilized or Starting Capital= 64,720 millions USD),are known, then it's (t= Tax Rate planned), is:

$$t = 1-[E-U]/([1-d]\{O-S'i[1+s]\})$$
$$= 1-[70,249-64,720]$$
$$/([1-0.3000]\{11,798-48,851$$
$$*0.0100[1+0.0500]\})$$
$$= 30.00\%$$

Rule-3864:
If both ($= Sales of Past Year= 48,851 millions USD), (O= Operational Surplus= 11,798 millions USD), (i= I/S= Interest Portion= 1.00%), (E= Equity or Capital= 70,249 millions USD), (s= [S/S']-1= Sales Growth= 5.00%), (t= T/B= Tax Rate= 30.00%), and (U= Utilized or Starting Capital= 64,720 millions USD),are known, then it's (d= Dividend Portion or Payout planned), is:

$$d = 1-[E-U]/([1-t]\{O-S'i[1+s]\})$$
$$= 1-[70,249-64,720]/([1-0.3000]$$
$$\{11,798-48,851*0.0100$$
$$[1+0.0500]\})$$
$$= 30.00\%$$

Rule-3865:
If both ($= Sales of Past Year= 48,851 millions USD), (O= Operational Surplus= 11,798 millions USD), (i= I/S= Interest Portion= 1.00%), (E= Equity or Capital= 70,249 millions USD), (s= [S/S']-1= Sales Growth= 5.00%), (t= T/B= Tax Rate= 30.00%), and (d= D/A= Dividend Portion or Payout= 30.00%), are known, then it's (U= Utilized or Starting Capital planned), is:

U= E-[1-d][1-t]{O-$'i[1+s]}
= 70,249-[1-0.3000][1-0.3000]{11,798
-48,851*0.0100[1+0.0500]}
= 64,720 millions USD

Rule-3866:
If both (U= Utilized or Starting Capital= 64,720 millions USD), (I= Interest Expense= 513 millions USD), (F= Fixed Cost= 29,750 millions USD), (M= Margin of Contribution= 41,548 millions USD), (t= T/B= Tax Rate= 30.00%), and (d= D/A=Dividend Portion or Payout= 30.00%), are known, then it's (E= Equity or Capital planned), is:

E= U+[1-d][1-t][M-F-I]
= 64,720+[1-0.3000][1-0.3000][41,548
-29,750-513]
= 70,249 millions USD

Rule-3867:
If both (**U**= Utilized or Starting Capital= 64,720 millions USD), (**I**= Interest Expense= 513 millions USD), (**F**= Fixed Cost= 29,750 millions USD), (**E**= Equity or Capital= 70,249 millions USD), (**t**= T/B= Tax Rate= 30.00%), and (**d**= D/A= Dividend Portion or Payout= 30.00%), are known, then it's (**M**= Margin of Contribution planned), is:
$$M = F-I+[E-U]/\{[1-d][1-t]\}$$
$$= 29{,}750+513+[70{,}249-64{,}720]$$
$$/\{[1-0.3000][1-0.3000]\}$$
$$= 41{,}548 \text{ millions USD}$$

Rule-3868:
If both (**U**= Utilized or Starting Capital= 64,720 millions USD), (**I**= Interest Expense= 513 millions USD), (**M**= Margin of Contribution= 41,548 millions USD), (**E**= Equity or Capital= 70,249 millions USD), (**t**= T/B= Tax Rate= 30.00%), and (**d**= D/A= Dividend Portion or Payout= 30.00%), are known, then it's (**F**= Fixed Cost planned), is:
$$F = M-I-[E-U]/\{[1-d][1-t]\}$$
$$= 41{,}548-513-[70{,}249-64{,}720]$$
$$/\{[1-0.3000][1-0.3000]\}$$
$$= 29{,}750 \text{ millions USD}$$

Rule-3869:
If both (**U**= Utilized or Starting Capital= 64,720 millions USD), (**F**= Fixed Cost= 29,750 millions USD), (**M**= Margin of Contribution= 41,548 millions USD), (**E**= Equity or Capital= 70,249 millions USD), (**t**= T/B= Tax Rate= 30.00%), and (**d**= D/A= Dividend Portion or Payout= 30.00%), are known, then it's (**I**= Interest Expense planned), is:
 I= **M-F-[E-U]**/{[1-**d**][1-**t**]}
  = 41,548-29,750-[70,249-64,720]
   /{[1-0.3000][1-0.3000]}
  = 513 millions USD

Rule-3870:
If both (**U**= Utilized or Starting Capital= 64,720 millions USD), (**F**= Fixed Cost= 29,750 millions USD), (**M**= Margin of Contribution= 41,548 millions USD), (**E**= Equity or Capital= 70,249 millions USD), (**I**= Interest Expense= 513 millions USD), and (**d**= D/A= Dividend Portion or Payout= 30.00%), are known, then it's (**t**= Tax Rate planned), is:
 t= 1-[**E-U**]/{[1-**d**][**M-F-I**]}
  = 1-[70,249-64,720]/{[1-0.3000]
   [41,548-29,750-513]}
  = 30.00%

Rule-3871:
If both (**U**= Utilized or Starting Capital= 64,720 millions USD), (**F**= Fixed Cost= 29,750 millions USD), (**M**= Margin of Contribution= 41,548 millions USD), (**E**= Equity or Capital= 70,249 millions USD), (**I**= Interest Expense= 513 millions USD), and (**t**= T/B= Tax Rate= 30.00%), are known, then it's (**d**= Dividend Portion or Payout planned), is:
 d= 1-[**E**-**U**]/{[1-**t**][**M**-**F**-**I**]}
   = 1-[70,249-64,720]/{[1-0.3000]
      [41,548-29,750-513]}
   = 30.00%

Rule-3872:
If both (**d**= D/A= Dividend Portion or Payout= 30.00%), (**F**= Fixed Cost= 29,750 millions USD), (**M**= Margin of Contribution= 41,548 millions USD), (**E**= Equity or Capital= 70,249 millions USD), (**I**= Interest Expense= 513 millions USD), and (**t**= T/B= Tax Rate= 30.00%), are known, then it's (**U**= Utilized or Starting Capital planned), is:
 U= **E**-[1-**d**][1-**t**][**M**-**F**-**I**]
   = 70,249-[1-0.3000][1-0.3000]
      [41,548-29,750-513]
   = 64,720 millions USD

Rule-3873:
If both (**d**= D/A= Dividend Portion or Payout= 30.00%), (**F**= Fixed Cost= 29,750 millions USD), (**M**= Margin of Contribution= 41,548 millions USD), (**U**= Utilized or Starting Capital= 64,720 millions USD), (**$**= Sales or Revenues= 51,294 millions USD), (**i**= I/S= Interest Portion= 1.00%), and (**t**= T/B= Tax Rate= 30.00%), are known, then it's (**E**= Equity or Capital planned), is:

$$E = U+[1-d][1-t][M-F-\$i]$$
$$= 64{,}720+[1-0.3000][1-0.3000][41{,}548 -29{,}750-51{,}294*0.0100]$$
$$= 70{,}249 \text{ millions USD}$$

Rule-3874:
If both (**d**= D/A= Dividend Portion or Payout= 30.00%), (**F**= Fixed Cost= 29,750 millions USD), (**E**= Equity or Capital= 70,249 millions USD), (**U**= Utilized or Starting Capital= 64,720 millions USD), (**$**= Sales or Revenues= 51,294 millions USD), (**i**= I/S= Interest Portion= 1.00%), and (**t**= T/B= Tax Rate= 30.00%), are known, then it's (**M**= Margin of Contribution planned), is:

$$M = F+\$i+[E-U]/\{[1-d][1-t]\}$$
$$= 29{,}750+51{,}294*0.0100+[70{,}249 -64{,}720]/\{[1-0.3000][1-0.3000]\}$$
$$= 41{,}548 \text{ millions USD}$$

Rule-3875:
If both (**d**= D/A= Dividend Portion or Payout= 30.00%), (**M**= Margin of Contribution= 41,548 millions USD), (**E**= Equity or Capital= 70,249 millions USD), (**U**= Utilized or Starting Capital= 64,720 millions USD), (**$**= Sales or Revenues= 51,294 millions USD), (**i**= I/S= Interest Portion= 1.00%), and (**t**= T/B= Tax Rate= 30.00%), are known, then it's (**F**= Fixed Cost planned), is:

$$F = M - \$i - [E-U]/\{[1-d][1-t]\}$$
$$= 41{,}548 - 51{,}294*0.0100 - [70{,}249 - 64{,}720]/\{[1-0.3000][1-0.3000]\}$$
$$= \underline{29{,}750} \text{ millions USD}$$

Rule-3876:
If both (**d**= D/A= Dividend Portion or Payout= 30.00%), (**M**= Margin of Contribution= 41,548 millions USD), (**E**= Equity or Capital= 70,249 millions USD), (**U**– Utilized or Starting Capital= 64,720 millions USD), (**F**= Fixed Cost= 29,750 millions USD), (**i**= I/S= Interest Portion= 1.00%), and (**t**= T/B= Tax Rate= 30.00%), are known, then it's (**$**= Sales or Revenues planned), is:

$$\$ = (M - F - [E-U]/\{[1-d][1-t]\})/i$$
$$= (41{,}548 - 29{,}750 - [70{,}249 - 64{,}720] /\{[1-0.3000][1-0.3000]\}) /0.0100$$
$$= \underline{51{,}294} \text{ millions USD}$$

Rule-3877:
If both (**d**= D/A= Dividend Portion or Payout= 30.00%), (**M**= Margin of Contribution= 41,548 millions USD), (**E**= Equity or Capital= 70,249 64,720 millions USD), (**F**= Fixed Cost= 29,750 millions USD), (**S**= Sales or Revenues= 51,294 millions USD), and (**t**= T/B= Tax Rate= 30.00%), are known, then it's (**i**= Interest Portion planned), is:

$$\mathbf{i} = (\mathbf{M}\text{-}\mathbf{F}\text{-}[\mathbf{E}\text{-}\mathbf{U}]/\{[1\text{-}\mathbf{d}][1\text{-}\mathbf{t}]\})/\mathbf{S}$$
$$= (41,548\text{-}29,750\text{-}[70,249\text{-}64,720]/\{[1\text{-}0.3000][1\text{-}0.3000]\})/51,294$$
$$= \underline{1.00\%}$$

Rule-3878:
If both (**d**= D/A= Dividend Portion or Payout= 30.00%), (**M**= Margin of Contribution= 41,548 millions USD), (**E**= Equity or Capital= 70,249 millions USD), (**U**= Utilized or Starting Capital= 64,720 millions USD), (**F**= Fixed Cost= 29,750 millions USD), (**S**= Sales or Revenues= 51,294 millions USD), and (**i**= I/S= Interest Portion= 1.00%), are known, then it's (**t**= Tax Rate planned), is:

$$\mathbf{t} = 1\text{-}[\mathbf{E}\text{-}\mathbf{U}]/\{[1\text{-}\mathbf{d}][\mathbf{M}\text{-}\mathbf{F}\text{-}\mathbf{Si}]\}$$
$$= 1\text{-}[70,249\text{-}64,720]/\{[1\text{-}0.3000]\{41,548\text{-}29,750\text{-}51,294*0.0100\}$$
$$= \underline{30.00\%}$$

Rule-3879:
If both (**t**= T/B= Tax Rate= 30.00%), (**M**= Margin of Contribution= 41,548 millions USD), (**E**= Equity or Capital= 70,249 millions USD), (**U**= Utilized or Starting Capital= 64,720 millions USD), (**F**= Fixed Cost= 29,750 millions USD), (**$**= Sales or Revenues= 51,294 millions USD), and (**i**= I/S= Interest Portion= 1.00%), are known, then it's (**d**= Dividend Portion or Payout planned), is:

$$d = 1-[\text{E-U}]/\{[1\text{-}t][\text{M-F-Si}]\}$$
$$= 1-[70,249-64,720]/\{[1-0.3000]$$
$$\{41,548-29,750-51,294*0.0100\}$$
$$= 30.00\%$$

Rule-3880:
If both (**t**= T/B= Tax Rate= 30.00%), (**M**= Margin of Contribution= 41,548 millions USD), (**E**= Equity or Capital= 70,249 millions USD), (**d**= D/A= Dividend Portion or Payout= 30.00%), (**F**= Fixed Cost= 29,750 millions USD), (**$**= Sales or Revenues= 51,294 millions USD), and (**i**= I/S= Interest Portion= 1.00%), are known, then it's (**U**= Utilized or Starting Capital planned), is:

$$U = \text{E-}[1\text{-}d][1\text{-}t][\text{M-F-Si}]$$
$$= 70,249-[1-0.3000][1-0.3000][41,548$$
$$-29,750-51,294*0.0100]$$
$$= 64,720 \text{ millions USD}$$

Rule-3881:
If both (**t**= T/B= Tax Rate= 30.00%), (**M**= Margin of Contribution= 41,548 millions USD), (**U**= Utilized or Starting Capital= 64,720 millions USD), (**d**= D/A= Dividend Portion or Payout= 30.00%), (**F**= Fixed Cost= 29,750 millions USD), (**S'**= Sales of Past Year= 48,851 millions USD), (**s**= [S/S']-1= Sales Growth= 5.00%), and (**i**= I/S= Interest Portion= 1.00%), are known, then it's (**E**= Equity or Capital planned), is:

$$E = U+[1-d][1-t]\{M-F-S'i[1+s]\}$$
$$= 64,720+[1-0.3000][1-0.3000]\{41,548 -29,750-48,851*0.0100 [1+0.0500]\}$$
$$= 70,249 \text{ millions USD}$$

Rule-3882:

If both (**t**= T/B= Tax Rate= 30.00%), (**E**= Equity or Capital= 70,249 millions USD), (**U**= Utilized or Starting Capital= 64,720 millions USD), (**d**= D/A= Dividend Portion or Payout= 30.00%), (**F**= Fixed Cost= 29,750 millions USD), (**S'**= Sales of Past Year= 48,851 millions USD), (**s**= [S/S']-1= Sales Growth= 5.00%), and (**i**= I/S= Interest Portion= 1.00%), are known, then it's (**M**= Margin of Contribution planned), is:

$$M = F + S'i[1+s] + [E-U]/\{[1-d][1-t]\}$$
$$= 29{,}750 + 48{,}851 * 0.0100[1+0.0500]$$
$$+ [70{,}249 - 64{,}720]$$
$$/\{[1-0.3000][1-0.3000]\}$$
$$= 41{,}548 \text{ millions USD}$$

Rule-3883:
If both (**t**= T/B= Tax Rate= 30.00%), (**E**= Equity or Capital= 70,249 millions USD), (**U**= Utilized or Starting Capital= 64,720 millions USD), (**d**= D/A= Dividend Portion or Payout= 30.00%), (**M**= Margin of Contribution= 41,548 millions USD), (**S'**= Sales of Past Year= 48,851 millions USD), (**s**= [S/S']-1= Sales Growth= 5.00%), and (**i**= I/S= Interest Portion= 1.00%), are known, then it's (**F**= Fixed Cost planned), is:

$$F = M - S'i[1+s] - [E-U]/\{[1-d][1-t]\}$$
$$= 41,548 - 48,851 \cdot 0.0100[1+0.0500]$$
$$-[70,249 - 64,720]$$
$$/\{[1-0.3000][1-0.3000]\}$$
$$= 29,750 \text{ millions USD}$$

Rule-3884:
If both (**t**= T/B= Tax Rate= 30.00%), (**E**= Equity or Capital= 70,249 millions USD), (**U**= Utilized or Starting Capital= 64,720 millions USD), (**d**= D/A= Dividend Portion or Payout= 30.00%), (**M**= Margin of Contribution= 41,548 millions USD), (**F**= Fixed Cost= 29,750 millions USD), (**s**= [S/S']-1= Sales Growth= 5.00%), and (**i**= I/S= Interest Portion= 1.00%), are known, then it's (**S'**= Sales Past), must be:

$$S' = (M - F - [E-U]/\{[1-d][1-t]\})/\{i[1+s]\}$$
$$= (41,548 - 29,750 - [70,249 - 64,720]$$
$$/\{[1-0.3000][1-0.3000]\})$$
$$/\{0.0100[1+0.0500]\}$$
$$= 48,851 \text{ millions USD}$$

Rule-3885:
If both (**t**= T/B= Tax Rate= 30.00%), (**E**= Equity or Capital= 70,249 millions USD), (**U**= Utilized or Starting Capital= 64,720 millions USD), (**d**= D/A= Dividend Portion or Payout= 30.00%), (**M**= Margin of Contribution= 41,548 millions USD), (**F**= Fixed Cost= 29,750 millions USD), (**s**= [S/S']-1= Sales Growth= 5.00%), and (**S'**= Sales of Past Year= 48,851 millions USD), are known, then it's (**i**= Interest Portion planned), is:

$$i = (M-F-[E-U]/\{[1-d][1-t]\})/\{S'[1+s]\}$$
$$= (41{,}548 - 29{,}750 - [70{,}249 - 64{,}720]/\{[1-0.3000][1-0.3000]\})/\{48{,}851[1+0.0500]\}$$
$$= 1.00\%$$

Rule-3886:
If both (**t**= T/B= Tax Rate= 30.00%), (**E**= Equity or Capital= 70,249 millions USD), (**U**= Utilized or Starting Capital= 64,720 millions USD), (**d**= D/A= Dividend Portion or Payout= 30.00%), (**M**= Margin of Contribution= 41,548 millions USD), (**F**= Fixed Cost= 29,750 millions USD), (**i**= I/S= Interest Portion= 1.00%), and (**$**'= Sales of Past Year= 48,851 millions USD), are known, then it's (**s**= Sales Growth planned), is:

s= (**M-F-[E-U]**/{[1-**d**][1-**t**]})/[**$'i**]-1
   = (41,548-29,750-[70,249-64,720]
       /{[1-0.3000][1-0.3000]})
       /{48,851*0.0100]-1
   = 5.00%

Rule-3887:
If both ($= [S/S']-1=$ Sales Growth= 5.00%), (E= Equity or Capital= 70,249 millions USD), (U= Utilized or Starting Capital= 64,720 millions USD), (d= D/A= Dividend Portion or Payout= 30.00%), (M= Margin of Contribution= 41,548 millions USD), (F= Fixed Cost= 29,750 millions USD), (i= I/S= Interest Portion= 1.00%), and ($'= Sales of Past Year= 48,851 millions USD), are known, then it's (t= Tax Rate planned), is:

$t = 1-[E-U]/([1-d]\{M-F-S'i[1+s]\})$
$= 1-[70,249-64,720]/([1-0.3000]$
$\{41,548-29,750-48,851*0.0100$
$[1+0.0500]\})$
$= 30.00\%$

Rule-3888:
If both (s= [S/S']-1= Sales Growth= 5.00%), (E= Equity or Capital= 70,249 millions USD), (U= Utilized or Starting Capital= 64,720 millions USD), (t= T/B= Tax Rate= 30.00%), (M= Margin of Contribution= 41,548 millions USD), (F= Fixed Cost= 29,750 millions USD), (i= I/S= Interest Portion= 1.00%), and (S'= Sales of Past Year= 48,851 millions USD), are known, then it's (d= Dividend Portion or Payout planned), is:

d= 1-[E-U]/([1-t]{M-F-$S'i$[1+s]})
  = 1-[70,249-64,720]/([1-0.3000]
    {41,548-29,750-48,851
    *0.0100[1+0.0500]})
  = 30.00%

Rule-3889:
If both (s= [S/S']-1= Sales Growth= 5.00%), (E= Equity or Capital= 70,249 millions USD), (d= D/A= Dividend Portion or Payout= 30.00%), (t= T/B= Tax Rate= 30.00%), (M= Margin of Contribution= 41,548 millions USD), (F= Fixed Cost= 29,750 millions USD), (i= I/S= Interest Portion= 1.00%), and (S'= Sales of Past Year= 48,851 millions USD), are known, then it's (U= Dividend Portion or Payout planned), is:
$U = E\text{-}[1\text{-}d][1\text{-}t]\{M\text{-}F\text{-}S'i[1+s]\}$)
= 70,249- [1-0.3000][1-0.3000]{41,548
-29,750-48,851*0.0100
[1+0.0500]})
= 64,720 millions USD

Rule-3890:
If both (U= Utilized or Starting Capital= 64,720 millions USD), (d= D/A= Dividend Portion or Payout= 30.00%), (t= T/B= Tax Rate= 30.00%), (I= Interest Expense= 513 millions USD), (M= Margin of Contribution= 41,548 millions USD), (S= Sales or Revenues= 51,294 millions USD), and (f= F/S= Fixed Portion= 58.00%), are known, then it's (E= Equity or Capital planned), is:
E= U+[1-d][1-t][M-Sf-I]
= 64,720+ [1-0.3000][1-0.3000][41,548
-51,294*0.5800-513]
= 70,249 millions USD

Rule-3891:
If both (**U**= Utilized or Starting Capital= 64,720 millions USD), (**d**= D/A= Dividend Portion or Payout= 30.00%), (**t**= T/B= Tax Rate= 30.00%), (**I**= Interest Expense= 513 millions USD), (**E**= Equity or Capital= 70,249 millions USD), (**S**= Sales or Revenues= 51,294 millions USD), and (**f**= F/S=Fixed Portion= 58.00%), are known, then it's (**M**= Margin of Contribution planned), is:

$$M = Sf + I + [E-U]/\{[1-d][1-t]\}$$
$$= 51,294*0.5800+513+[70,249-64,720]/\{[1-0.3000][1-0.3000]\}$$
$$= 41,548 \text{ millions USD}$$

Rule-3892:
If both (**U**= Utilized or Starting Capital= 64,720 millions USD), (**d**= D/A= Dividend Portion or Payout= 30.00%), (**t**= T/B= Tax Rate= 30.00%), (**I**= Interest Expense= 513 millions USD), (**E**= Equity or Capital= 70,249 millions USD), (**M**= Margin of Contribution= 41,548 millions USD), and (**f**= F/S= Fixed Portion= 58.00%), are known, then it's (**S**= Sales or Revenues planned), is:

$$S = (M-I-[E-U]/\{[1-d][1-t]\})/f$$
$$= (41,548-513-[70,249-64,720]/\{[1-0.3000][1-0.3000]\})/0.5800$$
$$= 51,294 \text{ millions USD}$$

Rule-3893:
If both (**U**= Utilized or Starting Capital= 64,720 millions USD), (**d**= D/A= Dividend Portion or Payout= 30.00%), (**t**= T/B= Tax Rate= 30.00%), (**I**= Interest Expense= 513 millions USD), (**E**= Equity or Capital= 70,249 millions USD), (**M**= Margin of Contribution= 41,548 millions USD), and (**S**= Sales or Revenues= 51,294 millions USD), are known, then it's (**f**= Fixed Portion planned), is:

$$f = (M-I-[E-U]/\{[1-d][1-t]\})/S$$
$$= (41,548-513-[70,249-64,720]$$
$$/\{[1-0.3000][1-0.3000]\})$$
$$/51,294$$
$$= 58.00\%$$

Rule-3894:
If both (**U**= Utilized or Starting Capital= 64,720 millions USD), (**d**= D/A= Dividend Portion or Payout= 30.00%), (**t**= T/B= Tax Rate= 30.00%), (**f**= F/S=Fixed Portion= 58.00%), (**E**= Equity or Capital= 70,249 millions USD), (**M**= Margin of Contribution= 41,548 millions USD), and (**S**= Sales or Revenues= 51,294 millions USD), are known, then it's (**I**= Interest Expense planned), is:

$$I = M-Sf-[E-U]/\{[1-d][1-t]\}$$
$$= 41,548-51,294*0.5800-[70,249$$
$$-64,720]/\{[1-0.3000][1-0.3000]\}$$

$$= 513 \text{ millions USD}$$

Rule-3895:
If both (**U**= Utilized or Starting Capital= 64,720 millions USD), (**d**= D/A= Dividend Portion or Payout= 30.00%), (**I**= Interest Expense= 513 millions USD), (**f**= F/S=Fixed Portion= 58.00%), (**E**= Equity or Capital= 70,249 millions USD), (**M**= Margin of Contribution= 41,548 millions USD), and (**S**= Sales or Revenues= 51,294 millions USD), are known, then it's (**t**= Tax Rate planned), is:
$$t = 1-[\mathbf{E}-\mathbf{U}]/\{[1-\mathbf{d}][\mathbf{M}-\mathbf{Sf}-\mathbf{I}]$$
$$= 1-[70,249-64,720]/\{[1-0.3000]$$
$$[41,548-51,294*0.5800-513]$$
$$= \underline{30.00\%}$$

Rule-3896:
If both (**U**= Utilized or Starting Capital= 64,720 millions USD), (**t**= T/B= Tax Rate= 30.00%), (**I**= Interest Expense= 513 millions USD), (**f**= F/S= Fixed Portion= 58.00%), (**E**= Equity or Capital= 70,249 millions USD), (**M**= Margin of Contribution= 41,548 millions USD), and (**S**= Sales or Revenues= 51,294 millions USD), are known, then it's (**d**= Dividend Portion or Payout planned), is:
$$d = 1-[\mathbf{E}-\mathbf{U}]/\{[1-\mathbf{t}][\mathbf{M}-\mathbf{Sf}-\mathbf{I}]$$
$$= 1-[70,249-64,720]/\{[1-0.3000]$$
$$[41,548-51,294*0.5800-513]$$
$$= \underline{30.00\%}$$

Rule-3897:
If both (**d**= D/A= Dividend Portion or Payout= 30.00%), (**t**= T/B= Tax Rate= 30.00%), (**I**= Interest Expense= 513 millions USD), (**f**= F/S= Fixed Portion= 58.00%), (**E**= Equity or Capital= 70,249 millions USD), (**M**= Margin of Contribution= 41,548 millions USD), and (**$**= Sales or Revenues= 51,294 millions USD), are known, then it's (**U**= Utilized or Starting Capital planned), is:

$$U = E - [1-d][1-t][M-\$f-I]$$
$$= 70,249 - [1-0.3000][1-0.3000][41,548 - 51,294*0.5800 - 513]$$
$$= \underline{64,720} \text{ millions USD}$$

Rule-3898:
If both (**d**= D/A= Dividend Portion or Payout= 30.00%), (**t**= T/B= Tax Rate= 30.00%), (**i**= I/S= Interest Portion= 1.00%), (**f**= F/S= Fixed Portion= 58.00%), (**U**= Utilized or Starting Capital= 64,720 millions USD), (**M**= Margin of Contribution= 41,548 millions USD), and (**$**= Sales or Revenues= 51,294 millions USD), are known, then it's (**E**= Equity or Capital planned), is:

$$E = U + [1-d][1-t][M-\$f-\$i]$$
$$= U + [1-d][1-t]\{M-\$[f+i]\}$$
$$= 64,720 + [1-0.3000][1-0.3000][41,548 - 51,294[0.5800+0.0100]]$$
$$= \underline{70,249} \text{ millions USD}$$

Rule-3899:
If both (**d**= D/A= Dividend Portion or Payout= 30.00%), (**t**= T/B= Tax Rate= 30.00%), (**i**= I/S= Interest Portion= 1.00%), (**f**= F/S= Fixed Portion= 58.00%), (**U**= Utilized or Starting Capital= 64,720 millions USD), (**E**= Equity or Capital= 70,249 millions USD), and (**S**= Sales or Revenues= 51,294 millions USD), are known, then it's (**M**= Margin of Contribution planned), is:

$$M = S[f+i]+[E-U]/\{[1-d][1-t]\}$$
$$= 51,294[0.5800+0.0100]+[70,249$$
$$-64,720]/\{[1-0.3000][1-0.3000]\}$$
$$= \underline{41,548} \text{ millions USD}$$

Rule-3900:
If both (**d**= D/A= Dividend Portion or Payout= 30.00%), (**t**= T/B= Tax Rate= 30.00%), (**i**= I/S= Interest Portion= 1.00%), (**S**= Sales or Revenues= 51,294 millions USD), (**U**= Utilized or Starting Capital= 64,720 millions USD), (**E**= Equity or Capital= 70,249 millions USD), and (**M**= Margin of Contribution= 41,548 millions USD), are known, then it's (**S**= Sales or Revenues planned), is:

$$S = (M-[E-U]/\{[1-d][1-t]\})/[f+i]$$
$$= (41,548-[70,249-64,720]/\{[1-0.3000]$$
$$[1-0.3000]\})/[0.5800+0.0100]$$
$$= \underline{51,294} \text{ millions USD}$$

Rule-3901:
If both (**d**= D/A= Dividend Portion or Payout= 30.00%), (**t**= T/B= Tax Rate= 30.00%), (**i**= I/S= Interest Portion= 1.00%), (**$**= Sales or Revenues= 51,294 millions USD), (**U**= Utilized or Starting Capital= 64,720 millions USD), (**E**= Equity or Capital= 70,249 millions USD), and (**M**= Margin of Contribution= 41,548 millions USD),are known, then it's (**f**= Fixed Portion planned), is:

$$f = (M-[E-U]/\{[1-d][1-t]\})/S - i$$
$$= (41,548-[70,249-64,720]/\{[1-0.3000][1-0.3000]\})/51,294-0.0100]$$
$$= 58.00\%$$

Rule-3902:
If both (**d**= D/A= Dividend Portion or Payout= 30.00%), (**t**= T/B= Tax Rate= 30.00%), (**f**= F/S= Fixed Portion= 58.00%), (**$**= Sales or Revenues= 51,294 millions USD), (**U**= Utilized or Starting Capital= 64,720 millions USD), (**E**= Equity or Capital= 70,249 millions USD), and (**M**= Margin of Contribution= 41,548 millions USD),are known, then it's (**i**= Interest Portion planned), is:

$$i = (M-[E-U]/\{[1-d][1-t]\})/S - f$$
$$= (41,548-[70,249-64,720]/\{[1-0.3000][1-0.3000]\})/51,294-0.5800]$$
$$= 1.00\%$$

Rule-3903:
If both (**d**= D/A=cDividend Portion or Payout= 30.00%), (**i**= I/S= Interest Portion= 1.00%), (**f**= F/S=xFixed Portion= 58.00%), (**$**= Sales or Revenues= 51,294 millions USD), (**U**= Utilized or Starting Capital= 64,720 millions USD), (**E**= Equity or Capital= 70,249 millions USD), and (**M**= Margin of Contribution= 41,548 millions USD),are known, then it's (**t**= Tax Rate planned), is:

$$\begin{aligned}\mathbf{t} &= 1\text{-}[\mathbf{E}\text{-}\mathbf{U}]/([1\text{-}\mathbf{d}]\{\mathbf{M}\text{-}\mathbf{\$}[\mathbf{f}\text{+}\mathbf{i}]\}) \\ &= 1\text{-}[70{,}249\text{-}64{,}720]/([1\text{-}0.3000] \\ &\quad \{41{,}548\text{-}51{,}294 \\ &\quad \text{-}[0.5800\text{+}0.0100]\}) \\ &= 30.00\%\end{aligned}$$

Rule-3904:
If both (**t**= T/B= Tax Rate= 30.00%), (**i**= I/S= Interest Portion= 1.00%), (**f**= F/S= Fixed Portion= 58.00%), (**$**= Sales or Revenues= 51,294 millions USD), (**U**= Utilized or Starting Capital= 64,720 millions USD), (**E**= Equity or Capital= 70,249 millions USD), and (**M**= Margin of Contribution= 41,548 millions USD),are known, then it's (**d**= Dividend Portion or Payout planned), is:

$$\begin{aligned}\mathbf{d} &= 1\text{-}[\mathbf{E}\text{-}\mathbf{U}]/([1\text{-}\mathbf{t}]\{\mathbf{M}\text{-}\mathbf{\$}[\mathbf{f}\text{+}\mathbf{i}]\}) \\ &= 1\text{-}[70{,}249\text{-}64{,}720]/([1\text{-}0.3000] \\ &\quad \{41{,}548\text{-}51{,}294 \\ &\quad \text{-}[0.5800\text{+}0.0100]\}) \\ &= 30.00\%\end{aligned}$$

Rule-3905:
If both (**t**= T/B= Tax Rate= 30.00%), (**i**= I/S= Interest Portion= 1.00%), (**f**= F/S= Fixed Portion= 58.00%), (**$**= Sales or Revenues= 51,294 millions USD), (**d**= D/A= Dividend Portion or Payout= 30.00%), (**E**= Equity or Capital= 70,249 millions USD), and (**M**= Margin of Contribution= 41,548 millions USD),are known, then it's (**U**= Utilized or Starting Capital planned), is:

$U = E-[1-d][1-t]\{M-\$[f+i]\}$
  $= 70,249-[1-0.3000][1-0.3000]\{41,548$
    $-51,294-[0.5800+0.0100]\})$
  $= 64,720$ millions USD

Rule-3906:
If both (**t**= T/B= Tax Rate= 30.00%), (**i**= I/S= Interest Portion= 1.00%), (**$'**= Sales of Past Year= 48,851 millions USD), (**s**= [S/S']-1= Sales Growth= 5.00%), (**f**= F/S= Fixed Portion= 58.00%), (**$**= Sales or Revenues= 51,294 millions USD), (**d**= D/A= Dividend Portion or Payout= 30.00%), (**U**= Utilized or Starting Capital= 64,720 millions USD), and (**M**= Margin of Contribution= 41,548 millions USD),are known, then it's (**E**= Equity or Capital planned), is:

$E = U+[1-d][1-t]\{M-\$f-\$'i[1+s]\}$
  $= 64,720+[1-0.3000][1-0.3000]\{41,548$
    $-51,294*0.5800-48,851*0.0100$
    $[1+0.0500]\}$
  $= 70,249$ millions USD

Rule-3907:
If both (**t**= T/B= Tax Rate= 30.00%), (**i**= I/S= Interest Portion= 1.00%), (**$'**= Sales of Past Year= 48,851 millions USD), (**s**= [S/S']-1= Sales Growth= 5.00%), (**f**= F/S= Fixed Portion= 58.00%), (**$**= Sales or Revenues= 51,294 millions USD), (**d**= D/A= Dividend Portion or Payout= 30.00%), (**U**= Utilized or Starting Capital= 64,720 millions USD), and (**E**= Equity or Capital= 70,249 millions USD), are known, then it's (**M**= Margin of Contribution planned), is:

$$M = \$f + \$'i[1+s] + [E-U]\{[1-d][1-t]\}$$
$$= 51{,}294 * 0.5800 + 48{,}851 * 0.0100$$
$$[1+0.0500] + [70{,}249 - 64{,}720]$$
$$/\{[1-0.3000][1-0.3000]\}$$
$$= \underline{41{,}548} \text{ millions USD}$$

Rule-3908:
If both (**t**= T/B= Tax Rate= 30.00%), (**i**= I/S= Interest Portion= 1.00%), (**$'$**= Sales of Past Year= 48,851 millions USD), (**s**= [S/S']-1= Sales Growth= 5.00%), (**f**= F/S=Fixed Portion= 58.00%), (**M**= Margin of Contribution= 41,548 millions USD), (**d**= D/A= Dividend Portion or Payout= 30.00%), (**U**= Utilized or Starting Capital= 64,720 millions USD), and (**E**= Equity or Capital= 70,249 millions USD), are known, then it's (**$**= Sales or Revenues planned), is:

$= (**M**-**$'i**[1+**s**]-[**E**-**U**]{[1-**d**][1-**t**]})/**f**
  = (41,548-48,851*0.0100[1+0.0500]
    -[70,249-64,720]/{[1-0.3000]
    [1-0.3000]})/ 0.5800
  = 51,294 millions USD

Rule-3909:
If both (**t**= T/B= Tax Rate= 30.00%), (**i**= I/S= Interest Portion= 1.00%), (**S'**= Sales of Past Year= 48,851 millions USD), (**s**= [S/S']-1= Sales Growth= 5.00%), (**S**= Sales or Revenues= 51,294 millions USD), (**M**= Margin of Contribution= 41,548 millions USD), (**d**= D/A= Dividend Portion or Payout= 30.00%), (**U**= Utilized or Starting Capital= 64,720 millions USD), and (**E**= Equity or Capital= 70,249 millions USD), are known, then it's (**f**= Fixed Portion planned), is:

$f = (M-S'i[1+s]-[E-U]\{[1-d][1-t]\})/S$
$= (41,548-48,851*0.0100[1+0.0500]$
$\quad -[70,249-64,720]/\{[1-0.3000]$
$\quad [1-0.3000]\})/1,294$
$= 58.00\%$

Rule-3910:
If both (**t**= T/B= Tax Rate= 30.00%), (**i**= I/S= Interest Portion= 1.00%), (**f**= F/S= Fixed Portion= 58.00%), (**s**= [S/S']-1= Sales Growth= 5.00%), (**$**= Sales or Revenues= 51,294 millions USD), (**M**= Margin of Contribution= 41,548 millions USD), (**d**= D/A= Dividend Portion or Payout= 30.00%), (**U**= Utilized or Starting Capital= 64,720 millions USD), and (**E**= Equity or Capital= 70,249 millions USD), are known, then it's (**$'**= Sales Past), must be:

$'= (**M-$f-[E-U]**{[1-**d**][1-**t**]})/{**i**[1+**s**]}
  = (41,548-51,294 *0.5800-[70,249
      -64,720]/{[1-0.3000]
      [1-0.3000]})
      /{ 0.0100[1+0.0500]}
  = 48,851 millions USD

Rule-3911:
If both (**t**= T/B= Tax Rate= 30.00%), (**$'**= Sales of Past Year= 48,851 millions USD), (**f**= F/S= Fixed Portion= 58.00%), (**s**= [S/S']-1= Sales Growth= 5.00%), (**$**= Sales or Revenues= 51,294 millions USD), (**M**= Margin of Contribution= 41,548 millions USD), (**d**= D/A= Dividend Portion or Payout= 30.00%), (**U**= Utilized or Starting Capital= 64,720 millions USD), and (**E**= Equity or Capital= 70,249 millions USD), are known, then it's (**i**= Interest Portion planned), is:

$$i = (M\text{-}\$f\text{-}[E\text{-}U]\{[1\text{-}d][1\text{-}t]\})/\{\$'[1+s]\}$$
$$= (41,548\text{-}51,294*0.5800\text{-}[70,249$$
$$ \text{-}64,720]/\{[1\text{-}0.3000]$$
$$ [1\text{-}0.3000]\})$$
$$ /\{48,851[1+0.0500]\}$$
$$= 1.00\%$$

Rule-3912:
If both (**t**= T/B= Tax Rate= 30.00%), (**$'**= Sales of Past Year= 48,851 millions USD), (**f**= F/S= Fixed Portion= 58.00%), (**i**= I/S= Interest Portion= 1.00%), (**$**= Sales or Revenues= 51,294 millions USD), (**M**= Margin of Contribution= 41,548 millions USD), (**d**= D/A= Dividend Portion or Payout= 30.00%), (**U**= Utilized or Starting Capital= 64,720 millions USD), and (**E**= Equity or Capital= 70,249 millions USD), are known, then it's (**s**= Sales Growth planned), is:

s= (**M-$f-[E-U]**{[1-**d**][1-**t**]})/[**$'i**]-1
= (41,548-51,294 *0.5800-[70,249
-64,720]/{[1-0.3000]
[1-0.3000]})
/[48,851*0.0100]-1
= 5.00%

Rule-3913:
If both (**s**= [S/S']-1= Sales Growth= 5.00%), (**S'**= Sales of Past Year= 48,851 millions USD), (**f**= F/S= Fixed Portion= 58.00%), (**i**= I/S= Interest Portion= 1.00%), (**S**= Sales or Revenues= 51,294 millions USD), (**M**= Margin of Contribution= 41,548 millions USD), (**d**= D/A= Dividend Portion or Payout= 30.00%), (**U**= Utilized or Starting Capital= 64,720 millions USD), and (**E**= Equity or Capital= 70,249 millions USD), are known, then it's (**t**= Tax Rate planned), is:

$$t = 1-[\mathbf{E}-\mathbf{U}]/([1-\mathbf{d}]\{\mathbf{M}-\mathbf{Sf}-\mathbf{S'i}[1+\mathbf{s}]\})$$
$$= 1-[70,249-64,720]/([1-0.3000]$$
$$\{41,548-51,294*0.5800$$
$$-48,851*0.0100[1+0.0500]\})$$
$$= 30.00\%$$

Rule-3914:
If both ($=$ [S/S']-1= Sales Growth= 5.00%), ($'=$ Sales of Past Year= 48,851 millions USD), ($f=$ F/S=Fixed Portion= 58.00%), ($i=$ I/S= Interest Portion= 1.00%), ($\$=$ Sales or Revenues= 51,294 millions USD), (**M**= Margin of Contribution= 41,548 millions USD), ($t=$ T/B= Tax Rate= 30.00%), (**U**= Utilized or Starting Capital= 64,720 millions USD), and (**E**= Equity or Capital= 70,249 millions USD), are known, then it's (**d**= Dividend Portion or Payout planned), is:

$$d = 1-[E-U]/([1-t]\{M-\$f-\$'i[1+\$]\})$$
$$= 1-[70,249-64,720]/([1-0.3000]$$
$$\{41,548-51,294*0.5800$$
$$-48,851*0.0100[1+0.0500]\})$$
$$= 30.00\%$$

Rule-3915:
If both (s= [S/S']-1= Sales Growth= 5.00%), (S'= Sales of Past Year= 48,851 millions USD), (f= F/S= Fixed Portion= 58.00%), (i= I/S= Interest Portion= 1.00%), (S= Sales or Revenues= 51,294 millions USD), (M= Margin of Contribution= 41,548 millions USD), (t= T/B= Tax Rate= 30.00%), (d= D/A= Dividend Portion or Payout= 30.00%), and (E= Equity or Capital= 70,249 millions USD), are known, then it's (U= Utilized or Starting Capital planned), is:

U= E-[1-d][1-t]{M-Sf-$S'i$[1+s]}
= 70,249-[1-0.3000] [1-0.3000]{41,548
-51,294 *0.5800-
48,851*0.0100[1+0.0500]}
= 64,720 millions USD

Rule-3916:
If both (**s**= [S/S']-1= Sales Growth= 5.00%), (**S'**= Sales of Past Year= 48,851 millions USD), (**f**= F/S= Fixed Portion= 58.00%), (**I**= Interest Expense= 513 millions USD), (**M**= Margin of Contribution= 41,548 millions USD), (**t**= T/B= Tax Rate= 30.00%), (**d**= D/A= Dividend Portion or Payout= 30.00%), and (**U**= Utilized or Starting Capital= 64,720 millions USD), are known, then it's (**E**= Equity or Capital planned), is:

$$E = U + [1-d][1-t]\{M-I-S'f[1+s]\}$$
$$= 64{,}720 + [1-0.3000][1-0.3000]$$
$$\{41{,}548 - 513 - 48{,}851$$
$$*0.5800[1+0.0500]\}$$
$$= \underline{70{,}249} \text{ millions USD}$$

Rule-3917:
If both (s= [S/S']-1= Sales Growth= 5.00%), (S'= Sales of Past Year= 48,851 millions USD), (f= F/S= Fixed Portion= 58.00%), (I= Interest Expense= 513 millions USD), (E= Equity or Capital= 70,249 millions USD), (t= T/B= Tax Rate= 30.00%), (d= D/A= Dividend Portion or Payout= 30.00%), and (U= Utilized or Starting Capital= 64,720 millions USD), are known, then it's (M= Margin of Contribution planned), is:

$M = S'f[1+s]+I+[E-U]/\{[1-d][1-t]\}$
   = 48,851*0.5800[1+0.0500]+513
      +[70,249-64,720]/{[1-0.3000]
      [1-0.3000]}
   = 41,548 millions USD

Rule-3918:
If both (s= [S/S']-1= Sales Growth= 5.00%), (M= Margin of Contribution= 41,548 millions USD), (f= F/S= Fixed Portion= 58.00%), (I= Interest Expense= 513 millions USD),(E= Equity or Capital= 70,249 millions USD), (t= T/B= Tax Rate= 30.00%), (d= D/A= Dividend Portion or Payout= 30.00%), and (U= Utilized or Starting Capital= 64,720 millions USD), are known, then it's (S'= Sales Past), must be:

$S' = (M-I-[E-U]/\{[1-d][1-t]\})/\{f[1+s]\}$
    = (41,548-513-[70,249-64,720]
       /{[1-0.3000] [1-0.3000]})
       /{0.5800[1+0.0500]}
    = 48,851 millions USD

Rule-3919:
If both (s= [S/S']-1= Sales Growth= 5.00%), (M= Margin of Contribution= 41,548 millions USD), (S'= Sales of Past Year= 48,851 millions USD), (I= Interest Expense= 513 millions USD),(E= Equity or Capital= 70,249 millions USD), (t= T/B= Tax Rate= 30.00%), (d= D/A= Dividend Portion or Payout= 30.00%), and (U= Utilized or Starting Capital= 64,720 millions USD), are known, then it's (f= Fixed Portion planned), is:

f= (M-I-[E-U]/{[1-d][1-t]})/{S'[1+s]}
= (41,548-513-[70,249-64,720]
/{[1-0.3000][1-0.3000]})
/{48,851[1+0.0500]}
= 58.00%

Rule-3920:
If both (**f**= F/S= Fixed Portion= 58.00%), (**M**= Margin of Contribution= 41,548 millions USD), (**S'**= Sales of Past Year= 48,851 millions USD), (**I**= Interest Expense= 513 millions USD),(**E**= Equity or Capital= 70,249 millions USD), (**t**= T/B= Tax Rate= 30.00%), (**d**= D/A= Dividend Portion or Payout= 30.00%), and (**U**= Utilized or Starting Capital= 64,720 millions USD), are known, then it's (**s**= Sales Growth planned), is:

$$s = (M-I-[E-U]/\{[1-d][1-t]\})/[S'f]-1$$
$$= (41,548-513-[70,249-64,720]/\{[1-0.3000][1-0.3000]\})/[48,851*0.5800]-1$$
$$= 5.00\%$$

Rule-3921:
If both (**f**= F/S= Fixed Portion= 58.00%), (**M**= Margin of Contribution= 41,548 millions USD), (**S'**= Sales of Past Year= 48,851 millions USD), (**s**= [S/S']-1= Sales Growth= 5.00%), (**E**= Equity or Capital= 70,249 millions USD), (**t**= T/B= Tax Rate= 30.00%), (**d**= D/A= Dividend Portion or Payout= 30.00%), and (**U**= Utilized or Starting Capital= 64,720 millions USD), are known, then it's (**I**= Interest Expense planned), is:

$$I = M - S'f[1+s] - [E-U]/\{[1-d][1-t]\}$$
$$= 41{,}548 - 48{,}851 * 0.5800[1+0.0500]$$
$$\quad - [70{,}249 - 64{,}720]/\{[1-0.3000][1-0.3000]\}$$
$$= 513 \text{ millions USD}$$

Rule-3922:
If both (**f**= F/S= Fixed Portion= 58.00%), (**M**= Margin of Contribution= 41,548 millions USD), (**S'**= Sales of Past Year= 48,851 millions USD), (**s**= [S/S']-1= Sales Growth= 5.00%), (**E**= Equity or Capital= 70,249 millions USD), (**I**= Interest Expense= 513 millions USD), (**d**= D/A=Dividend Portion or Payout= 30.00%), and (**U**= Utilized or Starting Capital= 64,720 millions USD), are known, then it's (**t**= Tax Rate planned), is:

$$\begin{aligned}
\mathbf{t} &= 1-[\mathbf{E}\text{-}\mathbf{U}]/([1\text{-}\mathbf{d}][1\text{-}\mathbf{t}]\{\mathbf{M}\text{-}\mathbf{S'f}[1+\mathbf{s}]\text{-}\mathbf{I}\}) \\
&= 1-[70,249\text{-}64,720]/([1\text{-}0.3000] \\
&\quad \{41,548\text{-}48,851*0.5800 \\
&\quad [1+0.0500]\text{-}513\}) \\
&= \underline{30.00\%}
\end{aligned}$$

Rule-3923:

If both (**f**= F/S= Fixed Portion= 58.00%), (**M**= Margin of Contribution= 41,548 millions USD), (**S'**= Sales of Past Year= 48,851 millions USD), (**s**= [S/S']-1= Sales Growth= 5.00%), (**E**= Equity or Capital= 70,249 millions USD), (**I**= Interest Expense= 513 millions USD), (**t**= T/B= Tax Rate= 30.00%), and (**U**= Utilized or Starting Capital= 64,720 millions USD), are known, then it's (**d**= Dividend Portion or Payout planned), is:

$$d = 1-[\mathbf{E}-\mathbf{U}]/([1-\mathbf{t}][1-\mathbf{t}]\{\mathbf{M}-\mathbf{S'f}[1+\mathbf{s}]-\mathbf{I}\})$$
$$= 1-[70,249-64,720]/([1-0.3000]$$
$$\{41,548-48,851*0.5800$$
$$[1+0.0500]-513\})$$
$$= 30.00\%$$

Rule-3924:
If both (**f**= F/S= Fixed Portion= 58.00%), (**M**= Margin of Contribution= 41,548 millions USD), (**S'**= Sales of Past Year= 48,851 millions USD), (**s**= [S/S']-1= Sales Growth= 5.00%), (**E**= Equity or Capital= 70,249 millions USD), (**I**= Interest Expense= 513 millions USD), (**t**= T/B= Tax Rate= 30.00%), and (**d**= D/A= Dividend Portion or Payout= 30.00%), are known, then it's (**U**= Utilized or Starting Capital planned), is:

$$U = E-[1-d][1-t]\{M-S'f[1+s]-I\}$$
$$= 70,249-[1-0.3000][1-0.3000]\{41,548 -48,851*0.5800[1+0.0500]-513\}$$

$$= 64,720 \text{ millions USD}$$

Rule-3925:
If both (**f**= F/S= Fixed Portion= 58.00%), (**M**= Margin of Contribution= 41,548 millions USD), (**S'**= Sales of Past Year= 48,851 millions USD), (**s**= [S/S']-1= Sales Growth= 5.00%), (**U**= Utilized or Starting Capital= 64,720 millions USD), (**S**= Sales or Revenues= 51,294 millions USD), (**i**= I/S= Interest Portion= 1.00%), (**t**= T/B= Tax Rate= 30.00%), and (**d**= D/A= Dividend Portion or Payout= 30.00%), are known, then it's (**E**= Equity or Capital planned), is:

$$E = U+[1-d][1-t]\{M-S'f[1+s]-Si\}$$
$$= 64{,}720+[1-0.3000][1-0.3000]\{41{,}548$$
$$-48{,}851*0.5800[1+0.0500]$$
$$-51{,}294*0.0100\}$$
$$= 70{,}249 \text{ millions USD}$$

Rule-3926:
If both (**f**= F/S= Fixed Portion= 58.00%), (**E**= Equity or Capital= 70,249 millions USD), (**$'**= Sales of Past Year= 48,851 millions USD), (**s**= [S/S']-1= Sales Growth= 5.00%), (**U**= Utilized or Starting Capital= 64,720 millions USD), (**$**= Sales or Revenues= 51,294 millions USD), (**i**= I/S= Interest Portion= 1.00%), (**t**= T/B= Tax Rate= 30.00%), and (**d**= D/A= Dividend Portion or Payout= 30.00%), are known, then it's (**M**= Margin of Contribution planned), is:

$$M = \$'f[1+s]+\$i+[E-U]/\{[1-d][1-t]\}$$
$$= 48,851*0.5800[1+0.0500]+51,294$$
$$*0.0100+[70,249-64,720]$$
$$/\{[1-0.3000][1-0.3000]\}$$
$$= 41,548 \text{ millions USD}$$

Rule-3927:
If both (**f**= F/S= Fixed Portion= 58.00%), (**E**= Equity or Capital= 70,249 millions USD), (**M**= Margin of Contribution= 41,548 millions USD), (**s**= [S/S']-1= Sales Growth= 5.00%), (**U**= Utilized or Starting Capital= 64,720 millions USD), (**$**= Sales or Revenues= 51,294 millions USD), (**i**= I/S=Interest Portion=1.00%), (**t**= T/B= Tax Rate= 30.00%), and (**d**= D/A= Dividend Portion or Payout= 30.00%), are known, then it's (**$'**= Sales Past), must be:

$'= (**M-$i-[E-U]**/{[1-**d**][1-**t**]})/{**f**[1+**s**]}
  = (41,548-51,294*0.0100-[70,249
    -64,720]/{[1-0.3000]
    [1-0.3000]})
    /{0.5800[1+0.0500]}
  = 48,851 millions USD

Rule-3928:
If both ($'= Sales of Past Year= 48,851 millions USD), (E= Equity or Capital= 70,249 millions USD), (M= Margin of Contribution= 41,548 millions USD), (s= [S/S']-1= Sales Growth= 5.00%), (U= Utilized or Starting Capital= 64,720 millions USD), ($= Sales or Revenues= 51,294 millions USD), (i= I/S= Interest Portion=1.00%), (t= T/B= Tax Rate= 30.00%), and (d= D/A= Dividend Portion or Payout= 30.00%), are known, then it's (f= Fixed Portion planned), is:

f= (M-$i-[E-U]/{[1-d][1-t]})/{$'[1+s]}
= (41,548-51,294*0.0100-[70,249
   -64,720]/{[1-0.3000]
   [1-0.3000]})
   /{48,851[1+0.0500]}
= 58.00%

Rule-3929:
If both ($'=$ Sales of Past Year= 48,851 millions USD), (**E**= Equity or Capital= 70,249 millions USD), (**M**= Margin of Contribution= 41,548 millions USD), (**f**= F/S= Fixed Portion= 58.00%), (**U**= Utilized or Starting Capital= 64,720 millions USD), ($=$ Sales or Revenues= 51,294 millions USD), (**i**= I/S= Interest Portion=1.00%), (**t**= T/B= Tax Rate= 30.00%), and (**d**= D/A= Dividend Portion or Payout= 30.00%), are known, then it's (**s**= Sales Growth planned), is:

$s =$ (**M-$i-[E-U]**/{[1-**d**][1-**t**]})/[$'**f**]-1
 = (41,548-51,294*0.0100-[70,249
   -64,720]/{[1-0.3000]
   [1-0.3000]})
   /[48,851*0.5800]-1
 = 5.00%

Rule-3930:
If both ($'= Sales of Past Year= 48,851 millions USD), (E= Equity or Capital= 70,249 millions USD), (M= Margin of Contribution= 41,548 millions USD), (f= F/S= Fixed Portion= 58.00%), (U= Utilized or Starting Capital= 64,720 millions USD), (s= [S/S']-1= Sales Growth= 5.00%), (i= I/S= Interest Portion=1.00%), (t= T/B= Tax Rate= 30.00%), and (d= D/A= Dividend Portion or Payout= 30.00%), are known, then it's ($= Sales or Revenues planned), is:

$= (M-$'f[1+s]-[E-U]/{[1-d][1-t]})/i
 = (41,548-48,851*0.5800][1+0.0500]
   -[70,249-64,720]/{[1-0.3000]
   [1-0.3000]}) /0.0100
 = 51,294 millions USD

Rule-3931:
If both ($'$= Sales of Past Year= 48,851 millions USD), (**E**= Equity or Capital= 70,249 millions USD), (**M**= Margin of Contribution= 41,548 millions USD), (**f**= F/S= Fixed Portion= 58.00%), (**U**= Utilized or Starting Capital= 64,720 millions USD), (s= [S/S']-1= Sales Growth= 5.00%), (**$**= Sales or Revenues= 51,294 millions USD), (**t**= T/B= Tax Rate= 30.00%), and (**d**= D/A= Dividend Portion or Payout= 30.00%), are known, then it's (**i**= Interest Portion planned), is:

$\mathsf{i}= (\mathbf{M}\text{-}\mathbf{\$'f}[1+\mathsf{s}]\text{-}[\mathbf{E}\text{-}\mathbf{U}]/\{[1\text{-}\mathbf{d}][1\text{-}\mathbf{t}]\})/\mathbf{\$}$
$= (41,548\text{-}48,851*0.5800][1+0.0500]$
$\quad\quad\quad\text{-}[70,249\text{-}64,720]/\{[1\text{-}0.3000]$
$\quad\quad\quad [1\text{-}0.3000]\})/51,294$
$=\underline{1.00}\%$

Rule-3932:

If both (**S'**= Sales of Past Year= 48,851 millions USD), (**E**= Equity or Capital= 70,249 millions USD), (**M**= Margin of Contribution= 41,548 millions USD), (**f**= F/S=Fixed Portion= 58.00%), (**U**= Utilized or Starting Capital= 64,720 millions USD), (**s**= [S/S']-1= Sales Growth= 5.00%), (**S**= Sales or Revenues= 51,294 millions USD), (**i**= I/S= Interest Portion= 1.00%), and (**d**= D/A= Dividend Portion or Payout= 30.00%), are known, then it's (**t**= Tax Rate planned), is:

t = 1-[**E-U**]/([1-**d**]{**M-S'f**[1+**s**]-**Si**})
   = 1-[70,249-64,720]/([1-0.3000]
       {41,548-48,851*0.5800]
       [1+0.0500]-51,294*0.0100})
   = 30.00%

Rule-3933:
If both ($'$= Sales of Past Year= 48,851 millions USD), (E= Equity or Capital= 70,249 millions USD), (M= Margin of Contribution= 41,548 millions USD), (f= F/S= Fixed Portion= 58.00%), (U= Utilized or Starting Capital= 64,720 millions USD), (s= [S/S']-1= Sales Growth= 5.00%), (S= Sales or Revenues= 51,294 millions USD), (i= I/S= Interest Portion= 1.00%), and (t= T/B= Tax Rate= 30.00%), are known, then it's (d= Dividend Portion or Payout planned), is:

d= 1-[E-U]/([1-t]{M-$S'f$[1+s]-Si})
  = 1-[70,249-64,720]/([1-0.3000]
          {41,548-48,851*0.5800]
          [1+0.0500]-51,294*0.0100})
  = 30.00%

Rule-3934:
If both ($'= Sales of Past Year= 48,851 millions USD), (E= Equity or Capital= 70,249 millions USD), (M= Margin of Contribution= 41,548 millions USD), (f= F/S= Fixed Portion= 58.00%), (d= D/A= Dividend Portion or Payout= 30.00%), (s= [S/S']-1= Sales Growth= 5.00%), ($= Sales or Revenues= 51,294 millions USD), (i= I/S= Interest Portion= 1.00%), and (t= T/B= Tax Rate= 30.00%), are known, then it's (U= Utilized or Starting Capital planned), is:

U= E-[1-d][1-t]{M-$'f[1+s]-$i}
= 70,249-[1-0.3000][1-0.3000]{41,548
-48,851*0.5800][1+0.0500]
-51,294*0.0100}
= 64,720 millions USD

Rule-3935:
If both (S'= Sales of Past Year= 48,851 millions USD), (U= Utilized or Starting Capital= 64,720 millions USD), (M= Margin of Contribution= 41,548 millions USD), (f= F/S= Fixed Portion= 58.00%), (d= D/A= Dividend Portion or Payout= 30.00%), (s= [S/S']-1= Sales Growth= 5.00%), (i= I/S= Interest Portion= 1.00%), and (t= T/B= Tax Rate= 30.00%), are known, then it's (E= Equity or Capital planned), is:

$$E = U+[1-d][1-t]\{M-S'f[1+s]-S'i[1+s]\}$$
$$= U+[1-d][1-t]\{M-S'[1+s][f+i]\}$$
$$= 64{,}720+[1-0.3000][1-0.3000]$$
$$\{41{,}548-48{,}851[1+0.0500]$$
$$[0.5800+0.0100]\}$$
$$= 70{,}249 \text{ millions USD}$$

<u>Rule-3936</u>:
If both ($**$'**= Sales of Past Year= <u>48,851</u> millions USD), (**U**= Utilized or Starting Capital= <u>64,720</u> millions USD), (**E**= Equity or Capital= <u>70,249</u> millions USD), (**f**= F/S= Fixed Portion= <u>58.00</u>%), (**d**= D/A= Dividend Portion or Payout= 30.00%), (**s**= [S/S']-1= Sales Growth= <u>5.00</u>%), (**i**= I/S= Interest Portion= <u>1.00</u>%), and (**t**= T/B= Tax Rate= <u>30.00</u>%), are known, then it's (**M**= Margin of Contribution planned), is:

$$M = \$'[1+s][f+i]+[E-U]/\{[1-d][1-t]\}$$
$$= 48,851[1+0.0500][0.5800+0.0100]$$
$$\qquad +[70,249-64,720]/\{[1-0.3000]$$
$$\qquad [1-0.3000]\}$$
$$= \underline{41,548} \text{ millions USD}$$

Rule-3937:
If both (**M**= Margin of Contribution= 41,548 millions USD), (**U**= Utilized or Starting Capital= 64,720 millions USD), (**E**= Equity or Capital= 70,249 millions USD), (**f**= F/S=Fixed Portion= 58.00%), (**d**= D/A= Dividend Portion or Payout= 30.00%), (**s**= [S/S']-1= Sales Growth= 5.00%), (**i**= I/S= Interest Portion= 1.00%), and (**t**= T/B= Tax Rate= 30.00%), are known, then it's (**S'**= Sales Past), must be:

$$S' = (M-[E-U]/\{[1-d][1-t]\})/\{[1+s][f+i]\}$$
$$= (41{,}548-[70{,}249-64{,}720]$$
$$/\{[1-0.3000][1-0.3000]\})$$
$$/\{[1+0.0500][0.5800+0.0100]\}$$
$$= 48{,}851 \text{ millions USD}$$

Rule-3938:
If both (**M**= Margin of Contribution= 41,548 millions USD), (**U**= Utilized or Starting Capital= 64,720 millions USD), (**E**= Equity or Capital= 70,249 millions USD), (**f**= F/S= Fixed Portion= 58.00%), (**d**= D/A= Dividend Portion or Payout= 30.00%), (**$'**= Sales of Past Year= 48,851 millions USD), (**i**= I/S= Interest Portion= 1.00%), and (**t**= T/B= Tax Rate= 30.00%), are known, then it's (**$**= Sales Growth planned), is:

$$\$ = (M-[E-U]/\{[1-d][1-t]\})/\{\$'[f+i]\}-1$$
$$= (41,548-[70,249-64,720]$$
$$/\{[1-0.3000][1-0.3000]\})$$
$$/\{48,851[0.5800+0.0100]\}-1$$
$$= 5.00\%$$

Rule-3939:
If both (**M**= Margin of Contribution= 41,548 millions USD), (**U**= Utilized or Starting Capital= 64,720 millions USD), (**E**= Equity or Capital= 70,249 millions USD), (**s**= [S/S']-1= Sales Growth= 5.00%), (**d**= D/A= Dividend Portion or Payout= 30.00%), (**S'**= Sales of Past Year= 48,851 millions USD), (**i**= I/S= Interest Portion= 1.00%), and (**t**= T/B= Tax Rate= 30.00%), are known, then it's (**f**= Fixed Portion planned), is:

f= (**M**-[**E**-**U**]/{[1-**d**][1-**t**]})/{**S'**[1+**s**]}-i
  = (41,548-[70,249-64,720]
      /{[1-0.3000][1-0.3000]})
      /{48,851[1+0.0500]}-0.0100
  = 58.00%

Rule-3940:
If both (**M**= Margin of Contribution= 41,548 millions USD), (**U**= Utilized or Starting Capital= 64,720 millions USD), (**E**= Equity or Capital= 70,249 millions USD), (**s**= [S/S']-1= Sales Growth= 5.00%), (**d**= D/A= Dividend Portion or Payout= 30.00%), (**S'**= Sales of Past Year= 48,851 millions USD), (**f**= F/S= Fixed Portion= 58.00%), and (**t**= T/B= Tax Rate= 30.00%), are known, then it's (**i**= Interest Portion planned), is:

$$i = (M-[E-U]/\{[1-d][1-t]\})/\{S'[1+s]\} - f$$
$$= (41,548-[70,249-64,720]$$
$$/\{[1-0.3000][1-0.3000]\})$$
$$/\{48,851[1+0.0500]\} - 0.5800$$
$$= 1.00\%$$

Rule-3941:
If both (**M**= Margin of Contribution= 41,548 millions USD), (**U**= Utilized or Starting Capital= 64,720 millions USD), (**E**= Equity or Capital= 70,249 millions USD), (**s**= [S/S']-1= Sales Growth= 5.00%), (**d**= D/A= Dividend Portion or Payout= 30.00%), (**S'**= Sales of Past Year= 48,851 millions USD), (**f**= F/S= Fixed Portion= 58.00%), and (**i**= I/S= Interest Portion= 1.00%), are known, then it's (**t**= Tax Rate planned), is:

$$t = 1-[\mathbf{E}-\mathbf{U}]/([1-\mathbf{d}]\{\mathbf{M}-\mathbf{S'}[1+\mathbf{s}][\mathbf{f}+\mathbf{i}]\})$$
$$= (41,548-[70,249-64,720]$$
$$/([1-0.3000]\{41,548-48,851$$
$$[1+0.0500][0.5800+0.0100]\})$$
$$= 30.00\%$$

Rule-3942:
If both (**M**= Margin of Contribution= 41,548 millions USD), (**U**= Utilized or Starting Capital= 64,720 millions USD), (**E**= Equity or Capital= 70,249 millions USD), (**s**= [S/S']-1= Sales Growth= 5.00%), (**t**= T/B= Tax Rate= 30.00%), (**S'**= Sales of Past Year= 48,851 millions USD), (**f**= F/S= Fixed Portion= 58.00%), and (**i**= I/S= Interest Portion= 1.00%), are known, then it's (**d**= Dividend Portion or Payout planned), is:

$$d = 1-[E-U]/([1-t]\{M-S'[1+s][f+i]\})$$
$$= (41,548-[70,249-64,720]$$
$$/([1-0.3000]\{41,548-48,851$$
$$[1+0.0500][0.5800+0.0100]\})$$
$$= 30.00\%$$

Rule-3943:
If both (**M**= Margin of Contribution= 41,548 millions USD), (**d**= D/A= Dividend Portion or Payout= 30.00%), (**E**= Equity or Capital= 70,249 millions USD), (**s**= [S/S']-1= Sales Growth= 5.00%), (**t**= T/B= Tax Rate= 30.00%), (**S'**= Sales of Past Year= 48,851 millions USD), (**f**= F/S= Fixed Portion= 58.00%), and (**i**= I/S= Interest Portion= 1.00%), are known, then it's (**U**= Utilized or Starting Capital planned), is:

$$U = E-[1-d][1-t]\{M-S'[1+s][f+i]\}$$
$$= 70,249-[1-0.3000][1-0.3000]$$
$$\{41,548-48,851[1+0.0500]$$
$$[0.5800+0.0100]\})$$
$$= 64,720 \text{ millions USD}$$

Rule-3944:
If both ($= Sales or Revenues= 51,294 millions USD), (V= Variable Cost= 9,746 millions USD), (F= Fixed Cost= 29,750 millions USD), (d= D/A= Dividend Portion or Payout= 30.00%), (U= Utilized or Starting Capital= 64,720 millions USD), (t= T/B= Tax Rate= 30.00%), and (I= Interest Expense= 513 millions USD), are known, then it's (E= Equity or Capital planned), is:

$$E = U+[1-d][1-t][\$-V-F-I]$$
$$= 64,720+[1-0.3000][1-0.3000][51,294 -9,746-29,750-513]$$
$$= 70,249 \text{ millions USD}$$

Rule-3945:
If both (E= Equity or Capital= 70,249 millions USD), (V= Variable Cost= 9,746 millions USD), (F= Fixed Cost= 29,750 millions USD), (d= D/A= Dividend Portion or Payout= 30.00%), (U= Utilized or Starting Capital= 64,720 millions USD), (t= T/B= Tax Rate= 30.00%), and (I= Interest Expense= 513 millions USD), are known, then it's ($= Sales or Revenues planned), is:

$$\$ = V+F+I+[E-U]/\{[1-d][1-t]\}$$
$$= 9,746+29,750+513+[70,249-64,720] /\{[1-0.3000][1-0.3000]\}$$
$$= 51,294 \text{ millions USD}$$

Rule-3946:
If both (**E**= Equity or Capital= 70,249 millions USD), (**$**= Sales or Revenues= 51,294 millions USD), (**F**= Fixed Cost= 29,750 millions USD), (**d**= D/A= Dividend Portion or Payout= 30.00%), (**U**= Utilized or Starting Capital= 64,720 millions USD), (**t**= T/B= Tax Rate= 30.00%), and (**I**= Interest Expense= 513 millions USD), are known, then it's (**V**= Variable Cost planned), is:

$$V = \$-F-I-[E-U]/\{[1-d][1-t]\}$$
$$= 51{,}294-29{,}750-513-[70{,}249-64{,}720]/\{[1-0.3000][1-0.3000]\}$$
$$= 9{,}746 \text{ millions USD}$$

Rule-3947:
If both (**E**= Equity or Capital= 70,249 millions USD), (**$**= Sales or Revenues= 51,294 millions USD), (**V**= Variable Cost= 9,746 millions USD), (**d**= D/A= Dividend Portion or Payout= 30.00%), (**U**= Utilized or Starting Capital= 64,720 millions USD), (**t**= T/B= Tax Rate= 30.00%), and (**I**= Interest Expense= 513 millions USD), are known, then it's (**F**= Variable Cost planned), is:

$$V = \$-V-I-[E-U]/\{[1-d][1-t]\}$$
$$= 51{,}294-9{,}746-513-[70{,}249-64{,}720]/\{[1-0.3000][1-0.3000]\}$$
$$= 29{,}750 \text{ millions USD}$$

Rule-3948:
If both (**E**= Equity or Capital= 70,249 millions USD), (**$**= Sales or Revenues= 51,294 millions USD), (**V**= Variable Cost= 9,746 millions USD), (**d**= D/A= Dividend Portion or Payout= 30.00%), (**U**= Utilized or Starting Capital= 64,720 millions USD), (**t**= T/B= Tax Rate= 30.00%), and (**F**= Fixed Cost= 29,750 millions USD), are known, then it's (**I**= Interest Expense planned), is:

$I = \$-V-F-[E-U]/\{[1-d][1-t]\}$
$= 51,294-9,746-29,750-[70,249$
$-64,720]/\{[1-0.3000][1-0.3000]\}$
$= \underline{513}$ millions USD

Rule-3949:
If both (**E**= Equity or Capital= 70,249 millions USD), (**$**= Sales or Revenues= 51,294 millions USD), (**V**= Variable Cost= 9,746 millions USD), (**d**= D/A= Dividend Portion or Payout= 30.00%), (**U**= Utilized or Starting Capital= 64,720 millions USD), (**I**= Interest Expense= 513 millions USD), and (**F**= Fixed Cost= 29,750 millions USD), are known, then it's (**t**= Tax Rate planned), is:

$t = 1-[E-U]/\{[1-d][\$-V-F-I]\}$
$= 1-[70,249-64,720]/\{[1-0.3000]$
$\{51,294-9,746-29,750-513]\}$
$= \underline{30.00\%}$

Rule-3950:
If both (**E**= Equity or Capital= 70,249 millions USD), (**$**= Sales or Revenues= 51,294 millions USD), (**V**= Variable Cost= 9,746 millions USD), (**t**= T/B= Tax Rate= 30.00%), (**U**= Utilized or Starting Capital= 64,720 millions USD), (**I**= Interest Expense= 513 millions USD), and (**F**= Fixed Cost= 29,750 millions USD), are known, then it's (**d**= Dividend Portion or Payout planned), is:

$$d = 1-[\mathbf{E}-\mathbf{U}]/\{[1-\mathbf{t}][\mathbf{\$}-\mathbf{V}-\mathbf{F}-\mathbf{I}]\}$$
$$= 1-[70,249-64,720]/\{[1-0.3000]\{51,294-9,746-29,750-513]\}$$
$$= 30.00\%$$

Rule-3951:
If both (**E**= Equity or Capital= 70,249 millions USD), (**$**= Sales or Revenues= 51,294 millions USD), (**V**= Variable Cost= 9,746 millions USD), (**t**= T/B= Tax Rate= 30.00%), (**d**= D/A= Dividend Portion or Payout= 30.00%), (**I**= Interest Expense= 513 millions USD), and (**F**= Fixed Cost= 29,750 millions USD), are known, then it's (**U**= Utilized or Starting Capital planned), is:

$$\mathbf{U} = \mathbf{E}-[1-\mathbf{d}][1-\mathbf{t}][\mathbf{\$}-\mathbf{V}-\mathbf{F}-\mathbf{I}]$$
$$= 70,249-[1-0.3000][1-0.3000]\{51,294-9,746-29,750-513]\}$$
$$= 64,720 \text{ millions USD}$$

Rule-3952:
If both (**U**= Utilized or Starting Capital= 64,720 millions USD), (**$**= Sales or Revenues= 51,294 millions USD), (**V**= Variable Cost= 9,746 millions USD), (**t**= T/B= Tax Rate= 30.00%), (**d**= D/A= Dividend Portion or Payout= 30.00%), (**i**= I/S= Interest Portion= 1.00%), and (**F**= Fixed Cost= 29,750 millions USD), are known, then it's (**E**= Equity or Capital planned), is:

$$\begin{aligned}
\mathbf{E} &= \mathbf{U} + [1-\mathbf{d}][1-\mathbf{t}][\mathbf{\$}-\mathbf{V}-\mathbf{F}-\mathbf{\$i}] \\
&= \mathbf{U} + [1-\mathbf{d}][1-\mathbf{t}]\{\mathbf{\$}[1-\mathbf{i}]-\mathbf{V}-\mathbf{F}\} \\
&= 64{,}720 + [1-0.3000][1-0.3000]\{51{,}294 \\
&\quad [1-0.0100]-9{,}746-29{,}750]\} \\
&= 70{,}249 \text{ millions USD}
\end{aligned}$$

Rule-3953:
If both (**U**= Utilized or Starting Capital= 64,720 millions USD), (**E**= Equity or Capital= 70,249 millions USD), (**V**= Variable Cost= 9,746 millions USD), (**t**= T/B= Tax Rate= 30.00%), (**d**= D/A= Dividend Portion or Payout= 30.00%), (**i**= I/S= Interest Portion= 1.00%), and (**F**= Fixed Cost= 29,750 millions USD), are known, then it's (**$**= Sales or Revenues planned), is:

$$\begin{aligned}
\mathbf{\$} &= (\mathbf{V}+\mathbf{F}+[\mathbf{E}-\mathbf{U}]/\{[1-\mathbf{d}][1-\mathbf{t}]\})/[1-\mathbf{i}] \\
&= (9{,}746+29{,}750+[70{,}249-64{,}720] \\
&\quad +[1-0.3000][1-0.3000]\}) \\
&\quad /\{[1-0.0100]\} \\
&= 51{,}294 \text{ millions USD}
\end{aligned}$$

Rule-3954:
If both (**U**= Utilized or Starting Capital= 64,720 millions USD), (**E**= Equity or Capital= 70,249 millions USD), (**V**= Variable Cost= 9,746 millions USD), (**t**= T/B= Tax Rate= 30.00%), (**d**= D/A= Dividend Portion or Payout= 30.00%), (**$**= Sales or Revenues= 51,294 millions USD), and (**F**= Fixed Cost= 29,750 millions USD), are known, then it's (**i**= Interest Portion planned), is:

$$i = 1-(\mathbf{V}+\mathbf{F}+[\mathbf{E}\text{-}\mathbf{U}]/\{[1\text{-}\mathbf{d}][1\text{-}\mathbf{t}]\})/\mathbf{\$}$$
$$= (9,746+29,750+[70,249-64,720]+[1-0.3000][1-0.3000]\})/51,294$$
$$= \underline{1.00\%}$$

Rule-3955:
If both (**U**= Utilized or Starting Capital= 64,720 millions USD), (**E**= Equity or Capital= 70,249 millions USD), (**i**= I/S= Interest Portion= 1.00%), (**t**= T/B= Tax Rate= 30.00%), (**d**= D/A= Dividend Portion or Payout= 30.00%), (**$**= Sales or Revenues= 51,294 millions USD), and (**F**= Fixed Cost= 29,750 millions USD), are known, then it's (**V**= Variable Cost planned), is:

$$V = \$[1\text{-}i]\text{-}F\text{-}[E\text{-}U]/\{[1\text{-}d][1\text{-}t]\}$$
$$= 51{,}294[1\text{-}0.0100]\text{-}29{,}750\text{-}[70{,}249\text{-}64{,}720]/\{[1\text{-}0.3000][1\text{-}0.3000]\}$$

$$= 9{,}746 \text{ millions USD}$$

Rule-3956:
If both (**U**= Utilized or Starting Capital= 64,720 millions USD), (**E**= Equity or Capital= 70,249 millions USD), (**i**= I/S= Interest Portion= 1.00%), (**t**= T/B= Tax Rate= 30.00%), (**d**= D/A= Dividend Portion or Payout= 30.00%), (**$**= Sales or Revenues= 51,294 millions USD), and (**V**= Variable Cost= 9,746 millions USD), are known, then it's (**F**= Fixed Cost planned), is:

$$F = \$[1\text{-}i]\text{-}V\text{-}[E\text{-}U]/\{[1\text{-}d][1\text{-}t]\}$$
$$= 51{,}294[1\text{-}0.0100]\text{-}9{,}746\text{-}[70{,}249\text{-}64{,}720]/\{[1\text{-}0.3000][1\text{-}0.3000]\}$$

$$= 29{,}750 \text{ millions USD}$$

Rule-3957:
If both (**U**= Utilized or Starting Capital= 64,720 millions USD), (**E**= Equity or Capital= 70,249 millions USD), (**i**= I/S= Interest Portion= 1.00%), (**F**= Fixed Cost= 29,750 millions USD), (**d**= D/A= Dividend Portion or Payout= 30.00%), (**$**= Sales or Revenues= 51,294 millions USD), and (**V**= Variable Cost= 9,746 millions USD), are known, then it's (**t**= Tax Rate planned), is:

$$t = 1-[\textbf{E}-\textbf{U}]/([1-\textbf{d}]\{\textbf{\$}[1-\textbf{i}]-\textbf{V}-\textbf{F}\})$$
$$= 1-[70,249-64,720]/([1-0.3000]\{51,294[1-0.0100]-9,746-29,750\})$$
$$= 30.00\%$$

Rule-3958:
If both (**U**= Utilized or Starting Capital= 64,720 millions USD), (**E**= Equity or Capital= 70,249 millions USD), (**i**= I/S= Interest Portion= 1.00%), (**F**= Fixed Cost= 29,750 millions USD), (**t**= T/B= Tax Rate= 30.00%), (**$**= Sales or Revenues= 51,294 millions USD), and (**V**= Variable Cost= 9,746 millions USD), are known, then it's (**d**= Dividend Portion or Payout planned), is:

$$d = 1-[\textbf{E}-\textbf{U}]/([1-\textbf{t}]\{\textbf{\$}[1-\textbf{i}]-\textbf{V}-\textbf{F}\})$$
$$= 1-[70,249-64,720]/([1-0.3000]\{51,294[1-0.0100]-9,746-29,750\})$$
$$= 30.00\%$$

Rule-3959:
If both (**d**= D/A= Dividend Portion or Payout= 30.00%), (**E**= Equity or Capital= 70,249 millions USD), (**i**= I/S= Interest Portion= 1.00%), (**F**= Fixed Cost= 29,750 millions USD), (**t**= T/B= Tax Rate= 30.00%), (**$**= Sales or Revenues= 51,294 millions USD), and (**V**= Variable Cost= 9,746 millions USD), are known, then it's (**U**= Utilized or Starting Capital planned), is:
$$U = E-[1-d][1-t]\{\$[1-i]-V-F\})$$
$$= 70,249-[1-0.3000][1-0.3000]\{51,294$$
$$[1-0.0100]-9,746-29,750\})$$
$$= 64,720 \text{ millions USD}$$

Rule-3960:
If both (**d**= D/A= Dividend Portion or Payout= 30.00%), (**$'**= Sales of Past Year= 48,851 millions USD), (**s**= [S/S']-1= Sales Growth= 5.00%), (**U**= Utilized or Starting Capital= 64,720 millions USD), (**i**= I/S= Interest Portion= 1.00%), (**F**= Fixed Cost= 29,750 millions USD), (**t**= T/B= Tax Rate= 30.00%), (**$**= Sales or Revenues= 51,294 millions USD), and (**V**= Variable Cost= 9,746 millions USD), are known, then it's (**E**= Equity or Capital planned), is:
$$E = U+[1-d][1-t]\{\$-V-F-\$'i[1+s]\}$$
$$= 64,720+[1-0.3000][1-0.3000]\{51,294$$
$$-9,746-29,750-48,851*0.0100$$
$$[1+0.0500]\}$$
$$= 70,249 \text{ millions USD}$$

Rule-3961:
If both (**d**= D/A= Dividend Portion or Payout= 30.00%), (**S'**= Sales of Past Year= 48,851 millions USD), (**s**= [S/S']-1= Sales Growth= 5.00%), (**U**= Utilized or Starting Capital= 64,720 millions USD), (**i**= I/S= Interest Portion= 1.00%), (**F**= Fixed Cost= 29,750 millions USD), (**t**= T/B= Tax Rate= 30.00%), (**E**= Equity or Capital= 70,249 millions USD), and (**V**= Variable Cost= 9,746 millions USD), are known, then it's (**S**= Sales or Revenues planned), is:

$S = V+F+S'i[1+s]+[E-U]/\{[1-d][1-t]\}$
$= 9,746+29,750+48,851*0.0100$
$[1+0.0500]+[70,249-64,720]$
$/\{[1-0.3000][1-0.3000]\}$
$= 51,294$ millions USD

Rule-3962:
If both (**d**= D/A= Dividend Portion or Payout= 30.00%), (**$'**= Sales of Past Year= 48,851 millions USD), (**s**= [S/S']-1= Sales Growth= 5.00%), (**U**= Utilized or Starting Capital= 64,720 millions USD), (**i**= I/S= Interest Portion= 1.00%), (**F**= Fixed Cost= 29,750 millions USD), (**t**= T/B= Tax Rate= 30.00%), (**E**= Equity or Capital= 70,249 millions USD), and (**$**= Sales or Revenues= 51,294 millions USD), are known, then it's (**V**= Variable Cost planned), is:

$$V = \$-F-\$'i[1+s]-[E-U]/\{[1-d][1-t]\}$$
$$= 51,294-29,750-48,851*0.0100$$
$$[1+0.0500]-[70,249-64,720]$$
$$/\{[1-0.3000][1-0.3000]\}$$
$$= 9,746 \text{ millions USD}$$

Rule-3963:
If both (**d**= D/A= Dividend Portion or Payout= 30.00%), (**$'**= Sales of Past Year= 48,851 millions USD), (**s**= [S/S']-1= Sales Growth= 5.00%), (**U**= Utilized or Starting Capital= 64,720 millions USD), (**i**= I/S= Interest Portion= 1.00%), (**V**= Variable Cost= 9,746 millions USD), (**t**= T/B= Tax Rate= 30.00%), (**E**= Equity or Capital= 70,249 millions USD), and (**$**= Sales or Revenues= 51,294 millions USD), are known, then it's (**F**= Fixed Cost planned), is:

$$F = \$-V-\$'i[1+s]-[E-U]/\{[1-d][1-t]\}$$
$$= 51{,}294-9{,}746-48{,}851*0.0100$$
$$[1+0.0500]-[70{,}249-64{,}720]$$
$$/\{[1-0.3000][1-0.3000]\}$$
$$= 29{,}750 \text{ millions USD}$$

Rule-3964:
If both (**d**= D/A= Dividend Portion or Payout= 30.00%), (**F**= Fixed Cost= 29,750 millions USD), (**s**= [S/S']-1= Sales Growth= 5.00%), (**U**= Utilized or Starting Capital= 64,720 millions USD), (**i**= I/S= Interest Portion= 1.00%), (**V**= Variable Cost= 9,746 millions USD), (**t**= T/B= Tax Rate= 30.00%), (**E**= Equity or Capital= 70,249 millions USD), and (**S**= Sales or Revenues= 51,294 millions USD), are known, then it's (**S'**= Sales Past), must be:

$S' = (S-V-F-[E-U]/\{[1-d][1-t]\})/\{i[1+s]\}$
 = (51,294-9,746-29,750-[70,249
 -64,720]/{[1-0.3000]
 [1-0.3000]})
 /{ 0.0100[1+0.0500]}
 = 48,851 millions USD

Rule-3965:
If both (**d**= D/A= Dividend Portion or Payout= 30.00%), (**F**= Fixed Cost= 29,750 millions USD), (**s**= [S/S']-1= Sales Growth= 5.00%), (**U**= Utilized or Starting Capital= 64,720 millions USD), (**$'**= Sales of Past Year= 48,851 millions USD), (**V**= Variable Cost= 9,746 millions USD), (**t**= T/B= Tax Rate= 30.00%), (**E**= Equity or Capital= 70,249 millions USD), and (**$**= Sales or Revenues= 51,294 millions USD), are known, then it's (**i**= Interest Portion planned), is:

$$i = (\$-V-F-[E-U]/\{[1-d][1-t]\})/\{\$'[1+s]\}$$
$$= (51,294-9,746-29,750-[70,249-64,720]/\{[1-0.3000][1-0.3000]\})/\{48,851[1+0.0500]\}$$
$$= 1.00\%$$

Rule-3966:
If both (**d**= D/A= Dividend Portion or Payout= 30.00%), (**F**= Fixed Cost= 29,750 millions USD), (**i**= I/S= Interest Portion= 1.00%), (**U**= Utilized or Starting Capital= 64,720 millions USD), (**$'**= Sales of Past Year= 48,851 millions USD), (**V**= Variable Cost= 9,746 millions USD), (**t**= T/B= Tax Rate= 30.00%), (**E**= Equity or Capital= 70,249 millions USD), and (**$**= Sales or Revenues= 51,294 millions USD), are known, then it's (**s**= Sales Growth planned), is:

s= ($-**V**-**F**-[**E**-**U**]/{[1-**d**][1-**t**]})/[$'**i**]-1
= (51,294-9,746-29,750-[70,249
-64,720]/{[1-0.3000]
[1-0.3000]})
/[48,851*0.0100]-1
= 5.00%

Rule-3967:
If both (**d**= D/A= Dividend Portion or Payout= 30.00%), (**F**= Fixed Cost= 29,750 millions USD), (**i**= I/S= Interest Portion= 1.00%), (**U**= Utilized or Starting Capital= 64,720 millions USD), (**S'**= Sales of Past Year= 48,851 millions USD), (**V**= Variable Cost= 9,746 millions USD), (**s**= [S/S']-1= Sales Growth= 5.00%), (**E**= Equity or Capital= 70,249 millions USD), and (**S**= Sales or Revenues= 51,294 millions USD), are known, then it's (**t**= Tax Rate planned), is:

t= 1-[**E-U**]/([1-**d**]{**S-V-F-S'i**[1+**s**]})
 = 1-[70,249-64,720]/([1-0.3000]
     {51,294-9,746-29,750
     -48,851*0.0100[1+0.0500]})
 = 30.00%

Rule-3968:
If both (**t**= T/B= Tax Rate= 30.00%), (**F**= Fixed Cost= 29,750 millions USD), (**i**= I/S= Interest Portion= 1.00%), (**U**= Utilized or Starting Capital= 64,720 millions USD), (**$'**= Sales of Past Year= 48,851 millions USD), (**V**= Variable Cost= 9,746 millions USD), (**s**= [S/S']-1= Sales Growth= 5.00%), (**E**= Equity or Capital= 70,249 millions USD), and (**$**= Sales or Revenues= 51,294 millions USD), are known, then it's (**d**= Dividend Portion or Payout planned), is:

$$d = 1-[\textbf{E-U}]/([1-\textbf{t}]\{\textbf{\$-V-F-\$'i}[1+\textbf{s}]\})$$
$$= 1-[70,249-64,720]/([1-0.3000]$$
$$\{51,294-9,746-29,750$$
$$-48,851*0.0100[1+0.0500]\})$$
$$= 30.00\%$$

Rule-3969:
If both (**t**= T/B= Tax Rate= 30.00%), (**F**= Fixed Cost= 29,750 millions USD), (**i**= I/S= Interest Portion= 1.00%), (**d**= D/A= Dividend Portion or Payout= 30.00%), (**$'**= Sales of Past Year= 48,851 millions USD), (**V**= Variable Cost= 9,746 millions USD), (**s**= [S/S']-1= Sales Growth= 5.00%), (**E**= Equity or Capital= 70,249 millions USD), and (**$**= Sales or Revenues= 51,294 millions USD), are known, then it's (**U**= Utilized or Starting Capital planned), is:

$U = E - [1-d][1-t]\{\$-V-F-\$'i[1+s]\}$)
   = 70,249-[1-0.3000][1-0.3000]
     {51,294-9,746-29,750-
      48,851*0.0100[1+0.0500]})
   = 64,720 millions USD

Rule-3970:
If both (**t**= T/B= Tax Rate= 30.00%), (**f**= F/S= Fixed Portion= 58.00%), (**d**= D/A= Dividend Portion or Payout= 30.00%), (**I**= Interest Expense= 513 millions USD), (**V**= Variable Cost= 9,746 millions USD), (**U**= Utilized or Starting Capital= 64,720 millions USD), and (**S**= Sales or Revenues= 51,294 millions USD), are known, then it's (**E**= Equity or Capital planned), is:

$$\begin{aligned}\mathbf{E} &= \mathbf{U}+[1-\mathbf{d}][1-\mathbf{t}]\{\mathbf{S}-\mathbf{V}-\mathbf{Sf}-\mathbf{I}\} \\ &= \mathbf{U}+[1-\mathbf{d}][1-\mathbf{t}]\{\mathbf{S}[1-\mathbf{f}]-\mathbf{V}-\mathbf{I}\} \\ &= 64{,}720+[1-0.3000][1-0.3000] \\ &\quad \{51{,}294[1-0.5800]-9{,}746-513\} \\ &= 70{,}249 \text{ millions USD}\end{aligned}$$

Rule-3971:
If both (**t**= T/B= Tax Rate= 30.00%), (**f**= F/S= Fixed Portion= 58.00%), (**d**= D/A= Dividend Portion or Payout= 30.00%), (**I**= Interest Expense= 513 millions USD), (**V**= Variable Cost= 9,746 millions USD), (**U**= Utilized or Starting Capital= 64,720 millions USD), and (**E**= Equity or Capital= 70,249 millions USD), are known, then it's (**E**= Equity or Capital planned), is:

$$\begin{aligned}\mathbf{S} &= (\mathbf{V}+\mathbf{I}+[\mathbf{E}-\mathbf{U}]/\{[1-\mathbf{d}][1-\mathbf{t}]\})/[1-\mathbf{f}] \\ &= (9{,}746+513+[70{,}249-64{,}720] \\ &\quad /\{[1-0.3000][1-0.3000]\}) \\ &\quad /[1-0.5800] \\ &= 51{,}294 \text{ millions USD}\end{aligned}$$

Rule-3972:
If both (**t**= T/B= Tax Rate= 30.00%), (**$**= Sales or Revenues= 51,294 millions USD), (**d**= D/A= Dividend Portion or Payout= 30.00%), (**I**= Interest Expense= 513 millions USD), (**V**= Variable Cost= 9,746 millions USD), (**U**= Utilized or Starting Capital= 64,720 millions USD), and (**E**= Equity or Capital= 70,249 millions USD), are known, then it's (**f**= Fixed Portion planned), is:

$$\mathbf{f} = 1-(\mathbf{V}+\mathbf{I}+[\mathbf{E}-\mathbf{U}]/\{[1-\mathbf{d}][1-\mathbf{t}]\})/\mathbf{\$}$$
$$= 1-(9{,}746+513+[70{,}249-64{,}720]$$
$$/\{[1-0.3000][1-0.3000]\})$$
$$/51{,}294$$
$$= 58.00\%$$

Rule-3973:
If both (**t**= T/B= Tax Rate= 30.00%), (**$**= Sales or Revenues= 51,294 millions USD), (**d**= D/A= Dividend Portion or Payout= 30.00%), (**I**= Interest Expense= 513 millions USD), (**f**= F/S= Fixed Portion= 58.00%), (**U**= Utilized or Starting Capital= 64,720 millions USD), and (**E**= Equity or Capital= 70,249 millions USD), are known, then it's (**V**= Fixed Portion planned), is:

$$\mathbf{V} = \mathbf{\$}[1-\mathbf{f}]-\mathbf{I}-[\mathbf{E}-\mathbf{U}]/\{[1-\mathbf{d}][1-\mathbf{t}]\}$$
$$= 51{,}294[1-0.5800]-513-[70{,}249$$
$$-64{,}720]/\{[1-0.3000][1-0.3000]\}$$

$$= 9{,}746 \text{ millions USD}$$

Rule-3974:
If both (**t**= T/B= Tax Rate= 30.00%), (**$**= Sales or Revenues= 51,294 millions USD), (**d**= D/A= Dividend Portion or Payout= 30.00%), (**V**= Variable Cost= 9,746 millions USD), (**f**= F/S= Fixed Portion= 58.00%), (**U**= Utilized or Starting Capital= 64,720 millions USD), and (**E**= Equity or Capital= 70,249 millions USD), are known, then it's (**I**= Interest Expense planned), is:

I= **$**[1-**f**]-**V**-[**E**-**U**]/{[1-**d**][1-**t**]}
= 51,294[1-0.5800]-9,746-[70,249
-64,720]/{[1-0.3000][1-0.3000]}

= 513 millions USD

Rule-3975:
If both (**I**= Interest Expense= 513 millions USD), (**$**= Sales or Revenues= 51,294 millions USD), (**d**= D/A= Dividend Portion or Payout= 30.00%), (**V**= Variable Cost= 9,746 millions USD), (**f**= F/S= Fixed Portion= 58.00%), (**U**= Utilized or Starting Capital= 64,720 millions USD), and (**E**= Equity or Capital= 70,249 millions USD), are known, then it's (**t**= Tax Rate planned), is:

t= 1-[**E**-**U**]/([1-**d**]{**$**[1-**f**]-**V**-**I**})
= 1-[70,249-64,720]/([1-0.3000
{51,294[1-0.5800]-9,746-513})
= 30.00%

Rule-3976:
If both (**I**= Interest Expense= 513 millions USD), (**$**= Sales or Revenues= 51,294 millions USD), (**t**= T/B= Tax Rate= 30.00%), (**V**= Variable Cost= 9,746 millions USD), (**f**= F/S= Fixed Portion= 58.00%), (**U**= Utilized or Starting Capital= 64,720 millions USD), and (**E**= Equity or Capital= 70,249 millions USD), are known, then it's (**d**= Dividend Portion or Payout planned), is:

$$d = 1-[\mathbf{E}-\mathbf{U}]/([1-\mathbf{t}]\{\mathbf{\$}[1-\mathbf{f}]-\mathbf{V}-\mathbf{I}\})$$
$$= 1-[70{,}249-64{,}720]/([1-0.3000]\{51{,}294[1-0.5800]-9{,}746-513\})$$
$$= 30.00\%$$

Rule-3977:
If both (**I**= Interest Expense= 513 millions USD), (**$**= Sales or Revenues= 51,294 millions USD), (**t**= T/B= Tax Rate= 30.00%), (**V**= Variable Cost= 9,746 millions USD), (**f**= F/S= Fixed Portion= 58.00%), (**d**= D/A= Dividend Portion or Payout= 30.00%), and (**E**= Equity or Capital= 70,249 millions USD), are known, then it's (**U**= Utilized or Starting Capital planned), is:

$$\mathbf{U} = \mathbf{E}-[1-\mathbf{d}][1-\mathbf{t}]\{\mathbf{\$}[1-\mathbf{f}]-\mathbf{V}-\mathbf{I}\}$$
$$= 70{,}249-[1-0.3000][1-0.3000\{51{,}294[1-0.5800]-9{,}746-513\}$$
$$= 64{,}720 \text{ millions USD}$$

Rule-3978:

If both (**i**= I/S= Interest Portion= 1.00%), (**$**= Sales or Revenues= 51,294 millions USD), (**t**= T/B= Tax Rate= 30.00%), (**V**= Variable Cost= 9,746 millions USD), (**f**= F/S= Fixed Portion= 58.00%), (**d**= D/A= Dividend Portion or Payout= 30.00%), and (**U**= Utilized or Starting Capital= 64,720 millions USD), are known, then it's (**E**= Equity or Capital planned), is:

$$E = U+[1-d][1-t][\$-V-\$f-\$i]$$
$$= U+[1-d][1-t]\{\$[1-f-i]-V\}$$
$$= 64{,}720+[1-0.3000][1-0.3000]$$
$$\{51{,}294[1-0.5800-0.0100]$$
$$-9{,}746\}$$
$$= 70{,}249 \text{ millions USD}$$

Rule-3979:

If both (**i**= I/S= Interest Portion= 1.00%), (**E**= Equity or Capital= 70,249 millions USD), (**t**= T/B= Tax Rate= 30.00%), (**V**= Variable Cost= 9,746 millions USD), (**f**= F/S= Fixed Portion= 58.00%), (**d**= D/A= Dividend Portion or Payout= 30.00%), and (**U**= Utilized or Starting Capital= 64,720 millions USD), are known, then it's (**$**= Sales or Revenues planned), is:

$$\$ = (V+[E-U]/\{[1-d][1-t]\})/[1-f-i]$$
$$= (9{,}746+[70{,}249-64{,}720]/\{[1-0.3000]$$
$$[1-0.3000]\})/[1-0.5800-0.0100]$$

$$= 51{,}294 \text{ millions USD}$$

Rule-3980:
If both (**i**= I/S= Interest Portion= 1.00%), (**E**= Equity or Capital= 70,249 millions USD), (**t**= T/B= Tax Rate= 30.00%), (**V**= Variable Cost= 9,746 millions USD), (**S**= Sales or Revenues= 51,294 millions USD), (**d**= D/A= Dividend Portion or Payout= 30.00%), and (**U**= Utilized or Starting Capital= 64,720 millions USD), are known, then it's (**f**= Fixed Portion planned), is:

$$f = 1-i-(V+[E-U]/\{[1-d][1-t]\})/S$$
$$= 1-0.0100-(9,746+[70,249-64,720]$$
$$/\{[1-0.3000][1-0.3000]\})$$
$$/51,294$$
$$= 58.00\%$$

Rule-3981:
If both (**f**= F/S= Fixed Portion= 58.00%), (**E**= Equity or Capital= 70,249 millions USD), (**t**= T/B= Tax Rate= 30.00%), (**V**= Variable Cost= 9,746 millions USD), (**S**= Sales or Revenues= 51,294 millions USD), (**d**= D/A= Dividend Portion or Payout= 30.00%), and (**U**= Utilized or Starting Capital= 64,720 millions USD), are known, then it's (**i**= Interest Portion planned), is:

$$i = 1-f-(V+[E-U]/\{[1-d][1-t]\})/S$$
$$= 1-0.5800-(9,746+[70,249-64,720]$$
$$/\{[1-0.3000][1-0.3000]\})$$
$$/51,294$$
$$= 1.00\%$$

Rule-3982:
If both (**f**= F/S= Fixed Portion= 58.00%), (**E**= Equity or Capital= 70,249 millions USD), (**t**= T/B= Tax Rate= 30.00%), (**i**= I/S= Interest Portion= 1.00%), (**S**= Sales or Revenues= 51,294 millions USD), (**d**= D/A= Dividend Portion or Payout= 30.00%), and (**U**= Utilized or Starting Capital= 64,720 millions USD), are known, then it's (**V**= Variable Cost planned), is:

$V = S[1-f-i]-[E-U]/\{[1-d][1-t]\}$
$= 51,294[1-0.5800-0.0100]-[70,249 -64,720]/\{[1-0.3000][1-0.3000]\}$

$= 9,746$ millions USD

Rule-3983:
If both (**f**= F/S= Fixed Portion= 58.00%), (**E**= Equity or Capital= 70,249 millions USD), (**V**= Variable Cost= 9,746 millions USD), (**i**= I/S= Interest Portion= 1.00%), (**S**= Sales or Revenues= 51,294 millions USD), (**d**= D/A= Dividend Portion or Payout= 30.00%), and (**U**= Utilized or Starting Capital= 64,720 millions USD), are known, then it's (**t**= Tax Rate planned), is:

$t = 1-[E-U]/([1-d]\{S[1-f-i]-V\}$
$= 1-[70,249-64,720]/([1-0.3000] \{51,294[1-0.5800-0.0100] -9,746\})$
$= 30.00\%$

Rule-3984:
If both (**f**= F/S= Fixed Portion= 58.00%), (**E**= Equity or Capital= 70,249 millions USD), (**V**= Variable Cost= 9,746 millions USD), (**i**= I/S= Interest Portion= 1.00%), (**$**= Sales or Revenues= 51,294 millions USD), (**t**= T/B= Tax Rate= 30.00%), and (**U**= Utilized or Starting Capital= 64,720 millions USD), are known, then it's (**d**= Dividend Portion or Payout planned), is:
$$d = 1-[E-U]/([1-t]\{\$[1-f-i]-V\})$$
$$= 1-[70,249-64,720]/([1-0.3000]\{51,294[1-0.5800-0.0100]-9,746\})$$
$$= 30.00\%$$

Rule-3985:
If both (**f**= F/S= Fixed Portion= 58.00%), (**E**= Equity or Capital= 70,249 millions USD), (**V**= Variable Cost= 9,746 millions USD), (**i**= I/S= Interest Portion= 1.00%), (**$**= Sales or Revenues= 51,294 millions USD), (**t**= T/B= Tax Rate= 30.00%), and (**d**= D/A= Dividend Portion or Payout= 30.00%), are known, then it's (**U**= Utilized or Starting Capital planned), is:
$$U = E-[1-d][1-t]\{\$[1-f-i]-V\}$$
$$= 70,249-[1-0.3000][1-0.3000]\{51,294[1-0.5800-0.0100]-9,746\}$$
$$= 64,720 \text{ millions USD}$$

Rule-3986:
If both (**f**= F/S= Fixed Portion= 58.00%), (**$'**= Sales of Past Year= 48,851 millions USD), (**s**= [S/S']-1= Sales Growth= 5.00%), (**U**= Utilized or Starting Capital= 64,720 millions USD), (**V**= Variable Cost= 9,746 millions USD), (**i**= I/S= Interest Portion= 1.00%), (**$**= Sales or Revenues= 51,294 millions USD), (**t**= T/B= Tax Rate= 30.00%), and (**d**= D/A= Dividend Portion or Payout= 30.00%), are known, then it's (**E**= Equity or Capital planned), is:

$$E = U+[1-d][1-t]\{\$-V-\$f-\$'i[1+s]\}$$
$$= U+[1-d][1-t]\{\$[1-f]-V-\$'i[1+s]\}$$
$$= 64{,}720+[1-0.3000][1-0.3000]\{51{,}294$$
$$[1-0.5800]-9{,}746-48{,}851$$
$$*0.0100[1+0.0500]\}$$
$$= 70{,}249 \text{ millions USD}$$

Rule-3987:
If both (**f**= F/S= Fixed Portion= 58.00%), (**$'**= Sales of Past Year= 48,851 millions USD), (**s**= [S/S']-1= Sales Growth= 5.00%), (**U**= Utilized or Starting Capital= 64,720 millions USD), (**V**= Variable Cost= 9,746 millions USD), (**i**= I/S= Interest Portion= 1.00%), (**E**= Equity or Capital= 70,249 millions USD), (**t**= T/B= Tax Rate= 30.00%), and (**d**= D/A=Dividend Portion or Payout= 30.00%), are known, then it's (**$**= Sales or Revenues planned), is:

$= (**V**+**$'i**[1+**s**]+[**E**-**U**]/{[1-**d**][1-**t**]})/[1-**f**]
   = (9,746+48,851*0.0100[1+0.0500]
     +[70,249-64,720]/{[1-0.3000]
     [1-0.3000]})/ [1-0.5800]
   = 51,294 millions USD

Rule-3988:
If both (**f**= F/S= Fixed Portion= 58.00%), (**$'**= Sales of Past Year= 48,851 millions USD), (**s**= [S/S']-1= Sales Growth= 5.00%), (**U**= Utilized or Starting Capital= 64,720 millions USD), (**$**= Sales or Revenues= 51,294 millions USD), (**i**= I/S= Interest Portion= 1.00%), (**E**= Equity or Capital= 70,249 millions USD), (**t**= T/B= Tax Rate= 30.00%), and (**d**= D/A= Dividend Portion or Payout= 30.00%), are known, then it's (**V**= Variable Cost planned), is:

$$V = \$[1-f] - \$'i[1+s] - [E-U]/\{[1-d][1-t]\}$$
$$= 51{,}294[1-0.5800] - 48{,}851*0.0100$$
$$[1+0.0500] - [70{,}249-64{,}720]$$
$$/\{[1-0.3000][1-0.3000]\}$$
$$= 9{,}746 \text{ millions USD}$$

Rule-3989:
If both (**f**= F/S= Fixed Portion= 58.00%), (**V**= Variable Cost= 9,746 millions USD), (**s**= [S/S']-1= Sales Growth= 5.00%), (**U**= Utilized or Starting Capital= 64,720 millions USD), (**$**= Sales or Revenues= 51,294 millions USD), (**i**= I/S= Interest Portion= 1.00%), (**E**= Equity or Capital= 70,249 millions USD), (**t**= T/B= Tax Rate= 30.00%), and (**d**= D/A= Dividend Portion or Payout= 30.00%), are known, then it's (**$'**= Sales Past), must be:

$'= ($[1-**f**]-**V**-[**E**-**U**]/{[1-**d**][1-**t**]})/{**i**[1+**s**]}
   = (51,294[1-0.5800] -9,746-[70,249
        -64,720]/{[1-0.3000]
        [1-0.3000]})
        /{0.0100[1+0.0500]}
   = 48,851 millions USD

Rule-3990:
If both ($'= Sales of Past Year= 48,851 millions USD), (V= Variable Cost= 9,746 millions USD), ($= [S/S']-1= Sales Growth= 5.00%), (U= Utilized or Starting Capital= 64,720 millions USD), ($= Sales or Revenues= 51,294 millions USD), (i= I/S= Interest Portion= 1.00%), (E= Equity or Capital= 70,249 millions USD), (t= T/B= Tax Rate= 30.00%), and (d= D/A= Dividend Portion or Payout= 30.00%), are known, then it's (f= Fixed Portion planned), is:

$f = 1-(\$'i[1+s]+V+[E-U]/\{[1-d][1-t]\})/\$$
$= 1-(48,851*0.0100[1+0.0500]$
$+9,746+[70,249-64,720]$
$/\{[1-0.3000][1-0.3000]\})$
$/51,294$
$= \underline{58.00}\%$

Rule-3991:
If both ($= Sales of Past Year= 48,851 millions USD), (V= Variable Cost= 9,746 millions USD), (f= F/S= Fixed Portion= 58.00%), (U= Utilized or Starting Capital= 64,720 millions USD), ($= Sales or Revenues= 51,294 millions USD), (i= I/S= Interest Portion= 1.00%), (E= Equity or Capital= 70,249 millions USD), (t= T/B= Tax Rate= 30.00%), and (d= D/A= Dividend Portion or Payout= 30.00%), are known, then it's (s= Sales Growth planned), is:

s= ($[1-f]-V-[E-U]/{[1-d][1-t]})/[$'i]-1
= 1-(51,294[1-0.5800]-9,746-[70,249
-64,720]/{[1-0.3000]
[1-0.3000]})
/[48,851*0.0100]-1
= 5.00%

Rule-3992:
If both ($'= Sales of Past Year= 48,851 millions USD), (V= Variable Cost= 9,746 millions USD), (f= F/S= Fixed Portion= 58.00%), (U= Utilized or Starting Capital= 64,720 millions USD), ($= Sales or Revenues= 51,294 millions USD), (s= [S/S']-1= Sales Growth= 5.00%), (E= Equity or Capital= 70,249 millions USD), (t= T/B= Tax Rate= 30.00%), and (d= D/A= Dividend Portion or Payout= 30.00%), are known, then it's (i= Interest Portion planned), is:

i= ($[1-f]-V-[E-U]/{[1-d][1-t]})/{$'[1+s]}
= (51,294[1-0.5800] -9,746-[70,249 -64,720]/{1-0.3000]
[1-0.3000]})
/{48,851[1+0.0500]}
= 1.00%

Rule-3993:
If both ($= Sales of Past Year= 48,851 millions USD), (V= Variable Cost= 9,746 millions USD), (f= F/S= Fixed Portion= 58.00%), (U= Utilized or Starting Capital= 64,720 millions USD), ($= Sales or Revenues= 51,294 millions USD), (s= [S/S']-1= Sales Growth= 5.00%), (E= Equity or Capital= 70,249 millions USD), (i= I/S= Interest Portion= 1.00%), and (d= D/A= Dividend Portion or Payout= 30.00%), are known, then it's (t= Tax Rate planned), is:

$$t = 1-[E-U]/([1-d]\{\$[1-f]-V-\$'[1+s]\})$$
$$= 1-[70,249-64,720]/([1-0.3000]$$
$$\{51,294[1-0.5800]-9,746$$
$$-48,851[1+0.0500]\})$$
$$= 30.00\%$$

Rule-3994:
If both ($'= Sales of Past Year= 48,851 millions USD), (V= Variable Cost= 9,746 millions USD), (f= F/S= Fixed Portion= 58.00%), (U= Utilized or Starting Capital= 64,720 millions USD), ($= Sales or Revenues= 51,294 millions USD), (s= [S/S']-1= Sales Growth= 5.00%), (E= Equity or Capital= 70,249 millions USD), (i= I/S= Interest Portion= 1.00%), and (t= T/B= Tax Rate= 30.00%), are known, then it's (d= Dividend Portion or Payout planned), is:

$d = 1-[E-U]/([1-t]\{\$[1-f]-V-\$'[1+s]\})$
$= 1-[70,249-64,720]/([1-0.3000]$
$\{51,294[1-0.5800]-9,746-$
$48,851[1+0.0500]\})$
$= 30.00\%$

Rule-3995:
If both ($'= Sales of Past Year= 48,851 millions USD), (V= Variable Cost= 9,746 millions USD), (f= F/S= Fixed Portion= 58.00%), (d= D/A= Dividend Portion or Payout= 30.00%), ($= Sales or Revenues= 51,294 millions USD), (s= [S/S']-1= Sales Growth= 5.00%), (E= Equity or Capital= 70,249 millions USD), (i= I/S= Interest Portion= 1.00%), and (t= T/B= Tax Rate= 30.00%), are known, then it's (U= Utilized or Starting Capital planned), is:

U= E-[1-d][1-t]{$[1-f]-V-$'[1+s]}
 = 70,249-[1-0.3000][1-0.3000]
   {51,294[1-0.5800]-9,746
   -48,851[1+0.0500]}
 = 64,720 millions USD

Rule-3996:
If both ($\$'$= Sales of Past Year= 48,851 millions USD), (V= Variable Cost= 9,746 millions USD), (f= F/S= Fixed Portion= 58.00%), (d= D/A= Dividend Portion or Payout= 30.00%), ($\$$= Sales or Revenues= 51,294 millions USD), (s= [S/S']-1= Sales Growth= 5.00%), (U= Utilized or Starting Capital= 64,720 millions USD), (I= Interest Expense= 513 millions USD),and (t= T/B= Tax Rate= 30.00%), are known, then it's (E= Equity or Capital planned), is:

E= U+[1-d][1-t]{$\$$-V-I-$\$'$f[1+s]}
   = 64,720+[1-0.3000][1-0.3000]
        {51,294-9,746-513-48,851
        *0.5800[1+0.0500]}
   = 70,249 millions USD

Rule-3997:
If both ($'= Sales of Past Year= 48,851 millions USD), (V= Variable Cost= 9,746 millions USD), (f= F/S= Fixed Portion= 58.00%), (d= D/A= Dividend Portion or Payout= 30.00%), (E= Equity or Capital= 70,249 millions USD), (s= [S/S']-1= Sales Growth= 5.00%), (U= Utilized or Starting Capital= 64,720 millions USD), (I= Interest Expense= 513 millions USD),and (t= T/B= Tax Rate= 30.00%), are known, then it's ($= Sales or Revenues planned), is:

$= V+I+$'f[1+s]+[E-U]/{[1-d][1-t]}
 = 9,746+513+48,851*0.5800
   [1+0.0500]+[70,249-64,720]
   /{[1-0.3000][1-0.3000]}
 = 51,294 millions USD

Rule-3998:
If both ($'$= Sales of Past Year= 48,851 millions USD), ($= Sales or Revenues= 51,294 millions USD), (f= F/S= Fixed Portion= 58.00%), (d= D/A= Dividend Portion or Payout= 30.00%), (E= Equity or Capital= 70,249 millions USD), (s= [S/S']-1= Sales Growth= 5.00%), (U= Utilized or Starting Capital= 64,720 millions USD), (I= Interest Expense= 513 millions USD),and (t= T/B= Tax Rate= 30.00%), are known, then it's (V= Variable Cost planned), is:

$V = S - I - S'f[1+s] - [E-U]/\{[1-d][1-t]\}$
$= 51,294 - 513 - 48,851*0.5800$
$\qquad [1+0.0500] - [70,249 - 64,720]$
$\qquad /\{[1-0.3000][1-0.3000]\}$
$= 9,746$ millions USD

Rule-3999:

If both (V= Variable Cost= 9,746 millions USD), ($\$$= Sales or Revenues= 51,294 millions USD), (f= F/S= Fixed Portion= 58.00%), (d= D/A= Dividend Portion or Payout= 30.00%), (E= Equity or Capital= 70,249 millions USD), (s= [S/S']-1= Sales Growth= 5.00%), (U= Utilized or Starting Capital= 64,720 millions USD), (I= Interest Expense= 513 millions USD), and (t= T/B= Tax Rate= 30.00%), are known, then it's ($\$'$= Sales Past), must be:

$\$' = (\$-V-I-[E-U]/\{[1-d][1-t]\})/\{f[1+s]\}$
= (51,294-9,746-513-[70,249-64,720]
/{[1-0.3000][1-0.3000]})
/{0.5800[1+0.0500]}
= 48,851 millions USD

Rule-4000:
If both (**V**= Variable Cost= 9,746 millions USD), (**$**= Sales or Revenues= 51,294 millions USD), (**$'**= Sales of Past Year= 48,851 millions USD), (**d**= D/A= Dividend Portion or Payout= 30.00%), (**E**= Equity or Capital= 70,249 millions USD), (**s**= [S/S']-1= Sales Growth= 5.00%), (**U**= Utilized or Starting Capital= 64,720 millions USD), (**I**= Interest Expense= 513 millions USD), and (**t**= T/B= Tax Rate= 30.00%), are known, then it's (**f**= Fixed Portion planned), is:

f= (**$**-**V**-**I**-[**E**-**U**]/{[1-**d**][1-**t**]})/{**$'**[1+**s**]}
= (51,294-9,746-513-[70,249-64,720]
/{[1-0.3000][1-0.3000]})
/{48,851[1+0.0500]}
= 58.00%

Finance Construction-4, *Tim Asikin, Steve Asikin, Indra Senihardja*

The Rest will be Continued in
The Series of Book

Finance Construction 5

Corporate IFRS-GAAP (B/S-I/S) Engineering
Technologies No. 4,001-5,000 of 111,111 Laws

At
http: //www.Amazon.Com

REFERENCES:

A. Books:

1) Amin, A.R. & Asikin, S.(2011a- 110 pages), *CEO Time Matrix: Productivity Management Skill that Increase CEO Effectiveness by 12400%*, Seattle (US): Amazon Createspace. http://www.amazon.com/CEO-Time-Matrix-Productivity-Effectiveness/dp/1461066069/ref=sr_1_1?s=books&ie=UTF8&qid=1387091880&sr=1-1&keywords=ceo+time+matrix. ISBN-13: 978-1461066064, ISBN-10: 1461066069

2) Amin, A.R. & Asikin, S. (2014- 76 pages), *No Urgency CEO: A High-Tech Book in Modern Management*, Seattle (US): Amazon Createspace. http://www.amazon.com/No-Urgency-CEO-Hi-Tech-Management/dp/1489509356/ref=sr_1_1?s=books&ie=UTF8&qid=1396440688&sr=1-1&keywords=no+urgency. ISBN-13: 978-1489509352, ISBN-10: 1489509356

3} Amin, A.R. & Asikin, S. (2011b- 292 pages), *The Celestial Management: Spiritual Wisdom for All Human Beings*, Seattle (US): Amazon Createspace. http://www.amazon.com/Celestial-Management-Islamic-Wisdom-Beings/dp/1456450700/ref=sr_1_1?s=books&ie=UTF8&qid=1387092036&sr=1-1&keywords=celestial+management. ISBN-13: 978-1456450700, ISBN-10: 1456450700

4) Asikin, S. (2009- 548 pages), *Arigato, Obrigado... (Thanks): Portuguese-Japan 1550-1647 Historic Novel*, Seattle (US): Amazon Createspace. http://www.amazon.com/Arigato-Obrigado-Thanks-Portuguese-Japan-1550-1647/dp/1453786368/ref=sr_1_1?s=books&ie=UTF8&qid=1396442162&sr=1-1&keywords=arigato. ISBN-13: 978-1453786369, ISBN-10: 1453786368

5) Asikin, S. (2013a- 824 pages), *Econometric Rex: 888 Marketing Promo 6666 Six Sigma Parametric Theories (of 10 Data)*, Seattle (US): Amazon Createspace. http://www.amazon.co.uk/Econometric-Rex-Marketing-Parametric-Theories/dp/1482745259, ISBN-13: 978-1482745252, ISBN-10: 1482745259

6) Asikin, S. (2010b- 224 pages), *FINANCE Architecture: VISUAL art, for Corporate Finance's Beauty, Safety and Simplicity*, Seattle (US): Amazon Createspace. http://www.amazon.co.uk/FINANCE-Architecture-Corporate-Finances-Simplicity/dp/1453829547/ref=sr_1_14?s=books&ie=UTF8&qid=1385371842&sr=1-14&keywords=finance+architecture. ISBN-13: 978-1453829547, ISBN-10: 1453829547

7) Asikin, S. (2010c- 150 pages), *Geometric FINANCE: Useful concepts that work better than Economic Noble Laureates'* Seattle (US): Amazon Createspace. http://www.amazon.co.uk/Magic-Geometric-FINANCE-concepts-Laureates/dp/1453790284/ref=tmm_pap_title_0, ISBN-13: 978-1453790281, ISBN-10: 1453790284

8) **Asikin, S.** (2010a- 290 pages), *Investment ALGEBRA: Strategic Profitable Comprehensive Business Finance Engineering Technology*, Seattle (US): Amazon Createspace. http://www.amazon.co.uk/Investment-ALGEBRA-Profitable-Comprehensive-Engineering/dp/1453856811/ref=tmm_pap_title_0?ie=UTF8&qid=1385371309&sr=1-1, ISBN-13: 978-1453856819, ISBN-10: 1453856811
9) **Asikin, S.** (2011a- 352 pages), *State Economics: Comprehensive Macro-Micro Economics' Simple Fiscal-Monetary Export-Import Accouting, Integrated Supply-Demand Managerial ... Mathematical Engineering Visual* . Seattle (US): Amazon Createspace. http://www.amazon.co.uk/STATE-ECONOMICS-Steve-Asikin-ebook/dp/B00G89Q5SS, http://www.amazon.co.uk State-Economics-Steve-Asikin-ebook/dp/B00BBPLQDI, ISBN-13: 978-1461066095, ISBN-10: 1461066093
10) **Asikin, S.** (2013b- 260 pages), *Technoeconomistat: Economic &Statistical Technology for Financial Planning*, Seattle (US): Amazon Createspace. http://www.amazon.co.uk/Technoeconomistat-Economic-Statistical-Technology-Financial/dp/1482304244, ISBN-13: 978-1482304244, ISBN-10: 1482304244
11) **Asikin, S.** (2011b- 540 pages), *Terima Kasih: Novel Internasional Sejarah Jepang-Portugis 1550-1647*, Seattle (US): Amazon Createspace. http://www.amazon.com/Terima-Kasih-Internasional-Jepang-Portugis-1550-1647/dp/146351106X/ref=sr_1_1?s=books&ie=UTF8&qid=1396442361&sr=1-1&keywords=terima, ISBN-13: 978-1463511067, ISBN-10: 146351106X
12) **Asikin, S.** (2010d- 294 pages), *TREASURY Planology: Strategic Profitable Comprehensive Business Finance Engineering Technology*, Seattle (US): Amazon Createspace. http://www.amazon.co.uk/TREASURY-PLANOLOGY-Steve-Asikin-ebook/dp/B00G88MTFW/ref=sr_1_1?s=books&ie=UTF8&qid=1385403690&sr=1-1&keywords=Treasury+Planology. ISBN-13: 978-1453891476, ISBN 10: 1453891471
13) **Asikin, S.** (2014a- 190 pages), *Calculus Accountancy: Leibnitz Newton Pacioli's Polynomial Quantitative Finance Risk Modelling Optimization*, Seattle (US): Amazon Createspace. http://www.amazon.com/Calculus-Accountancy-Polynomial-Quantitative-Optimization/dp/1495463028/ref=sr_1_1?s=books&ie=UTF8&qid=1396441338&sr=1-1&keywords=calculus+accountancy, ISBN-13: 978-1495463020, ISBN-10: 1495463028
14) **Asikin, S.**(2014b- 318 pages), *No Parametric: No Parametric: Hyperbolic Octahedron Central 3-Dimensional Orthogonal Risk Distribution Topology*, Seattle (US): Amazon Createspace. http://www.amazon.com/Parametric-Hyperbolic-Octahedron-3-Dimensional-Distribution/dp/1496089731/ref=sr_1_1?s=books&ie=UTF8&qid=1396441543&sr=1-1&keywords=No+Parametric, ISBN-13: 978-1496089731, ISBN-10: 1496089731

15) Asikin, S. (2014c- 138 pages), *Quick Help: Religion vs Real-Legion Philosophy*, Seattle (US): Amazon Createspace.
http:/www.amazon.com/Quick-Help-Religion-Real-Legion-Philosophy/dp/1497466520/ref=sr_1_2?s=books&ie=UTF8&qid=1396441740&sr=1-2&keywords=quick+help, ISBN-13: 978-1497466524, ISBN-10: 1497466520

16) Asikin, S. (2014d- 824 pages), *Anne of Denmark: King James I the Bible, Historic Plays Drama*, Seattle (US): Amazon Createspace.
http:/www.amazon.com/Anne-Denmark-James-Bible-Drama/dp/149299734X/ref=sr_1_1?s=books&ie=UTF8&qid=1396441857&sr=1-1&keywords=anne+asikin, ISBN-13: 978-1492997344, ISBN-10: 149299734X

17) Asikin, S. (2014e- 184 pages), *Constructive Spells: Positive Spiritual Mentality Personal Development*, Seattle (US): Amazon Createspace.
http:/www.amazon.com/Constructive-Spells-Spiritual-Mentality-Development/dp/1497498899/ref=sr_1_1?s=books&ie=UTF8&qid=1397664512&sr=1-1&keywords=constructive+spells; ISBN-13: 978-1497498891, ISBN-10: 1497498899

18) Asikin, S. (2014f- 466 pages), *Elizabeth of Russia: Another Virgin Queen Gloariana: A Romanov Plays Drama*, Seattle (US): Amazon Createspace.
http:/www.amazon.com/Elizabeth-RUSSIA-Another-Gloriana-Romanov/dp/1505542480/ref=sr_1_4?s=books&ie=UTF8&qid=1423888482&sr=1-4&keywords=elizabeth+russia; ISBN-13: 978-1505542486, ISBN-10: 1505542480

19) Asikin, S. (2014f- 708 pages), *Shinmen Munisai:Miyamoto Musashi's Father, A Sword Samurai Plays Drama*, Seattle (US): Amazon Createspace.
http:/www.amazon.com/Shinmen-Munisai-Miyamoto-Musashis-Samurai/dp/1507857799/ref=sr_1_1?s=books&ie=UTF8&qid=1428094280&sr=1-1&keywords=munisai; ISBN-13: 978-1507857793, ISBN-10: 1507857799

20) Asikin, S. (2015a-767 pages), *Math Fin Law-1: Mathematical Financial Law for Public Listed Firms Rule 1-4441*, Seattle (US): Amazon Createspace.
http://www.amazon.com/Math-Fin-Law-Mathematical-Financial-ebook/dp/B00WLNX3Y4/ref=asap_bc?ie=UTF8; ISBN-13: 978-1511792219, ISBN-10: 1511792213

21) Asikin, S. (2015b-813 pages), *Math Fin Law-2: Mathematical Financial Law for Public Listed Firms Rule 4442-8712*, Seattle (US): Amazon Createspace. http://www.amazon.com/Math-Fin-Law-Mathematical-No-4442-8712/dp/1512285080/ref=sr_1_1?s=books&ie=UTF8&qid=1432732199&sr=1-1&keywords=math+fin+law+2; ISBN-13: 978-1512285086, ISBN-10: 1512285080

22) Asikin, S. (2015c-813 pages), *Math Fin Law-3: Mathematical Financial Law for Public Listed Firms Rule 8713-12575*, Seattle (US): Amazon Createspace. http://www.amazon.com/Math-Fin-Law-Mathematical-8712-12575/dp/1514276763/ref=sr_1_1?s=books&ie=UTF8&qid=1433982149&sr=1-1&keywords=math+fin+law+3, ISBN-13: 978-1514276761, ISBN-10: 1514276763

23) **Asikin, S.** (2015d-813 pages), *Math Fin Law-4: Mathematical Financial Law for Public Listed Firms Rule 12576-16333,* Seattle (US): Amazon Createspace. http://www.amazon.com/Math-Fin-Law-Mathematical-12576-16333/dp/1514685132/ref=sr_1_1?s=books&ie=UTF8&qid=1435393628&sr=1-1&keywords=math+fin+law+4, ISBN-13: 978-1514685136, ISBN-10: 1514685132
24) **Asikin, S.** (2015d-816 pages), *Math Fin Law-5: Mathematical Financial Law for Public Listed Firms Rule 16334-19904,* Seattle (US): Amazon Createspace. http://www.amazon.com/Math-Fin-Law-Mathematical-No-16334-19904/dp/1515073009/ref=sr_1_1?s=books&ie=UTF8&qid=1437224693&sr=1-1&keywords=math+fin+law+5. ISBN-13: 978-1515073000, ISBN-10: 1515073009
25) **Asikin, S.** (2015d-728 pages), *Math Fin Law-6: Mathematical Financial Law for Public Listed Firms Rule 19905-23237,* Seattle (US): Amazon Createspace. https://www.amazon.com/Math-Fin-Law-Mathematical-No-19905-23238-ebook/dp/B0129ILVNK/ref=sr_1_2?ie=UTF8&qid=1478017624&sr=8-2&keywords=Math+Fin+Law+Asikin, ISBN-13: 978-1515158691, ISBN-10: 1515158691
26) **Asikin, S.** (2016a-728 pages), *Math Fin Law-7: Mathematical Financial Law for Public Listed Firms Rule 23238-27115,* Seattle (US): Amazon Createspace. https://www.amazon.com/Math-Fin-Law-Mathematical-No-23238-27115/dp/1508518521/ref=la_B00428V6JU_1_27?s=books, ISBN-13: 978-1508518525, ISBN-10: 1508518521
27) **Asikin, S.** (2016b-728 pages), *Math Fin Law-8: Mathematical Financial Law for Public Listed Firms Rule 27116-30330,* Seattle (US): Amazon Createspace. https://www.amazon.com/dp/1541091884, https://www.amazon.com/Math-Fin-Law-Mathematical-No-27116-30330/dp/1541091884/ref=la_B00428V6JU_1_27?s=books&ie=UTF8&qid=14793985386 sr=1-27&refinements=p_82%3AB0042RV6JU, ISDN-13: 978-1508518525, ISBN-10: 1508518521
28) **Asikin, S.** (2017a-828 pages), *Math Fin Law-9: Mathematical Financial Law for Public Listed Firms Rule 30330-33373,* Seattle (US): Amazon Createspace. https://www.amazon.com/Math-Fin-Law-Mathematical-30331-33373-ebook/dp/B01MRA8HZ0/ref=la_B00428V6JU_1_4?s=books&ie=UTF8&qid=1485921040&sr=1-4, ISBN-13: 978-1541092433, ISBN-10: 1541092430
29) **Asikin, S.** (2017b-823 pages), *Math Fin Law-10: Mathematical Financial Law for Public Listed Firms Rule 33374-36242,* Seattle (US): Amazon Createspace. https://www.amazon.com/s/ref=nb_sb_noss?url=search-alias%3Dstripbooks&field-keywords=math+fin+law+10, https://www.amazon.com/s/ref=nb_sb_noss?url=search-alias%3Dstripbooks&field-keywords=math+fin+law+10, ISBN-13: 978-1541092761, ISBN-10: 1541092767

Finance Construction-4, *Tim Asikin, Steve Asikin, Indra Senihardja*

30) **Asikin, S.** (2017c-823 pages), *Math Finance Law-11*: *Mathematical Financial Law for Public Listed Firms Rule 36243-39158*, Seattle (US): Amazon Createspace. https://www.amazon.com/s/ref=nb_sb_noss?url=search-alias%3Dstripbooks&field-keywords=math+finance+law+11, https://www.amazon.com/dp/1541215265/ref=sr_1_1?s=books&ie=UTF8&qid=1488029978&sr=1-1&keywords=math+finance+law, ISBN-13: 978-1541215269. ISBN-10: 1541215265

31) **Asikin, S.** (2017d-823 pages), *Math Finance Law-12*: *Mathematical Financial Law for Public Listed Firms Rule 30150-42152*, Seattle (US): Amazon Createspace. https://www.amazon.com/s/ref=nb_sb_noss?url=search-alias%3Dstripbooks&field-keywords=math+finance+law+12, https://www.amazon.com/Math-Finance-Law-Mathematical-39159-42152/dp/1541215516/ref=sr_1_2?s=books&ie=UTF8&qid=1490990342&sr=1-2&keywords=math+finance+law+12, ISBN-13: 978-1541215511, ISBN-10: 1541215516

32) **Asikin, S.** (2017e-824 pages), *Math Finance Law-13*: *Mathematical Financial Law for Public Listed Firms Rule 42153-44938*, Seattle (US): Amazon Createspace. https://www.amazon.com/dp/1542675219/ref=sr_1_1?s=books&ie=UTF8&qid=1491146227&sr=1-1&keywords=math+finance+law+13, ISBN-13: 978-1542675215, ISBN-10: 1542675219

33) **Asikin, S.** (2017f-824 pages), *Math Finance Law-14*: *Mathematical Financial Law for Public Listed Firms Rule 44939-47745*, Seattle (US): Amazon Createspace. https://www.amazon.com/dp/1542675219/ref=sr_1_1?s=books&ie=UTF8&qid=1491146227&sr=1-1&keywords=math+finance+law+14, ISBN-13: 978-1544628967, ISBN-10: 154462896X

34) **Asikin, S.** (2017g-810 pages), *Math Finance Law-15*: *Mathematical Financial Law for Public Listed Firms Rule 47746-50465*, Seattle (US): Amazon Createspace. https://www.amazon.com/dp/1545118647/ref=sr_1_1?s=books&ie=UTF8&qid=1493626425&sr=1-1&keywords=math+finance+law+15, ISBN-13: 978-1545118641, ISBN-10: 1545118647

35) **Asikin, S.** (2017h-816 pages), *Math Finance Law 16*: *Mathematical Financial Law Public Listed Firm Rule No.50,466-53,320*, Seattle (US), Amazon CreateSpace, https://www.amazon.com/dp/1545320179/ref=sr_1_1?s=books&ie=UTF8&qid=1495692695&sr=1-1&keywords=math+finance+law+16, ISBN-13: 978-1545320174, ISBN-10: 1545320179

Finance Construction-4, *Tim Asikin, Steve Asikin, Indra Senihardja*

36) **Asikin, S.** (2017i-809 pages), *Math Finance Law 17*: *Mathematical Financial Law Public Listed Firm Rule No.53,231-56,124*, Seattle (US), Amazon CreateSpace,
https://www.amazon.com/dp/1545320179/ref=sr_1_1?s=books&ie=UTF8&qid=14956926956 sr=1-1&keywords=math+finance+law+17, ISBN-13: 978-1545486474, ISBN-10: 1545486476

37) **Asikin, S.** (2017j-815 pages), *Math Finance Law 18*: *Mathematical Financial Law Public Listed Firm Rule No.56,125-58,942*, Seattle (US), Amazon CreateSpace,
https://www.amazon.com/dp/1545320179/ref=sr_1_1?s=books&ie=UTF8&qid=14956926956 sr=1-1&keywords=math+finance+law+18, ISBN-13: 978-1546912361, ISBN-10: 1546912363

38) **Asikin, S.** (2014-47 pages), *Doing Business in Comprehensive Interactive Interdisciplinary Micro-Macro Eco nomic International Accounting Architecture of Serial Well Known General Great Theories*, Proceedings in Finance and Risk Perspective 13, ACRN Oxford Publishing House, UK, p.197-243,
https://books.google.com/books?id=4e4KAwAAQBAJ&pg=PA2196lpg=PA2196dq=steve+asiki n+oxford&source=bl&ots=qVq9KyAmem&sig=ABdBliwuzUwRiqlsuLimuJ5ZdW4&hl=en&sa=X&s qi=2&ved=0ahUKEwjZkMrF7KjTAhUDSiYKHREDB-qQ6AEILjAC#v=onepage&q=steve%20asikin%20oxford&f=false, ISBN 978-3-9503518-1-1.

39) **Asikin, T.& Asikin, S.** (2013- 828 pages), *448 Poems: KARAOKE Song Book: 24 Hours Story of 30 Languages*, Seattle (US): Amazon Createspace. http://www.amazon.com/448-Poems-KARAOKE-Song-Book/dp/1480034363/ref=sr_1_1?s=books&ie=UTF8&qid=1396442034&sr=1-1&keywords=448+poems, ISBN-13: 978 1480034365, ISBN-10: 1480034363

40) **Asikin, T, Asikin, S, Senihardja, I.** (2017a- 508 pages), *Finance Construction 1*: *Corporate IFRS-GAAP (B/S-I/S) Engineering Technologies No. 1-1,000 of 111,111 Laws*, Seattle (US): Amazon Createspace. https://www.amazon.com/s/ref=nb_sb_noss?url=search-alias%3Dstripbooks&field-keywords=finance+construction+1+asikin. ISBN-13: 978-1544059525, ISBN-10: 1544059523

41) **Asikin, T, Asikin, S, Senihardja, I.** (2017b- 526 pages), *Finance Construction 2*: *Corporate IFRS-GAAP (B/S-I/S) Engineering Technologies No. 1,001-2,000 of 111,111 Laws*, Seattle (US): Amazon Createspace.
https://www.amazon.com/s/ref=nb_sb_noss?url=search-alias%3Dstripbooks&field-keywords=finance+construction+2+asikin. ISBN-13: 978-1542402415, ISBN-10: 1542402417

42) **Asikin, T, Asikin, S, Senihardja, I.** (2017c- 738 pages), *Finance Construction 3: Corporate IFRS-GAAP (B/S-I/S) Engineering Technologies No. 2,001-3,000 of 111,111 Laws*, Seattle (US): Amazon Createspace. https://www.amazon.com/s/ref=nb_sb_noss?url=search-alias%3Dstripbooks&field-keywords=finance+construction+3+asikin. ISBN-13: 978-1975999292, ISBN-10: 1975999290

43) **Brigham, EF & Gapenski, LC,** . (1988- Textbook), *Financial Management: Theory and Practice (Study Guide)*, NY (US): Dryden Press. http:/www.amazon.com/Financial-Management-Theory-Practice-Study/dp/0030125391/ref=sr_1_1?s=books&ie=UTF8&qid=1429380835&sr=1-1 ;

44) **Chandonet, RE,** . (2011-32 pages), *Balancesheet Management, Trends & Strategies*, NY (US): Sandler O'Neil. https://www.ncbankers.org/uploads/File/Meetings/Presentations/2011CONV-Chandonnet.pdf;

45) **Cornuejols G & Tutuncu, R,** . (2006-358 pages), *Optimization Methods in Finance*, Cambridge (UK): Cambridge University Press. http://www.math.ku.dk/~rolf/CT_FinOpt.pdf;

46) **Crook, M.** (2015-6 pages), *Optimization Methods in Finance*, UBS FS (Swiss): CIO WM Research. https://www.ubs.com/content/dam/WealthManagementAmericas/documents/investment-strategy-insights-2015-01-09-balance-sheet-optimization.pdf;

47) **Fee, Greg.** (2012-127 pages), *Catalan's Constant: (Ramanujan's Formula}: Catalan Constant to 300,000 digits*, Seattle (US): Amazon Kindle. http://www.Amazon.com/gp/aw/d/B0082Z5IEW/ref=mp_sa_1_4?ie=UTF8%Qqid=14843678188sr=8-4&pi=AC_SX236_SY340_QL65&keywords=ramanujan&dpPl=1&dpID=51FqIwDZpRL&ref=plSrch

48) **Gardenfors, Peter.** (2014) *The Geometry of Meaning: Semantics Based on Conceptual Spaces*. Cambridge, Mass.: MIT Press. ISBN 0-262-02678-3.

49) **Gardenfors, Peter.** (2000), *Conceptual Spaces: The Geometry of Thought*. Cambridge, Mass.: MIT Press. ISBN 0-262-07199-1.

50) **Kanigel, Robert.** (2013-465 pages), *The Man Who Knew Infinity:A Life of the Genius Ramanujan*, Washington (US): Washington Square Press&Amazon Kindle. http://www.Amazon.com/gp/aw/d/B008BW4VEGM/ref=pd_aw_sim351_1?ie=UTF8&psc=1&refRID=VYN345GVDBJ6ATDQ35WZ&dpID=9lZgpc4ETdL

51) **Limlingan, Victor S.** (1993-1 page), *The Limlingan Financial Model*, Manila: Asian Manager. http://limlingan.com/images/article.gif

52) **Newton, Isaac.** (2015-631 pages), *Principia:The Mathematical Principles of Natural Philosophy (Active Content)*, London (UK): RSM&Amazon Kindle.
http://www.Amazon.com/gp/aw/sitb/B01ISFJSJ0?ref=sib_dp_aw_kb_udp
53) **Nikisa, Evets.** (2014g- 828 pages), *Wonderful Live Spells: 144-Steps Recreative Nice Children Story, Succesful Development:* Seattle (US): Amazon Createspace. http:/www.amazon.com/WONDERFUL-Live-Spells-Recreative-Development/dp/1499248423/ref=sr_1_1?s=books&ie=UTF8&qid=1418698379&sr=1-1&keywords=wonderful+asikin, ISBN-13: 978-1499248425, ISBN-10: 1499248423
54) **Nielson, SS.** (2007- 430 pages), *Practical Financial Optimization: Decision Making for Financial Engineers:* NY (US): Wiley.
https://books.google.com/books/about/Practical_Financial_Optimization.html?id=XkwLAAACAAJ.
55) **Niswonger CR, Fess PE & Warren CE.** (1993-Textbook), *Accounting Principles:* Seattle (US): South Western Publishing Co.
http:/www.amazon.com/Accounting-Principles-Chapters-C-Rollin-Niswonger/dp/0538818603
56) **Omerod, P.** (1993-Textbook), (1994), *The Death of Ecoomics*, St Martin's Press, London, . http:/www.amazon.com/Accounting-Principles-Chapters-C-Rollin-Niswonger/dp/0538818603, ISBN-13: 978-0312134648, ISBN-10: 0312134649
57) **Paccioli, L. (1989-Textbook),** *Summa de Arithmetica Geometria Proportioni et Proportionalita*, Venice 1494.
http:/www.amazon.com/Arithmetica-Geometria-Proportioni-Proportionalita-Venice/dp/B00T4HXTIW/ref=sr_1_fkmr0_2?s=books&ie=UTF8&qid=1479597966&sr=1-2-fkmr0&keywords=summa+mhematica+arithmetica
58) **Persson, Torsten & Tabellini, Guido** (2000), *Political Economics – Explaining Economic Policy,* MIT Press.
https://mitpress.mit.edu/authors/torsten-persson
59) **Persson, Torsten & Tabellini, Guido** (2003), *The Economic Effects of Constitutions*, MIT Press. https://mitpress.mit.edu/authors/torsten-persson
60) **Persson, Torsten & Tabellini, Guido** (2000), *Monetary and Fiscal Policy –1 Credibility*, MIT Press. https://mitpress.mit.edu/authors/torsten-persson
61) **Persson, Torsten & Tabellini, Guido** (2003), *Monetary and Fiscal Policy –2 Politics*, MIT Press. https://mitpress.mit.edu/authors/torsten-persson
62) **Puts, J** (2012-102 pages), *Balancesheet Optimization Under Basel III*, Amsterdam (Netherlands): Vrije Universiteit.
http://www.math.vu.nl/~sbhulai/papers/thesis-puts.pdf
63) **Timmermans, K** (2012-30 pages), *Balancesheet Optimization Under Basel III*, Amsterdam (Netherlads): ING Investor Day.
https://www.ing.com/web/file?uuid=09c803d9-79a1-47ec-bfe5...owner...

64) **Weston JF, Copeland TE & Shastri, K.** (2004-Textbook), *Financial Theory and Corporate Policy:* NY (US): Addison Wesley.
http:/www.amazon.com/Financial-Corporate-Copeland-Kuldeep-Paperback/dp/B008YSUE4M/ref=sr_1_1?s=books&ie=UTF8&qid=1429381700&sr=1-1&keywords=weston+copeland

65) **Zenios SA.** (2002-350 pages), *Financial Optimization:* Cambridge (UK): Cambridge University Press.
https://books.google.com/books/about/Financial_Optimization.html?id=-FJdgjZ0qSAC .

Finance Construction-4, Tim Asikin, Steve Asikin, Indra Senihardja

B. Intellectual Properties:

1) **Indonesian Banker Certification 1993**, *Sertifikasi Bankir Indonesia 1993*;
2) **MATHFINPLAN** (Mathematical Financial Planning);
   Perencanaan Keuangan Matematis
3) **BANK MATHPLAN** (Bank's Mathematical Planning);
   Perencanaan Matematis Bank
4) **TECHNO-ECONOMISTAT** (Technology for Economical Statistics);
   Teknologi Statistik untuk Ekonomi
5) **FINANCIAL GEOMETRY** (IT Program); *Ilmu Geometri Keuangan*
6) **BAK DOBEL** (Car Gasoline Tank: Left and Right Side Fuel Fill-in).
   Tangki Mobil yang bisa diisi dari Kiri atau Kanan
7) **Mister GO-CHENG** (Cheap Food Franchising). *Waralaba Pangan Murah*

C. SCIENTIFIC PAPERS:

1) **Eichhornia Crassipes**, as Anti Polutant and Fertile Sediment, *Pemanfaatan Eceng Gondok untuk Menetralkan Polusi dan Pengurukan Subur pada Genangan Air*, LKIR, Lomba Karya Ilmiah Remaja, LIPI (1979)
2) **Archimedes Catesian Buoyancy Law**, *Penyimpangan Hukum Archimedes pada Model Pelampung Cartesian*, Lomba Karya Ilmiah Remaja, Depdikbud-RI (1980)
3) **SURAMADU**, Surabaya-Madura Inter Island Bridge, Design Supervisor, *Sumarsono Sumali Engineering Contractor*, Tambak-V, Tandes, Surabaya (1982),
4) **GAZOLINE Taxi-Pool Depot**, Jemur Handayani, Surabaya, *Sumarsono Sumali Engineering Contractor*, Tambak-V, Tandes, Surabaya (1982),
5) **SOEKARNO-HATTA**, Jakarta International Airport, Garuda Maintenance Facilities, *Junior Architect*, As Bulit Drawings, Coinstruction Management, Aerospatiale-Encona Engineering (1983-1985),
6) **ADISUCIPTO**, Yogya International Airport, *Junior Architect*, Preliminary Technical Drawings, Budiono Surasno-Encona Engineering (1986),
7) **BLOK-M PLAZA**, "New Garden Hall", Kebayoran Baru Supermall, *Preliminary Design Architect*, Danisworo-James Ferry- Jasa Ferry-Encona (1986),
8) **NORTH JAKARTA PUBLIC LIBRARY**, Semper, Tanjung Priok, *Chief Architect*, Agung Darma Sarana-Asep Hermawan (1987),
9) **The Banker's Trainer**: A Manual for The Training Head in Banking. *Profesi Bankir Pelatih, Tuntunan bagi Kepala Diklat Perbankan*, (1200 Pages, 1993),
10) **Indonesian Bankers Certification**, a Strategic Move for the National Banking Improvement. *Sertifikasi Bankir Indonesia, Langkah Strategis Pembenahan Perbankan Nasional*, (KOMPAS, 20 Juli 1993).
11) **Indonesian Rupiah's Exchange Rate**, on a Quiet but Steady Movement. *Nilai Tukar Rupiah, Diam-diam Meningkat Pesat*, (MANAJEMEN, Jan-Feb 1994).
12) **The Bank's Credit**, for the Best Practices. *Kredit Praktis* (200 pages, 1994),
13) **Down to Earth Credit Meeting Credo**, *Kiat Rapat Pemberian Kredit yang Realistik*. (MANAJEMEN, Sep-Oct 1994).
14) **SCHIPHOL**, Amsterdam International Airport, 1st Anthony Fokker Weg, Sloten, *Airport Engineer*, Samsung- Fokker-Beurlin, BV (1996),
15) **Ultrapeneurship: The Secret of Starting Business Without Capital**. Ultrapreneur: *Kiat Usaha Tanpa Modal*, (MANAJEMEN, Nov-Dec 1996),
16) **Recipe from the Corporate Raider's Kitchen**, *Menguak Dapur Corporate Raiders*, (MANAJEMEN, Jan-Feb 1997).
17) **The 10-Tips and Tricks of Engineered Conglomeration**. *Sepuluh Rekayasa Keuangan Konglomerat*, (MANAJEMEN, Mar-Apr 1997),
18) **The Management of Succession in Five Parts of the World**. *Manajemen Suksesi di Lima bagian Dunia*, (MANAJEMEN, Mei-Jun 1997),

19) **Auditing**: The Global Language for the 2003's Human Resources. *Audit : Bahasa Global untuk SDM 2003,* (MANAJEMEN, Jul-Aug 1997),
20) **The Elected Parliament 1997-2002**, and the Hope of Professionals. *DPR 1997-2002 dan Harapan Profesional,* (MANAJEMEN, Sep-Okt 1997),
21) **Disappearance of World's Soft Currencies**, and Proposition for Global Economic Thinking. *Lenyapnya Mata Uang Lunak Dunia dan Usulan Pemikiran Ekonomi Global,* (MANAJEMEN, Nov-Dec 1997),
22) **The Human Resources and Its Accountability**, in Critical Situation. *Akuntabilitas Sumberdaya Manusia di Masa Krisis,* (MANAJEMEN, Jan-Feb 1998), '
23) **The Investment Economic Model**, and the Failure of Growth Economic Theories. *Model Ekonomi Investasi dan Kegagalan Teori Ekonomi Pertumbuhan,* (MANAJEMEN, Nov 1998),
24) **The Exact Equilibrium of Capital Budgeting**, *Persamaan-persamaan akurat dalam Penganggaran Modal Perusahaan.* (MANAGEMENT EXPOSE, November 1998),
25) **The Corruptive Economic Model**, and Its Solution for Long Term Debt's Problem. *Model Ekonomi Korupsi dan Penyelesaian Utang Jangka Panjang,* (MANAJEMEN, Dec 1998),
26) **The Self Equity Budgeting**, and Operations Research in Investments. *Penganggaran Modal Sendiri dan Riset Operasi dalam Investasi,* (MANAJEMEN, Jan 1999),
27) **Actual Impact of Inflation**, and the Need of Geometric Approach in Statistic. *Pengaruh Riil Inflasi dan Perlunya Statistika Geometri,* (MANAJEMEN, Feb 1999),
28) **Multiplication of Money**, to Overcome the Crisis. *Melipat Gandakan Uang untuk Mengusir Krisis,* (MANAJEMEN, Apr 1999),
29) **The Interactive Mathematical Model**, for Multi Constraints Financial Budgeting, *Model Matematika Ekonomi Interaktif bagi Penganggaran Keuangan Multi Kendala.* (SCRIPTA ECONOMICA, Vol. 2, April 1999)
30) **Application of Ethics**, for the Management Accounting Practices. *Aplikasi Etika dalam Praktek Akuntansi Manajemen,* (MANAJEMEN, May 1999),
31) **The Curriculum of Magistrate of Management in Finance**: Its Needs and Practices. *Kurikulum Magister Manajemen Keuangan 1988-1998: Antara Kebutuhan dan Praktek,* (MANAJEMEN, Jun 1999),
32) **The Attractiveness**, Personal Power and Management. *Pesona Pribadi: "Personal Power" dan Manajemen,* (MANAJEMEN, Jul 1999),
33) **The Market Research**, for Magistrate Degree in Management. *Soal Jawab Riset Pasar untuk Pascasarjana Manajemen* (126 Pages, 1999),
34) **The Thesis Manual**, Question and Answer for Magistrate Degree in Management. *Soal Jawab Tugas Akhir untuk Pascasarjana Manajemen* (126 Pages, 1999),
35) **The Risk Management**, Question and Answers for Magistrate Degree in Management. *Soal Jawab Manajemen Resiko untuk Pascasarjana Manajemen* (126 Pages, 1999),

Finance Construction-4, Tim Asikin, Steve Asikin, Indra Senihardja

36) **The Credit Manual**, Question and Answers for Magistrate Degree in Management. *Soal Jawab Manajemen Kredit untuk Pascasarjana Manajemen* (126 Pages, 1999),
37) **Enhancing the Morality Structure**. *Solidkan Struktur Moralitas*, (MANAJEMEN, Oct 1999)
38) **Performance of PT. Astra Otoparts**, Plc, in its Press Releases and Public Financial Reports. *Kinerja PT. Astra Otoparts dalam Siaran Pers dan Laporan Keuangan Publik*, (MANAJEMEN, Oct 1999)
39) **The Management of Conflicts**, for the New Indonesian's "Rainbow Colored" Cabinet. *Manajemen Konflik dan Kabinet Pelangi*, (MANAJEMEN, Dec 1999),
40) **Technoeconometric**: An Engineering Elaboration to Matrix Multi Variate Econometrics, *Tekno Ekonometrik sebagai sumbangan teknologi untuk Ekonometrika Matrik Multi Variat.* (MANAGEMENT EXPOSE, November 1999),
41) **Technoeconometric** : A Mathematical Shortcut for Non Linear Bivariate Econometrics, *Tekno Ekonometrik sebagai Usulan Jalan Pintas Matematik untuk Ekonometrika non Linear Bivariat.* (SCRIPTA ECONOMICA, Vol. 2, No. 3, Desember 1999),
42) **Developing "New Image"**, for the New Indonesia. *Membangun Citra Indonesia Baru*, (MANAJEMEN, Jan 2000),
43) **Segregations of Duties**, the Needs of Both Indonesian Governments and Private Sectors. *Pemerintah dan Swasta Perlu Berbagi Tugas*, MANAJEMEN, Mar 2000),
44) **The 812.500 Financial Models**, for the Autonomous Country Governments. *Tentang 812.500 Model Keuangan Otonomi Daerah*, (MANAJEMEN, Mar 2000),
45) **The Forecast of 10 Most Attractive Share Values**, in the Jakarta Stock Exchange, April 2000. *Prakiraan Harga Sepuluh Saham Unggulan di Bursa Efek Jakarta, April 2000*, (MANAJEMEN, Mar 2000),
46) **Let Us Merchandise the Intelectual Properties**!. *Jual "Intelectual Property" Saja!* (MANAJEMEN, Apr 2000),
47) **The Forecast of ten Most Attractive Share Values**, in the Jakarta Stock Exchange May 2000. *Prakiraan Harga Sepuluh Saham Unggulan di Bursa Efek Jakarta Mei 2000*, (MANAJEMEN, Apr 2000),
48) **HEATHROW**, London International Airport, Second Runway, *Q-Basic Algorithmic Engineer*, Beasley-Krishnamoorthy, Imperial College, Cambridge, Post Doctoral Project (2000),
49) **Why the Banking Sector Should be Healthy**, Before Others?. *Mengapa Perbankan Harus Sehat?* (MANAJEMEN, Mei 2000),
50) **Recycling of Wastes as a Negative Costing**. *Memanfaatkan Sampah atau Negative Costing*, (MANAJEMEN, Agt 2000),
51) **The Guidance for Indonesia Corporate Restructuring**. *Kiat Restrukturisasi Perusahaan Indonesia*, (MANAJEMEN, Agt 2000),

Finance Construction-4, *Tim Asikin, Steve Asikin, Indra Senihardja*

52) **Redefinition of the Ethical Comprehension**. *Pemaknaan Kembali atas Pemahaman Etika*, (MANAJEMEN, Sep 2000),
53) **The Critical View the 1999's Indonesians Central Bank Financial Statement**. *Analisa Laporan Keuangan Bank Indonesia 1999*, (MANAJEMEN, Oct 2000),
54) **Becoming an "Achiever"**, not Just Common "Worker". *Bukan Sekadar "Kerja", Tapi Menghasilkan Karya!* (MANAJEMEN, Feb 2001),
55) **Ten Steps to Real Wealth!** *Sepuluh Langkah Menjadi Kaya,* (MANAJEMEN, Jun 2001),
56) **Creative Financial Breakthrough**, *Seminar Sehari tentang Terobosan Keuangan yang Kreatif,* (Gramedia-PPM dan MANAJEMEN, Jun 2001),
57) **The Marketing Practices of Financial Softwares**, for Publics Listed Corporation. *Praktek Pemasaran Program Perangkat Lunak untuk Keuangan Perusahaan Publik* (Univ Satyagama, Jakarta Juni 24, 2001).
58) **The Mathematical Financial Planning**, *Perancangan Keuangan Secara Eksak dengan Matematik.* (MANAJEMEN, Jul 2001),
59) **The Real Research and Development, not Rework and Destruction**. *Litbang yang Bukan "Sulit Berkembang",* (MANAJEMEN, Jul 2001),
60) **The Treatise on the Advantages and Disadvantages**, of Revolutionary Learning Methods. *Baik Buruknya: Revolusi Cara Belajar,* (MANAJEMEN, Jul 2001),
61) **Better Make an Indonesian "Corruption Prevention" Body**, not Just a "Corruption Watch". *Buat Indonesian Corruption Prevention (ICP), Bukan Indonesian Corruption Watch, ICW!* (MANAJEMEN, Okt 2001),
62) **Preparation of the Worlds Business Leaders**, for our Century and the Century After. *Mempersiapkan Manusia-manusia yang akan menjadi Khafilah Bisnis Dunia Abad ini dan Abad Mendatang.* (Univ Satyagama – Gregorio Araneta Univ, Jakarta, Oktober 27, 2001).
63) **The Good Corporate Governance**, is not just a Morale Condition. *Kepemerintahan Perusahaan yang Baik, Bukan Sekedar Moralitas,* (MANAJEMEN, Nov 2001),
64) **Corporate Healthiness and Cleanliness as a Challenge**. *Bersih, Sehat Perusahaan, sebuah Tantangan* (MANAJEMEN, Jan 2002),
65) **Developing the Cultural Base for Future Competitions**. *Budaya untuk Bersaing di Masa Depan* (MANAJEMEN, Jan 2002),
66) **"Catapult", "Bullet" or "Ballistic Missiles" Approach**, for Corporate Healthiness and Cleanliness. *Pola "Ketapel", "Peluru Biasa" atau "Rudal" untuk Keuangan Perusahaan yang Bersih Sehat.* (MANAJEMEN, Feb 2002),
67) **The Mathematical Invesment Solution Model**, for Indonesian Foreign Debt and Its Corruption, *Solusi Matematika Investasi bagi Penyelesaian Masalah Utang Luar Negeri dan Korupsi di Indonesia.* (MANAGEMENT EXPOSE, Mar 2002),

Finance Construction-4, *Tim Asikin, Steve Asikin, Indra Senihardja*

68) **The Attitude, Skills and Knowledge (ASK)**, in Educational System Approach. *Sikap, Keterampilan dan Pengetahuan Menurut Pendekatan Sistemik Pendidikan*, (MANAJEMEN, Apr 2002),
69) **Strategic Treatise on Regional Autonomy**, in Futurization, Digitalization, Securitization and Globalization Era in Accordance to the Awakening of the Mercantile Nations. *Kajian Strategik Otonomi Daerah : Era Futurisasi, Digitalisasi, Strukturisasi, dan Globalisasi dalam Kebangkitan Negara-Negara Dagang.* (Training for Provincial Government Regional Management, Indonesian Association of Provincial Governments, *Pelatihan Pengembangan Manajemen Pemerintah Daerah bagi Perangkat Pemerintah Propinsi, Angkatan II*, (APPSI), June 27, 2002.
70) **The Suggestion for the Doctoral Course**, for Islamic Financial Management System. *Usulan-usulan Perbaikan dalam Kurikulum Doktor Manajemen Keuangan Islam.* (December, 17-19, 2002, Canadian International Development Agency - CIDA, ADSGM, Universiti Utara Malaysia, Kedah, Da'arul Sintok).
71) **The Defensive Strategy for the Market Leader**, a Nokia Cellular Telephone's Case. *Strategi Bertahan Sang Pemimpin Pasar dengan Menggunakan Multi Merk, Kasus Telepon Seluler Nokia.* (March, 12, 2004, Univ Tarumanegara, Jakarta).

Finance Construction-4, *Tim Asikin, Steve Asikin, Indra Senihardja*

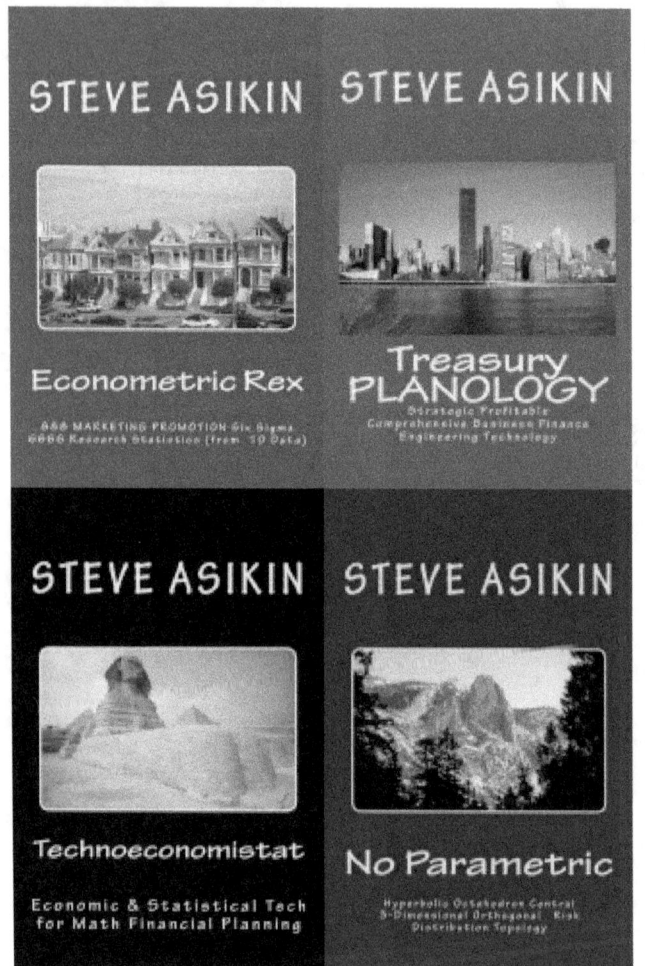

Finance Construction-4, *Tim Asikin, Steve Asikin, Indra Senihardja*

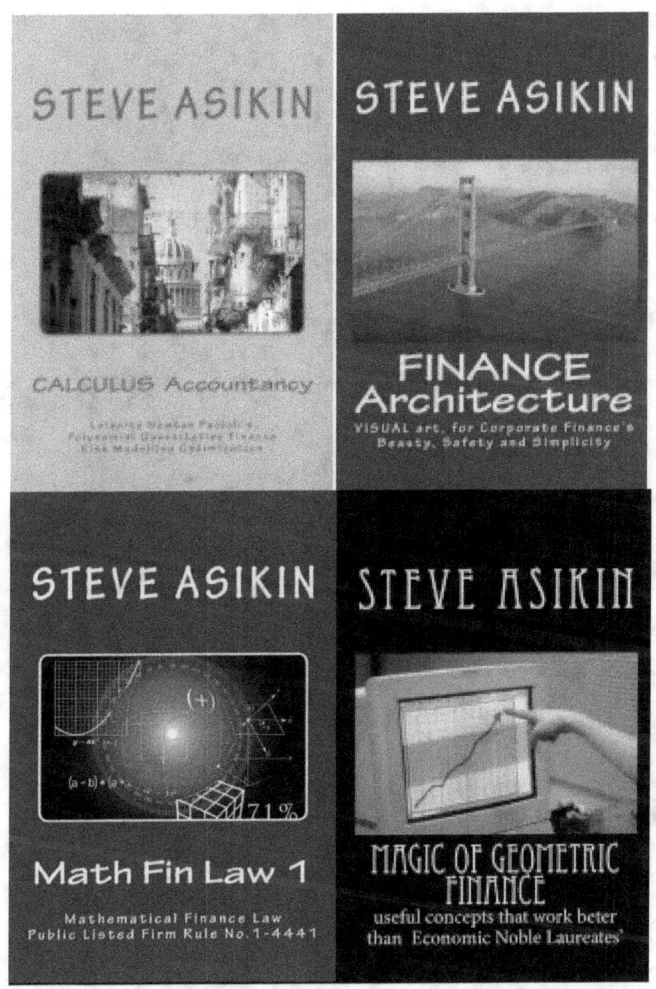

Finance Construction-4, *Tim Asikin, Steve Asikin, Indra Senihardja*

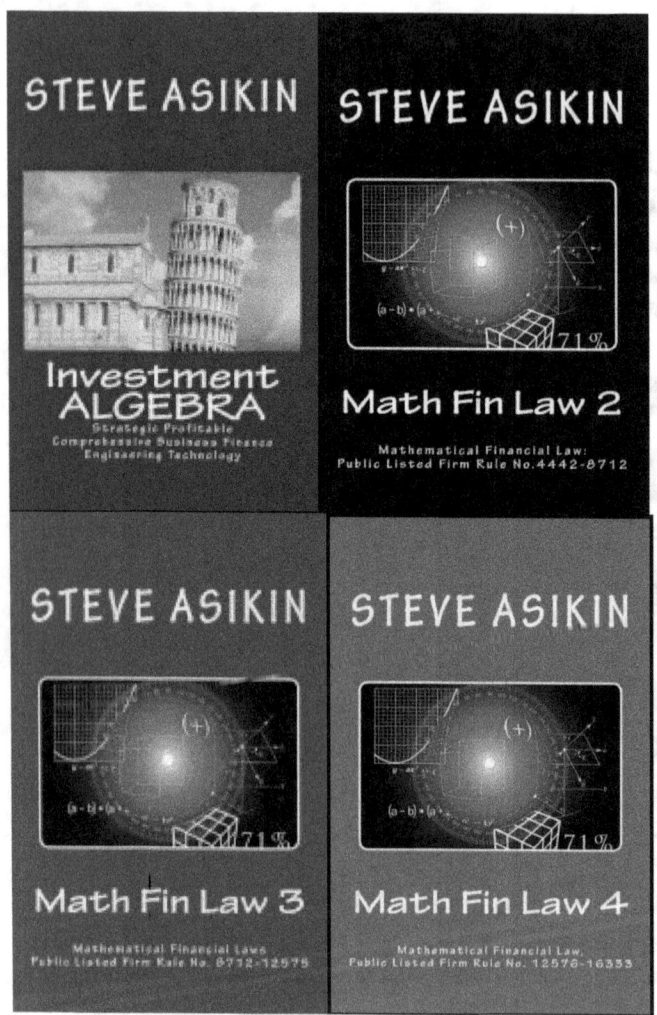

Finance Construction-4, *Tim Asikin, Steve Asikin, Indra Senihardja*

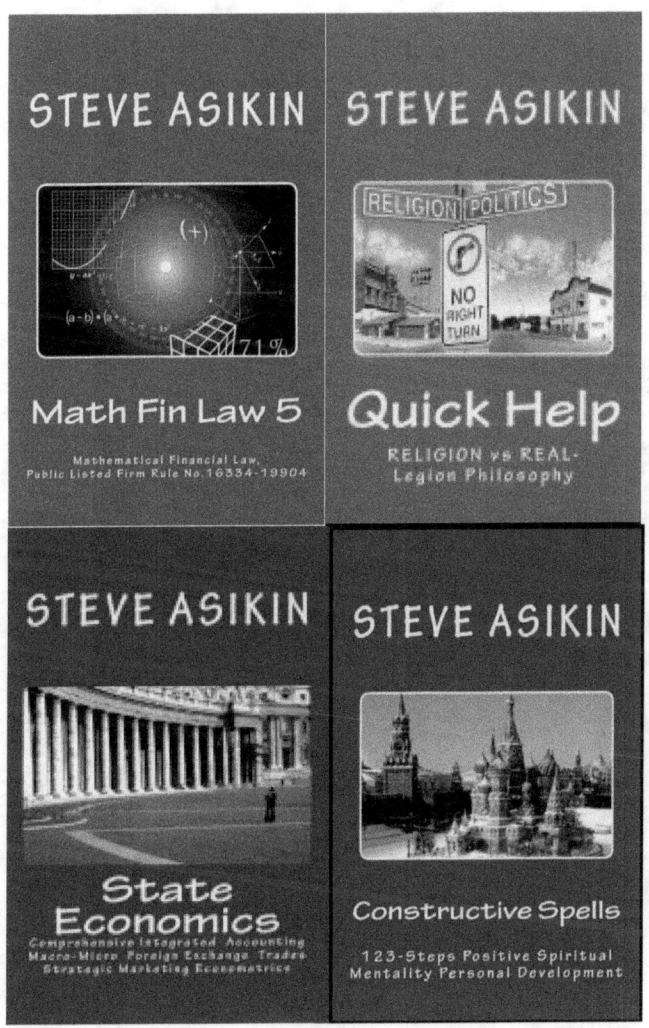

Finance Construction-4, *Tim Asikin, Steve Asikin, Indra Senihardja*

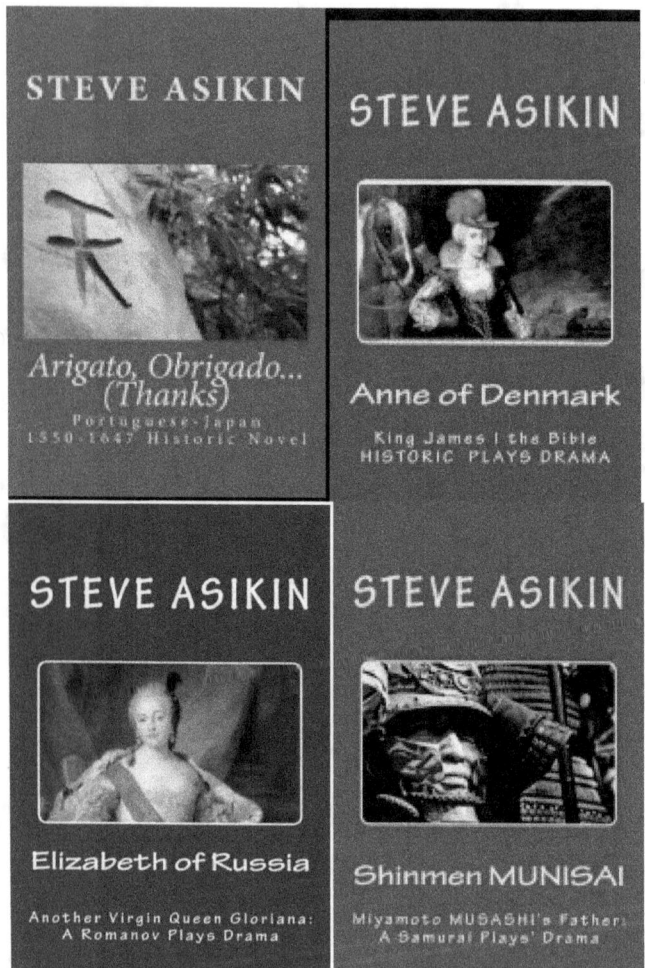

Finance Construction-4, *Tim Asikin, Steve Asikin, Indra Senihardja*

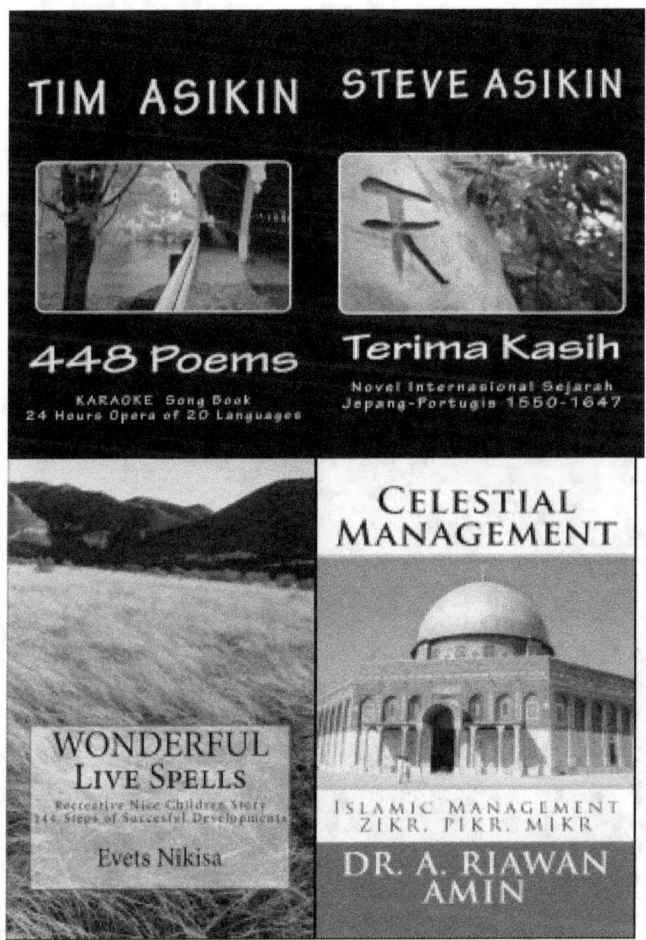

Finance Construction-4, *Tim Asikin, Steve Asikin, Indra Senihardja*

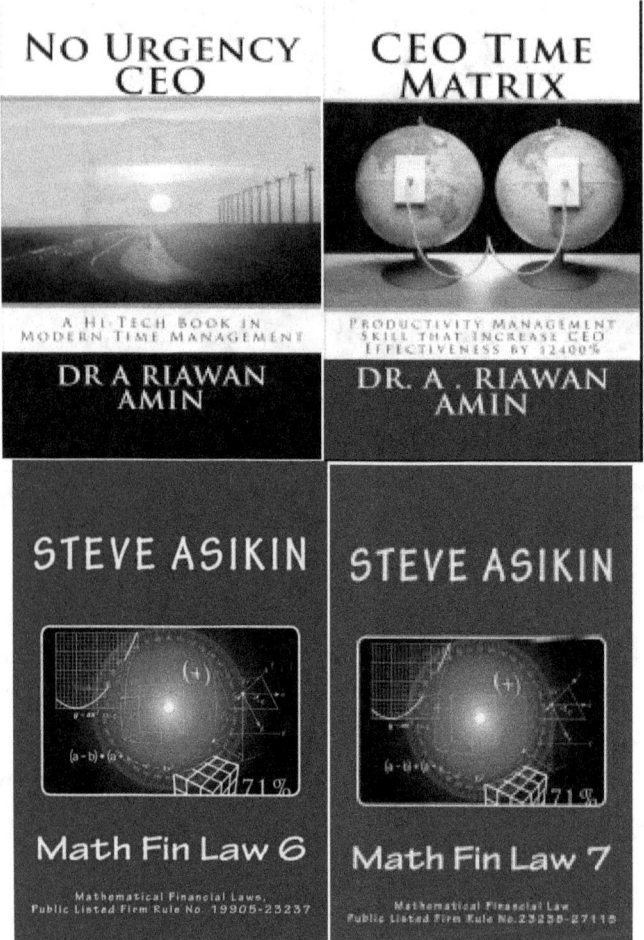

Finance Construction-4, *Tim Asikin, Steve Asikin, Indra Senihardja*

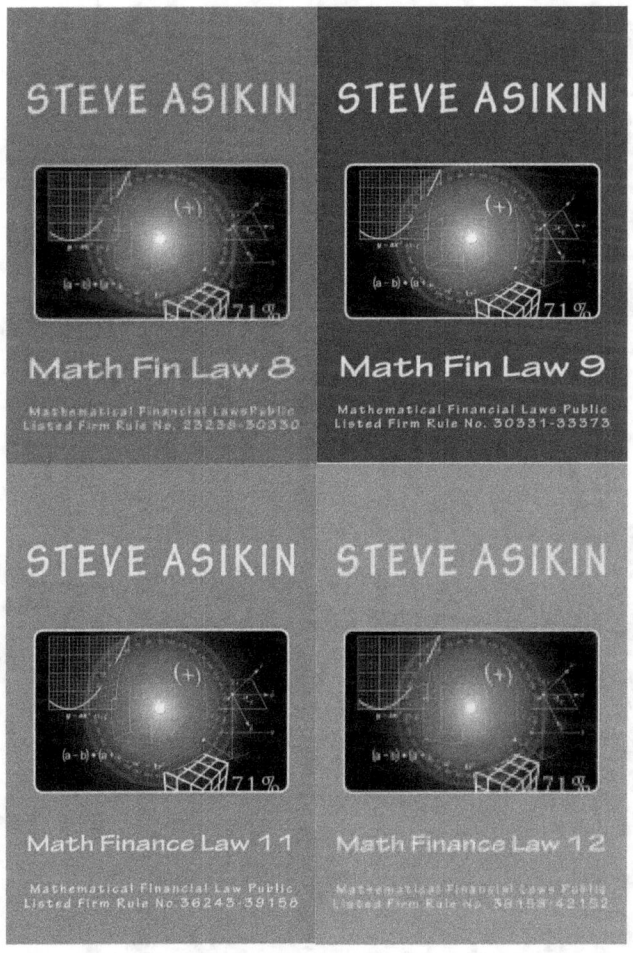

Finance Construction-4, *Tim Asikin, Steve Asikin, Indra Senihardja*

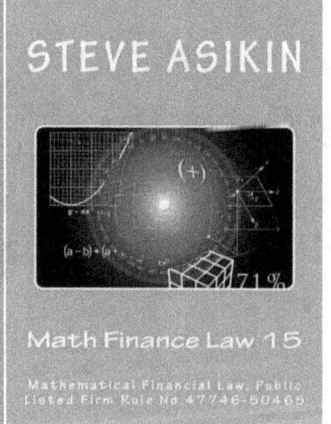

Finance Construction-4, *Tim Asikin, Steve Asikin, Indra Senihardja*

Finance Construction-4, *Tim Asikin, Steve Asikin, Indra Senihardja*

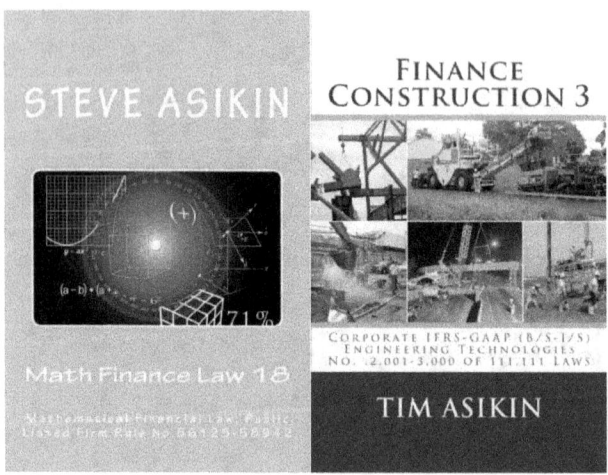

Law-3006:

If both (**I**= Interest Expense= 513 millions USD), (**y**= Y/D= Yielding Tax Rate of Dividend= 30.00%), (**H**= Home Taken Dividend= 1,659 millions USD), (**S**= Sales or Revenues= 51,294 millions USD), (**f**= F/S= Fixed Portion= 58.00%), (**S'**= Sales of Past Year= 48,851 millions USD), (**s**= [S/S']-1= Sales Growth= 5.00%), (**v**= V/S= Variable Portion= 19.00%), and (**t**= T/B= Tax Rate= 30.00%), are known, then it's (**d**= Dividend Portion or Payout planned) is:

$$\begin{aligned}
\mathbf{d} &= \mathbf{H}/([1-\mathbf{y}][1-\mathbf{t}]\{\mathbf{S}-\mathbf{S'v}[1+\mathbf{s}]-\mathbf{S'f}[1+\mathbf{s}]-\mathbf{I}\}) \\
&= \mathbf{H}/([1-\mathbf{y}][1-\mathbf{t}]\{\mathbf{S}-\mathbf{S'}[1+\mathbf{s}][\mathbf{v}+\mathbf{f}]-\mathbf{I}\} \\
&= 1,659/([1-0.3000][1-0.3000]\{51,294 \\
&\quad -48,851[1+0.0500][0.1900 \\
&\quad +0.5800]-513\}) \\
&= \underline{30.00\%}
\end{aligned}$$

Law-3007:

If both (**I**= Interest Expense= 513 millions USD), (**d**= D/A= Dividend Portion or Payout= 30.00%), (**H**= Home Taken Dividend= 1,659 millions USD), (**S**= Sales or Revenues= 51,294 millions USD), (**f**= F/S= Fixed Portion= 58.00%), (**S'**= Sales of Past Year= 48,851 millions USD), (**s**= [S/S']-1= Sales Growth= 5.00%), (**v**= V/S= Variable Portion= 19.00%), and (**t**= T/B= Tax Rate= 30.00%), are known, then it's (**y**= Yielding Tax Rate of Dividend planned) is:

$$\begin{aligned}
\mathbf{y} &= 1 - \mathbf{H}/(\mathbf{d}[1-\mathbf{t}]\{\mathbf{S}-\mathbf{S'v}[1+\mathbf{s}]-\mathbf{S'f}[1+\mathbf{s}]-\mathbf{I}\}) \\
&= \mathbf{H}/(\mathbf{d}[1-\mathbf{t}]\{\mathbf{S}-\mathbf{S'}[1+\mathbf{s}][\mathbf{v}+\mathbf{f}]-\mathbf{I}\}) \\
&= 1{,}659/([1-0.3000][1-0.3000]\{51{,}294 \\
&\quad -48{,}851[1+0.0500][0.1900 \\
&\quad +0.5800]-513\}) \\
&= 30.00\%
\end{aligned}$$

Law-3009:

If both (**I**= Interest Expense= 513 millions USD), (**d**= D/A= Dividend Portion or Payout= 30.00%), (**H**= Home Taken Dividend= 1,659 millions USD), (**t**= T/B= Tax Rate= 30.00%), (**f**= F/S= Fixed Portion= 58.00%), (**S'**= Sales of Past Year= 48,851 millions USD), (**s**= [S/S']-1= Sales Growth= 5.00%), (**v**= V/S= Variable Portion= 19.00%), and (**y**= Y/D= Yielding Tax Rate of Dividend= 30.00%), are known, then it's (**S**= Sales or Revenues planned) is:

$$S = I + S'[1+s][v+f] + H/\{d[1-y][1-t]\}$$
$$= 513 + 48,851[1+0.0500][0.1900 + 0.5800] + 1,659/\{[1-0.3000][1-0.3000]\}$$
$$= 51,294 \text{ millions USD}$$

Finance Construction-4, *Tim Asikin, Steve Asikin, Indra Senihardja*

Law-3008:
If both (**I**= Interest Expense= 513 millions USD), (**d**= D/A= Dividend Portion or Payout= 30.00%), (**H**= Home Taken Dividend= 1,659 millions USD), (**$**= Sales or Revenues= 51,294 millions USD), (**f**= F/S= Fixed Portion= 58.00%), (**$'**= Sales of Past Year= 48,851 millions USD), (**s**= [S/S']-1= Sales Growth= 5.00%), (**v**= V/S= Variable Portion= 19.00%), and (**y**= Y/D= Yielding Tax Rate of Dividend= 30.00%), are known, then it's (**t**= Tax Rate planned) is:

$$\begin{aligned}
\mathbf{t} &= 1-\mathbf{H}/(\mathbf{d}[1-\mathbf{y}]\{\$-\$'\mathbf{v}[1+\mathbf{s}]-\$'\mathbf{f}[1+\mathbf{s}]-\mathbf{I}\}) \\
&= 1-\mathbf{H}/(\mathbf{d}[1-\mathbf{y}]\{\$-\$'[1+\mathbf{s}][\mathbf{v}+\mathbf{f}]-\mathbf{I}\}) \\
&= 1{,}659/([1-0.3000][1-0.3000]\{51{,}294 \\
&\qquad -48{,}851[1+0.0500][0.1900 \\
&\qquad +0.5800]-513\}) \\
&= \underline{30.00\%}
\end{aligned}$$

www.ingramcontent.com/pod-product-compliance
Lightning Source LLC
Chambersburg PA
CBHW050148230526
45470CB00001B/4